Lecture Notes in Computer Science 9263

Commenced Publication in 1973
Founding and Former Series Editors:
Gerhard Goos, Juris Hartmanis, and Jan van Leeuwen

Editorial Board

More information about this series at http://www.springer.com/series/7409

Sanjay Madria · Takahiro Hara (Eds.)

Big Data Analytics and Knowledge Discovery

17th International Conference, DaWaK 2015
Valencia, Spain, September 1–4, 2015
Proceedings

 Springer

Editors
Sanjay Madria
University of Science and Technology
Rolla, MO
USA

Takahiro Hara
Osaka University
Osaka
Japan

ISSN 0302-9743 ISSN 1611-3349 (electronic)
Lecture Notes in Computer Science
ISBN 978-3-319-22728-3 ISBN 978-3-319-22729-0 (eBook)
DOI 10.1007/978-3-319-22729-0

Library of Congress Control Number: 2015945610

LNCS Sublibrary: SL3 – Information Systems and Applications, incl. Internet/Web, and HCI

Springer Cham Heidelberg New York Dordrecht London
© Springer International Publishing Switzerland 2015

Printed on acid-free paper

Springer International Publishing AG Switzerland is part of Springer Science+Business Media
(www.springer.com)

Preface

Big data are rapidly growing in all domains. Knowledge discovery using data analytics is important to several applications ranging from healthcare to manufacturing to smart city. The purpose of the DaWaK conference is to provide a forum for the exchange of ideas and experiences among theoreticians and practitioners who are involved in design, management, and implementation of big data management, analytics, and knowledge discovery solutions.

We received 90 good-quality submissions, from which 31 were selected for presentation and inclusion in the proceedings, after peer-review by at least three international experts in the area. The selected papers were included in the following sessions: Big Data Management, Stream Processing, Data Warehousing, Social Network Applications, Data Mining, and Big Data Applications.

Major credit for the quality of the track program goes to the authors who submitted high-quality papers and the reviewers, who under relatively short time and deadlines completed the reviews. We thank all the authors who contributed the papers and the reviewers who selected very high quality papers. We would like to thank all the members of the DEXA committee for their support and help; in particular, we are very thankful to Gabriela Wagner for her non-stop support. Finally, we would like to thank the local Organizing Committee for the wonderful arrangements and all the participants for attending the DaWaK conference and stimulating the technical discussions.

We hope that all participants enjoyed the conference and city of Valencia.

July 2015

Sanjay Madria
Takahiro Hara

Organization

Program Committee Co-chairs

Sanjay K. Madria   Missouri University of Science and Technology, USA
Takahiro Hara   Osaka University, Japan

Program Committee

Abelló, Alberto   Universitat Politecnica de Catalunya, Spain
Agrawal, Rajeev   North Carolina A&T State University, USA
Al-Kateb, Mohammed   Teradata Labs, USA
Amagasa, Toshiyuki   University of Tsukuba, Japan
Bach Pedersen, Torben   Aalborg University, Denmark
Baralis, Elena   Politecnico di Torino, Italy
Bellatreche, Ladjel   ENSMA, France
Ben Yahia, Sadok   Faculty of Sciences of Tunis, France
Bernardino, Jorge   ISEC - Polytechnic Institute of Coimbra, Portugal
Bhatnagar, Vasudha   Delhi University, India
Boukhalfa, Kamel   USTHB, Algeria
Boussaid, Omar   University of Lyon, France
Bressan, Stephane   National University of Singapore, Singapore
Buchmann, Erik   Karlsruhe Institute of Technology, Germany
Chakravarthy, Sharma   The University of Texas at Arlington, USA
Cremilleux, Bruno   Université de Caen, France
Cuzzocrea, Alfredo   University of Trieste, Italy
Davis, Karen   University of Cincinnati, USA
Dobra, Alin   University of Florida, USA
Dou, Dejing   University of Oregon, USA
Dyreson, Curtis   Utah State University, USA
Endres, Markus   University of Augsburg, Germany
Estivill-Castro, Vladimir Griffith University, Australia
Ezeife, Christie   University of Windsor, Canada
Furfaro, Filippo   University of Calabria, Italy
Furtado, Pedro   Universidade de Coimbra, Portugal
Garcia-Alvarado, Carlos   Pivotal Inc., USA
Goda, Kazuo   University of Tokyo, Japan
Golfarelli, Matteo   DISI - University of Bologna, Italy
Greco, Sergio   University of Calabria, Italy
Hara, Takahiro   Osaka University, Japan
Hoppner, Frank   Ostfalia University of Applied Sciences, Germany

Ishikawa, Yoshiharu	Nagoya University, Japan
Josep, Domingo-Ferrer	Rovira i Virgili University, Spain
Kalogeraki, Vana	Athens University of Economics and Business, Greece
Kim, Sang-Wook	Hanyang University, Republic of Korea
Koncilia, Christian	University of Klagenfurt, Austria
Lechtenboerger, Jens	Westfälische Wilhelms-Universität Münster, Germany
Lehner, Wolfgang	Dresden University of Technology, Germany
Leung, Carson K.	The University of Manitoba, Canada
Maabout, Sofian	University of Bordeaux, France
Madria, Sanjay Kumar	Missouri University of Science and Technology, USA
Marcel, Patrick	Université François Rabelais Tours, France
Mondal, Anirban	Xerox Research, India
Morimoto, Yasuhiko	Hiroshima University, Japan
Onizuka, Makoto	Osaka University, Japan
Papadopoulos, Apostolos	Aristotle University, Greece
Patel, Dhaval	Indian Institute of Technology Roorkee, India
Rao, Praveen	University of Missouri-Kansas City, USA
Ristanoski, Goce	National ICT, Australia
Rizzi, Stefano	University of Bologna, Italy
Sapino, Maria Luisa	Università degli Studi di Torino, Italy
Sattler, Kai-Uwe	Ilmenau University of Technology, Germany
Simitsis, Alkis	HP Labs, USA
Talia, Domenico	University of Calabria, Italy
Taniar, David	Monash University, Australia
Teste, Olivier	IRIT, University of Toulouse, France
Theodoratos, Dimitri	New Jersey Institute of Technology, USA
Vassiliadis, Panos	University of Ioannina, Greece
Weldemariam, Komminist	IBM Research Africa, Kenya
Wrembel, Robert	Poznan University of Technology, Poland
Zeitouni, Karine	Université de Versailles Saint-Quentin-en-Yvelines, France
Zhou, Bin	University of Maryland, Baltimore County, USA

External Reviewers

Vasil Slavov	Bloomberg L.P., USA
Anas Katib	University of Missouri-Kansas City, USA
Anamika Gupta	CBS College, University of Delhi, India
Rakhi Saxena	Deshbhandu College, University of Delhi, India
Sharanjit Kaur	AND College, University of Delhi, India
Manju Bharadwaj	Maitreyi College, University of Delhi, India
Fan Jiang	University of Manitoba, Canada
Syed Tanbeer	University of Manitoba, Canada
Luca Cagliero	Politecnico di Torino, Italy
Paolo Garza	Politecnico di Torino, Italy
Tsuyoshi Ozawa	NTT Software Innovation Center, Japan

Hiroaki Shiokawa	NTT Software Innovation Center, Japan
Tsubasa Takahashi	NEC Corporation, Japan
Chiemi Watanabe	University of Tsukuba, Japan
Cem Aksoy	New Jersey Institute of Technology, USA
Ananya Dass	New Jersey Institute of Technology, USA
Xiaoying Wu	Wuhan University, USA
Gerasimos Marketos	University of Piraeus, Greece
Bin Yang	Aalborg University, Denmark
Xike Xie	Aalborg University, Denmark
Christian Thomsen	Aalborg University, Denmark
Kim A. Jakobsen	Aalborg University, Denmark
Hiroyuki Yamada	The University of Tokyo, Japan
Sang-Chul Lee	Carnegie Mellon University, USA
Hao Wang	University of Oregon, USA
Fernando Gutierrez	University of Oregon, USA
Sabin Kafle	University of Oregon, USA
Nisansa de Silva	University of Oregon, USA
Ali Masri	University of Versailles Saint-Quentin and Vedecom Institute, France
Kiki Maulana	Monash University, Australia
Nicolas Béchet	Université de Bretagne Sud, France
Arnaud Soulet	Université Francois Rabelais, France
Mohammad Shamsul Arefin	Chittagong University of Engineering and Technology, Bangladesh
Mohammad Anisuzzaman Siddique	Hiroshima University, Japan
Petar Jovanovic	Universitat Politecnica de Catalunya, Spain
Besim Bilalli	Universitat Politecnica de Catalunya, Spain
Francesco Folino	ICAR-CNR, Italy
Luigi Pontieri	ICAR-CNR, Italy
Elio Masciari	ICAR-CNR, Italy
Andrea Tagarelli	UNICAL, Italy
Rahma Djiroune	USTHB, Algiers, Algeria
Ibtissem Frihi	Boumerdes University, Algeria
Ritu Chaturvedi	University of Windsor, Canada
Stéphane Jean	LIAS/ISAE-ENSMA, France
Brice Chardin	LIAS/ISAE-ENSMA, France
Selma Khouri	LIAS/ISAE-ENSMA, France
Zoé Faget	LIAS/ISAE-ENSMA, France
Selma Bouarar	LIAS/ISAE-ENSMA, France

Contents

Heterogeneous Networks and Data

Data Warehouses

Stream Processing

Applications of Big Data Analysis

Big Data

Similarity Measure and Clustering

Determining Query Readiness
for Structured Data

Farid Alborzi[1]([✉]), Rada Chirkova[1], Jon Doyle[1], and Yahya Fathi[2]

[1] Computer Science Department, North Carolina State University,
Raleigh, NC, USA
{falborz,rychirko,jon_doyle}@ncsu.edu
[2] Industrial and Systems Engineering Department,
North Carolina State University, Raleigh, NC, USA
fathi@ncsu.edu

Abstract. The outcomes and quality of organizational decisions depend on the characteristics of the data available for making the decisions and on the value of the data in the decision-making process. Toward enabling management of these aspects of data in analytics, we introduce and investigate *Data Readiness Level (DRL)*, a quantitative measure of the value of a piece of data at a given point in a processing flow. Our DRL proposal is a multidimensional measure that takes into account the relevance, completeness, and utility of data with respect to a given analysis task. This study provides a formalization of DRL in a structured-data scenario, and illustrates how knowledge of rules and facts, both within and outside the given data, can be used to identify those transformations of the data that improve its DRL.

Keywords: Big data quality · Big data analytics and user interfaces · Data readiness level · Data quality measurement · Data quality improvement

1 Introduction

Organizations around the world increasingly apply analytics to their data to enable and facilitate recognition of events of interest, prediction of future events, and prescription of needed actions. These activities are made possible by the information and knowledge that analytics extract from the input data. By influencing the nature and quality of the extracted knowledge, the input data impact the overall decision-making process. A way to view this is that every datum exhibits inherent qualities that contribute in different ways to the value of the datum in a specific decision-making process. Low value qualities indicate that the datum is not reliable and may lead to inferior decision making.

As an example, consider the task of linking (a) the information collected about passing cars at a toll booth by a license-plate recognition system (LPR), with (b) the Department of Motor Vehicles (DMV) information about the owners of the cars. Imperfect LPR readings of license-plate information at the toll booth,

S. Madria and T. Hara (Eds.): DaWaK 2015, LNCS 9263, pp. 3–14, 2015.
DOI: 10.1007/978-3-319-22729-0_1

as well as missing car or driver information in the DMV database would be indications of "low-value" qualities of the data, which may lead to failures in performing the linkage task. In this paper we address detecting and correcting data-quality problems such as those in this "Toll Booth" example.

Our focus in this paper is on the problem of determining whether the inherent quality of the available data meets, or at least can be improved to meet, the data-quality expectations of the decision makers that have access to the data. We address this problem by introducing methods for quantifying and improving the quality of data. Our first contribution is in introducing and investigating *data readiness level (DRL)*, a quantitative measure of the value of the quality of a piece of data at a given point in a processing flow. As such, the DRL represents a paradigm shift from the qualitative nature of traditional exploratory data analysis towards a rigorous metrics-based assessment of the quality of data in various states of readiness. The intuition for DRL is the distance between the information that can be extracted from (i.e., the "information content" [17] of) the given data for the given task, and the information content required by the task. Our second contribution is in introducing approaches for improving the value of data quality (i.e., DRL) of the data.

We use "relevance" and "completeness" dimensions to define DRL. In the above Toll Booth example, suppose 10 % of the driver-name information in the LPR data set is represented by null values (i.e., missing information). By the approach introduced in this paper, the "completeness" dimension of the corresponding DRL value of the data set would be 90 %. One potential DRL-improving solution proposed in this paper would replace these null values with the appropriate driver-name values from an external source. This would potentially increase the quality of these data values all the way to 100 % in that DRL dimension.

The foundational fact concerning data readiness is that data-readiness assessments rely on knowledge extracted from the data and, in turn, constitute knowledge about the data. We consider two major goals for DRL:

(1) Determine whether the given data have information of sufficient quality with respect to the given analysis task that is to be performed on the data.
(2) If this determination process returns a negative answer, identify ways to increase the information quality of the data with respect to the task.

Our Contributions: In this paper we address these two goals, by formalizing the data-quality measure DRL and approaches to improve its value. The study covers the core case where the data are relational, and tasks are carried out via relational (SQL) queries. Further, we assume that all the "content" problems in the data arise from missing (NULL) values. We refer to these assumptions as *the relational setting with null values.*

The rest of this paper is structured as follows. Section 2 defines the data-quality measure DRL for the relational setting. Section 3 introduces operators that improve the DRL value of the data for the tasks at hand. Section 4 presents a use case for the proposed approaches. Section 5 summarizes related work; Section 6 identifies som extensions of the relational setting.

Relation1 in database D1 at toll booth

Relation2 in D2 at NC DMV

row	plate	lane	CL	user	time	date
1	NMU45	3	1.00	NULL	09:20:16	07/04/2014
2	STA00	1	0.73	n1	09:20:03	07/04/2014
3	ABWD9	3	0.85	n2	09:19:53	07/04/2014
4	TRC19	4	1.00	n3	09:19:52	07/04/2014

LicNum	name	plate	state
11156	n2	ABWD9	VA
78922	n1	STA00	VA
58556	w1	NMU45	NC
82659	n3	TRC19	MD

Fig. 1. Table `Relation1` in database `D1`, and table `Relation2` in database `D2`. (`CL` stands for "confidence level," and `LicNum` for "driver license number.")

2 Formalizing Data Readiness Level

In this section we formalize the notion of data readiness level for the case where the data are structured using the relational data model, and tasks are carried out using relational (SQL) queries. Further, we assume that all the "content" problems in the data arise from missing (`NULL`) values. (We call this core case *the relational setting with null values,* and discuss extensions in Sect. 6.) This section provides the definitions and intuition, as well as illustrations via the Toll Booth example of Sect. 1.[1]

2.1 Toll Booth Example — Traffic Flow Identification

We begin by providing details on the Toll Booth example of Sect. 1; the example will be used to illustrate the concepts and approaches introduced in this paper. For the traffic-management department of a city, consider the *task* of obtaining the names of owners of those cars that are registered in North Carolina (NC) and that entered the city through a toll road on July 4, 2014. Suppose the toll road operates toll booths equipped with a license-plate recognition system (LPR). LPR identifies the number on a license plate with some possibility of error, with a confidence-level value that is either in the interval [0,1] or equals `NULL`.

Suppose the database `D1` at a toll booth includes a table `Relation1` obtained from the LPR system, and database `D2` at the NC DMV has table `Relation2`, see Fig. 1. The task at hand can be expressed over each of `D1` and `D2` as:

```
(Q1 over D1): SELECT user FROM Relation1
              WHERE state='NC' AND date='07/04/2014' AND CL>=0.8
(Q2 over D2): SELECT name FROM Relation2
              WHERE state='NC' AND date='07/04/2014' AND CL>=0.8
```

Please note that neither `Q1` nor `Q2` is to be posed directly over the respective database. Instead, we envision a user interface where the data first get checked for their data readiness level with respect to each query. The query is posed over the database only after the data readiness level of the data has been improved

---

[1] Due to the page limit, the details can be found in the online version [2] of this paper.

in a satisfactory way. In particular, the query Q1 is originally formulated correctly *with respect to the task at hand* independently of the available table, even though Q1 mentions the state attribute that is absent from Relation1. Thus, we envision that the query Q1 will not be rejected outright when posed directly on database D1. Rather, the DRL value of D1 can be improved by importing the state attribute, perhaps from database D2. Once the data readiness level of D1 is improved this way, Q1 will be syntactically correct for execution over D1.

2.2   Data Readiness Level: Intuition and Preliminaries

Intuitively, data-readiness judgments concern *utility* of the given data for the given task. In general, utility might depend on numerous factors. In our relational setting, we consider a very simple form of utility, where the readiness level of a data set with respect to (w.r.t.) a task is a tuple of "relevance" and "completeness" dimensions. Here, *relevance* represents how close the structure of the data is to the task requirements, and *completeness* represents the fraction of useful (non-NULL) values among the data values available for addressing the task. We provide a formalization of this two-dimensional DRL measure in Sect. 2.5. The definition there uses the formalizations of relevance (Sect. 2.3) and completeness (Sect. 2.4) of a given data set w.r.t. a given task.

As an illustration, the relevance of the database D1 of Sect. 2.1 is not "100 % satisfactory" w.r.t. the query Q1, as the state column mentioned in Q1 is absent from Relation1. Similarly, the completeness of D1 is not "100 % adequate" w.r.t. Q1, because the user name is NULL in the first row of Relation1.

We now begin introducing the DRL formalism, by specifying the notion of task in our core relational setting. We consider tasks carried out by SQL SELECT–FROM–WHERE queries without extra clauses. For our purposes, it is sufficient to consider the "signature" of each task. The *signature* of a SQL task Q is a triple

$$T(Q) = [S(Q), F(Q), W(Q)], \tag{1}$$

in which $S(Q)$ is the set of attributes in the SELECT clause of Q, $F(Q)$ is the set of relations in the FROM clause of Q, and $W(Q)$ is the set of attributes in the WHERE clause of Q. For instance, the task Q1 in our running example has the signature $T(Q1) = [\{user\}, \{Relation1\}, \{state, date, CL\}]$, and the task Q2 has the signature $T(Q2) = [\{name\}, \{Relation2\}, \{state, date, CL\}]$.

In addition to the signature for each task, we also consider the attributes available for the task in the data: The *available attributes* of a query Q, written $A(Q)$, is the set of all the attributes in all the relations in $F(Q)$. For the queries Q1 and Q2 of Sect. 2.1 we have $A(Q1) = \{row, plate, lane, CL, user, time, date\}$, and $A(Q2) = \{LicNum, name, plate, state\}$.

Finally, we assume that each user who poses a SQL task on the available data can also specify, for each relational attribute mentioned in the task, the nonnegative *relevance utility* ρ and *completeness utility* κ of the attribute for the task. (In the baseline case where the user does not specify individual utilities of attributes, we assume that along each of the relevance and completeness

Table 1. Relevance (ρ) and completeness (κ) utility values for task Q1 of Sect. 2.1.

a	user	state	date	CL
$\rho(a, \mathtt{Q1})$	4	1	4	2
$\kappa(a, \mathtt{Q1})$	1	2	4	3

dimensions of DRL, the utility of each attribute mentioned in the task is 1.) For instance, for our example of Sect. 2.1, possible utilities of the attributes mentioned in task Q1 are listed in Table 1.

2.3 The Relevance Dimension of DRL

Given a relational data set, a SQL task, and the relevance utilities of the task attributes (see Sect. 2.2), we combine these relevance utilities to obtain a measure of the relevance of the data set to the task.

Definition 1. *The DRL relevance R of a database D for a task with signature $T = [S, F, W]$ and available attributes A, is defined as the ratio of the weighted relevance utility of those task attributes that are available in the database, to the weighted relevance utility of all the attributes in the task:*

$$R(D, T) = \begin{cases} \dfrac{\sum_{a \in S \cap A} \rho(a, T) + \sum_{a \in W \cap A} \rho(a, T)}{\sum_{a \in S} \rho(a, T) + \sum_{a \in W} \rho(a, T)} & \text{if } F \text{ is in } D \\ 0 & \text{otherwise.} \end{cases} \quad (2)$$

That is, if all the relations in the FROM clause of the task are in the database, we use a weighted average as the relevance of the database to the task. Otherwise, the corresponding relevance value is zero.

Suppose that in our example of Sect. 2.1, the relevance (ρ) utilities of attributes for task Q1 are as given in Table 1. Then the relevance of D1 for Q1 is:

$$R(\mathtt{D1}, \mathtt{Q1}) = R(\mathtt{D1}, [\{\mathtt{user}\}, \{\mathtt{Relation1}\}, \{\mathtt{state}, \mathtt{date}, \mathtt{CL}\}])$$
$$= \frac{4 + 0 + 4 + 2}{4 + 1 + 4 + 2} = 0.91$$

2.4 The Completeness Dimension of DRL

In our definition of completeness of a database for a task, we use the following notion. For a relation r in database D, denote by $|r|$ the number of rows in r, and for an attribute a in r, denote by $|r(a \neq \mathtt{NULL})|$ the number of rows in r in which a non-NULL value appears for a. Then the *completeness degree* of attribute a w.r.t. relation r, denoted by $\phi(a, r)$, is defined as the ratio $|r(a \neq \mathtt{NULL})|/|r|$. (When either $a \notin r$ or $|r| = 0$, we define $\phi(a, r)$ to be 0.)

Now for a data set and a SQL task with completeness utilities of its attributes (see Sect. 2.2), we combine these utilities with the respective completeness degrees to obtain a measure of the completeness of the data set for the task.

Definition 2. *The* DRL *completeness* K *of a database* D *for a task with signature* $T = [S, F, W]$ *and available attributes* A, *is defined as the ratio of the weighted completeness utility of those task attributes that are available in the database to the weighted completeness utility of all the attributes in the task:*

$$K(D, T) = \begin{cases} \dfrac{\sum_{a \in S} \kappa(a, T)\phi(a, F) + \sum_{a \in W} \kappa(a, T)\phi(a, F)}{\sum_{a \in S \cap A} \kappa(a, T) + \sum_{a \in W \cap A} \kappa(a, T)} & \text{if } F \text{ in } D \\ 0 & \text{otherwise.} \end{cases} \quad (3)$$

That is, the DRL completeness of a database w.r.t. a task is computed as the sum of the completeness utilities of the task attributes weighted by the respective completeness degrees, normalized by the total possible (i.e., ideal) completeness.

Suppose that in our example of Sect. 2.1, the completeness (κ) utilities of attributes for task Q1 are as given in Table 1. We obtain that $\phi(\text{user}, \text{Relation1}) = \phi(\text{date}, \text{Relation1}) = \phi(\text{CL}, \text{Relation1}) = 1$, that $\phi(\text{state}, \text{Relation1}) = 0$, and that $\phi(\text{user}, \text{Relation1}) = 3/4$. Then the completeness of D1 for Q1 is

$$K(\text{D1}, \text{Q1}) = K(\text{D1}, [\{\text{user}\}, \{\text{Relation1}\}, \{\text{state}, \text{date}, \text{CL}\}])$$
$$= \frac{1.(3/4) + 2.(0) + 4.(1) + 3.(1)}{1.(1) + 2.(0) + 4.(1) + 3.(1)} = 0.97$$

2.5 Putting It Together: Data Readiness Level Tuples

As discussed in Sect. 2.2, we define the data readiness level as a tuple of relevance and completeness dimensions. Here, relevance represents the distance between the structure of the data and of the task, and completeness represents the availability of non-null values in the task-relevant attributes:

Definition 3. *The* data readiness level *of database* D *with respect to task* T *is a tuple of the relevance and completeness values of* D *w.r.t.* T:

$$DRL(D, T) = [R(D, T), K(D, T)]. \quad (4)$$

In our running example of Sect. 2.1, the DRL for database D1 for task Q1 is $DRL(\text{D1}, \text{Q1}) = [0.91, 0.97]$. By similar calculations, for database D2 and task Q2 we obtain that $DRL(\text{D2}, \text{Q2}) = [0.45, 1]$. (The intuition for the relevance dimension of this DRL value for D2 and Q2 comes from the observation that the structure of Relation2 in D2 does not match well the attributes mentioned in Q2. At the same time, Relation2 has no null values, which explains the "perfect score" of 1 in the completeness dimension of the DRL value for D2 and Q2.)

3 Improving Readiness Level of Data for Task at Hand

In Sect. 2 we formalized the notion of data readiness level for the relational setting with null values. In this section, we discuss actions that can be taken in this context, to improve the DRL value of a data set for a given task. Our contribution is in providing a taxonomy and descriptions of meta-operators, which we refer to as "data-readying operators." The distinction between our meta-operators and data-improvement approaches in the literature (e.g., data-cleaning approaches) is that we consider as explicit inputs to the data-improvement process not only the data, but also its DRL value w.r.t the *task* to be performed on the data, and a threshold on the desirable DRL value. As a result, to become applicable to the specific data set and task at hand, each meta-operator that we discuss is to be *instantiated* with both data-specific and task-specific parameter values, as well as potentially with knowledge that is (external to the data and task, and) available to the meta-operator for the data-improvement purpose.

3.1   Taxonomy of DRL-Improving Operators

In our relational setting with null values, we consider meta-operators that improve the DRL of the data by increasing the information content [17] of the data. The specific meaning here (same as in [17]) is that each operator "repairs" the NULL values in the data, by converting these values, to the extent possible, into non-NULL values that conform to the real world modeled by the data set.

We classify meta-operators aimed at improving the DRL of the data in this context by using three categories: null-value cleaning operators, resource-changing operators, and database integrating operators. Figure 2 summarizes these as part of a taxonomy of operators that increase the information content [17] of the data. Among these operator types, null-value cleaning operators increase the completeness dimension of the DRL, by looking for null values present in the data and using one of several possible techniques to repair the null values. Resource-changing operators aim to increase the relevance of the data for the task, by finding alternative resources (relations) with the task-appropriate structure. Finally, database integrating operators look for data sets that can be integrated with the given data to improve its relevance *and* completeness.

3.2   Illustration Using the Running Example

We now illustrate the use of null-values cleaning operators via our running example of Sect. 2.1. As discussed above, meta-operators in this class improve the value of the completeness dimension of the DRL of the given data set, by repairing (i.e., replacing by meaningful values) null values present in the data. Some other classes of our data-readying operators are illustrated in Sect. 4. For further details, please refer to the online version [2] of this paper.

Among the null-values cleaning operators that we consider in our taxonomy, one approach to repairing null values of an attribute in a relation consists of importing into the relation the appropriate non-null values of a matching attribute in another table. At the meta-operator level, this approach accepts as inputs the data set, the task information, and knowledge about which attributes in the available relations represent the same concept in the subject-matter ontology. An instantiation of this meta-operator within a particular DRL-improving process would accept specific values of all these inputs and would then proceed to replace all the NULL values in the attribute of interest by the non-NULL values taken from another relation. The non-NULL values used to repair the NULLs in

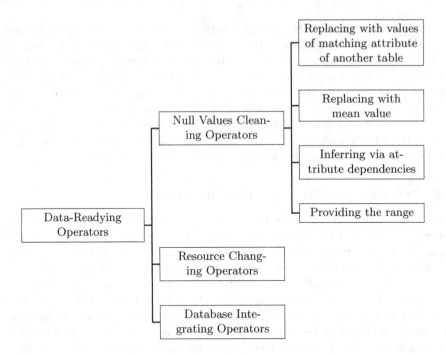

Fig. 2. Taxonomy of operators for increasing the information content of the data.

the attribute of interest would come from the same concept (i.e., attribute) and for the same real-world object (i.e., "appropriate" row in the relation) as those for the NULL value being replaced in the attribute of interest.

In our running example of Sect. 2.1, Relation1 includes attributes plate and user. Suppose these attributes are linked to the concepts *PlateNumber* and *UserName*, respectively, in the available knowledge base. Further, suppose that attributes plate and name of Relation2 are linked in the knowledge base to the same respective concepts. Then we say that plate in Relation1 "concept-matches" plate in Relation2, and user in Relation1 "concept-matches" name in Relation2. Observe that the plate attribute is an identifier in Relation2. Consequently, the instantiation of our meta-operator can replace the NULL values of the user attribute in Relation1 by those (non-NULL) values of the name attribute of Relation2 that correspond to the same value of the *PlateNumber* concept.

The output of applying this data-improvement operator to the user attribute of Relation1 in the Toll Booth example is demonstrated in Table 2.

4   Use Case: Marketing via Targeted Mailings

To further illustrate the meta-operators described in Sect. 3, we introduce the following use case that complements our example of Sect. 2.1. In this use case, the marketing department of a company decides to contact all the customers

Table 2. Repaired `NULL` values of the `user` attribute in `Relation1` of Sect. 2.1.

row	plate	lane	CL	user	time	date
1	NMU45	3	1.00	w1	09:20:16	07/01/2014
2	STA00	1	0.73	n1	09:20:03	07/01/2014
3	ABWD9	3	0.85	n2	09:19:53	07/01/2014
4	TRC19	4	1.00	n3	09:19:52	07/01/2014

who are over 20 years old and use the company's `Plan B`, to motivate them to switch to the company's `Plan A`. For this purpose, the company plans to use its database D3, with relations `Sales` and `Customer` as shown in Fig. 3.

Sales

id	custId	boughtPlan	date
1	103	Plan B	01/11/2015
2	102	Plan B	10/25/2014
3	104	Plan A	12/28/2014
4	102	Plan A	01/17/2014

Customer

id	name	address	currentPlan	age
101	David Smith	22nd St.	Plan A	NULL
102	Alfred Luck	20th St.	NULL	18
103	Daniel Bush	25th St.	Plan B	30
104	Goldy Elbetri	7th St.	NULL	NULL

Fig. 3. The `Sales` and `Customer` relations in database D3 for the Marketing use case.

The following task Q3, with signature [{name, address}, {Customer}, {age, currentPlan}], could be used to find the customers to be targeted in the mailing:

(Q3) SELECT name, address FROM Customer
    WHERE currentPlan = 'Plan B' AND age >= 20;

Since all the attributes mentioned in Q3 are in the data set D3, the relevance value of D3 for Q3 is 1. Assuming that the completeness utilities for Q3 are as in Table 3, the formulas in Sect. 2 yield $DRL(\text{D3}, \text{Q3}) = [1.0, 0.73]$.

Table 3. Completeness (κ) utilities for task Q3 in the Marketing use case.

a	name	address	currentPlan	age
$\kappa(a)$	1	4	4	2

Suppose the following facts, external to the data and the task, are known in this example: the values of the **age** attribute in the `Customer` relation are normally distributed; there are statistically enough observations with non-null values in that column of `Customer`; and most of the observations of customer age in `Customer` fall around the mean value. Then the null-value-cleaning operator

Table 4. Improved-quality `Customer` relation in the Marketing use case.

id	name	address	currentPlan	age
101	David Smith	22nd St	Plan A	24
102	Alfred Luck	20th St	Plan B	18
103	Daniel Bush	25th St	Plan B	30
104	Goldy Elbetri	7th St	Plan A	24

of Fig. 2 that replaces `NULL` values with the mean value of 24 can be applied to `age` in `Customer`. The results of improving the information content of the `age` attribute of the `Customer` relation are shown in Table 4.

Further, suppose that additional available knowledge prescribes replacing the `NULL` values of the attribute `currentPlan` in the `Customer` relation with the latest plan bought by the customer. The corresponding meta-operator, suitably instantiated, would look for tuples in the `Sales` relation that have the same value of `custId` as the value of `id` in the `Customer` relation. The meta-operator would then replace the associated `NULL` values of `currentPlan` in `Customer` with the information on the most recent transaction for the customer, which is represented by the tuple with the maximum value of `date` attribute. All this replacement knowledge can be encoded in a rule that is specific to the data set and task, but would be used by an instantiation of the generic meta-operator. The result of this meta-operator application to the `currentPlan` attribute in the `Customer` relation is shown in Table 4.

5   Related Work

Projects focused on defining and classifying data-quality dimensions have related the data format to syntactic criteria [9] and investigated the semantics of data values [13,14]. In measuring data quality, two directions have emerged: quality of conformance and quality of design. Quality of conformance aims to align an information system's existing data values and its design specifications, whereas in quality of design, checking the closeness of system specifications to the customer requirements is of interest [12]. Subjective data-quality assessments, such as those in quality of design, can be approached by distributing questionnaires among stakeholders [18]. In contrast, studies focused on objective measurements, such as quality of conformance, introduce and investigate descriptive metrics. The data-quality metrics introduced in the literature either assume fixed tasks in quantifying data quality, or do not consider tasks at all. In contrast, in this work we provide a framework for computing metrics of quality of the given data with respect to specific (but not fixed) data-processing tasks.

We formalize the notion of data relevance in a way similar to [15], where relevance is defined as "level of consistency between the data content and the area of interest of the user." While relevance has been addressed in several studies (e.g., [22,23]) as a data-quality dimension, to the best of our knowledge no metric has been provided to quantify it in the literature. The framework proposed in this current work includes a metric for computing the current level of data relevance and for measuring the quality of relevance-improving solutions.

Completeness is defined in [18] as a fraction of non-null values of an attribute. A number of approaches have been proposed in data management for addressing problems caused by null values in the data, e.g., [4,5,7,17]. Directions of studies of incomplete information [17] in the literature include approaches based on representation systems and certain answers [1,11]), logical theory [19,20], and programming semantics [3,16]. Unlike previous work, in this study we introduce a formalization where completeness of the data can be improved based on knowledge that is external to the data and task at hand.

Knowledge bases have been used in a variety of applications [8]. Notably, [6] introduced a platform that employs user-imposed rules to repair data, based on violations of the available data-cleaning rules. At the same time, [6] does not focus on evaluating quality of the data in presence of specific data-processing tasks. In this study we introduce a framework that measures the quality value of data in a task-specific way, and uses external knowledge to improve the quality value of the data for the task at hand.

6   Conclusion and Future Work

In this paper we studied the problem of quantifying the readiness of a relational data set to handle tasks carried out via SQL queries, in presence of missing information in the form of null values. We formalized this data-readiness problem and proposed approaches for evaluating the relevance and completeness aspects of our data readiness measure in presence of a given data-oriented task. In addition, we developed a taxonomy of "data-readying" meta-operators that improve the information content of the data for the task at hand, in presence of knowledge available about the data or the task. The proposed formalization can be extended to quantifying the data-quality dimensions (e.g., relevance and completeness) of a data set with respect to a collection of typical tasks posed to the system.

This initial study of the core case of data readiness involves several simplifying assumptions about the structure of data sets, tasks, and task-dependent utility of the data. Our future work aims to relax these assumptions, e.g., to extend the framework to allow subqueries in the tasks; to consider types of data-quality issues beyond null values; to enable treatment of more complex and realistic data-analysis scenarios and data-quality dimensions; and to consider data models beyond the relational model. Other extensions involve developing techniques to capture and express utility information and probabilistic properties of data-readying operators.

Acknowledgment. This material is based upon work supported in whole or in part with funding from the Laboratory for Analytic Sciences (LAS). Any opinions, findings, conclusions, or recommendations expressed in this material are those of the author(s) and do not necessarily reflect the views of the LAS and/or any agency or entity of the United States Government.

References

1. Abiteboul, S., Hull, R., Vianu, V.: Foundations of Databases. Addison-Wesley, San Diego (1995)
2. Alborzi, F., Chirkova, R., Doyle, J., Fathi, Y: Determining query readiness for structured data. Technical Report (which is not a publication) TR-2015-6, NCSU, 2015. http://www.csc.ncsu.edu/research/tech/reports.php
3. Buneman, P., Jung, A., Ohori, A.: Using power domains to generalize relational databases. TCS **91**(1), 23–55 (1991)
4. Codd, E.F.: Extending the database relational model to capture more meaning. ACM TODS **4**(4), 397–434 (1979)
5. Codd, E.F.: Understanding relations (installment #7). FDT - Bull. ACM SIGMOD **7**(3), 23–28 (1975)
6. Dallachiesa, M., Ebaid, A., Eldawy, A., Elmagarmid, A., Ilyas, I. F., Ouzzani, M., Tang, N.: NADEEF: a commodity data cleaning system. In: ACM SIGMOD, pp. 541–552 (2013)
7. Date, C.J.: Database in Depth - Relational Theory for Practitioners. OReilly, Sebastopol (2005)
8. Deshpande, O., Lamba, D.S., Tourn, M., Das, S., Subramaniam, S., Rajaraman, A., Doan, A.: Building, maintaining, and using knowledge bases: a report from the trenches. In: ACM SIGMOD, pp. 1209–1220 (2013)
9. Eppler, M.J.: Managing Information Quality: Increasing the Value of Information in Knowledge-intensive Products and Processes. Springer, Berlin (2006)
10. Gardyn, E.: A Data Quality Handbook for a Data Warehouse. Infrastructure. IQ, 267–290 (1997)
11. Grahne, G.: The Problem of Incomplete Information in Relational Databases. Springer, Berlin (1991)
12. Heinrich, B., Helfert, M.: Analyzing data quality investments in CRM: a model-based approach. In: Eighth International Conference on Information Quality, pp. 80–95 (2003)
13. Heinrich, B., Klier, M., Kaiser, M.: A procedure to develop metrics for currency and its application in CRM. J. Data Inf. Qual. **1**(1), 5 (2009)
14. Hinrichs, H.: Datenqualitatsmanagement in Data Warehouse-Systemen. Ph.D. thesis, Universitat Oldenburg (2002)
15. Kulikowski, J.L.: Data quality assessment. In: Ferraggine, V.E., Doorn, J.H., Rivero, L.C. (eds.) Handbook of Research on Innovations in Database Technologies and Applications, 378–384. Hershey, PA (2009)
16. Libkin, L.: A semantics-based approach to design of query languages for partial informatio. In: Thalheim, B., Libkin, L. (eds.) Semantics in Databases. LNCS, vol. 1358, pp. 170–208. Springer, Berlin (1995)
17. Libkin, L.: Incomplete data: what went wrong, and how to fix it. In: PODS, 1–13. ACM (2014)
18. Pipino, L.L., Lee, Y.W., Wang, R.Y.: Data quality assessment. Comm. ACM **45**(4), 211–218 (2002)
19. Reiter, R.: On closed world data bases. Logic Data Bases **33**, 55–76 (1977)
20. Reiter, R.: Towards a logical reconstruction of relational database theory. Conceptual Model. **33**, 191–233 (1982)
21. Teboul, J.: Managing Quality Dynamics. Prentice Hall, New York (1991)
22. Wand, Y., Wang, R.Y.: Anchoring data quality dimensions in ontological foundations. Commun. ACM **39**(11), 86–95 (1996)
23. Wang, R.Y., Strong, D.M.: Beyond accuracy: what data quality means to data consumers. J. Manage. Inf. Syst. **12**(4), 5–34 (1996)

Efficient Cluster Detection by Ordered Neighborhoods

Emin Aksehirli[1]([⊠]), Bart Goethals[1], and Emmanuel Müller[1,2]

[1] University of Antwerp, Antwerp, Belgium
{emin.aksehirli,bart.goethals}@uantwerpen.be
[2] Karlsruhe Institute of Technology, Karlsruhe, Germany
emmanuel.mueller@kit.edu

Abstract. Detecting cluster structures seems to be a simple task, i.e. separating similar from dissimilar objects. However, given today's complex data, (dis-)similarity measures and traditional clustering algorithms are not reliable in separating clusters from each other. For example, when too many dimensions are considered simultaneously, objects become unique and (dis-)similarity does not provide meaningful information to detect clusters anymore. While the (dis-)similarity measures might be meaningful for individual dimensions, algorithms fail to combine this information for cluster detection. In particular, it is the severe issue of a combinatorial search space that results in inefficient algorithms.

In this paper we propose a cluster detection method based on the *ordered neighborhoods*. By considering such ordered neighborhoods in each dimension individually, we derive properties that allow us to detect clustered objects in dimensions in linear time. Our algorithm exploits the ordered neighborhoods in order to find both the similar objects and the dimensions in which these objects show high similarity. Evaluation results show that our method is scalable with both database size and dimensionality and enhances cluster detection w.r.t. state-of-the-art clustering techniques.

1 Introduction

In the information era that we live in, there are a huge amount of data about almost anything. Institutions, both government and private, realized the importance of data and started to collect any kind of information on their business objects, hoping that they will derive useful knowledge out of it one day. Considering the amount, annotating and labeling the data is often infeasible. Therefore, unsupervised methods such as clustering are more suitable for knowledge extraction.

One of the challenging effects of this "data hoarding" is the increased number of attributes associated with each object. Unfortunately, due to the phenomenon of the so called *curse of dimensionality*, the similarity between objects becomes meaningless in high dimensional data spaces [5]. This means that the clusters cannot be separated from each other by (dis-)similarity assessment in high dimensional space, which in turn poses a serious challenge for traditional

© Springer International Publishing Switzerland 2015
S. Madria and T. Hara (Eds.): DaWaK 2015, LNCS 9263, pp. 15–27, 2015.
DOI: 10.1007/978-3-319-22729-0_2

clustering algorithms such as k-means or hierarchical clustering. Nevertheless, (dis-)similarity in the individual dimensions can still be exploited and clusters can be detected in combinations of these dimensions. The open challenge is how to exploit this (dis-)similarity information in individual dimensions for any combination of dimensions while not falling prey to the exponential nature of this combinatorial search space.

For example, nowadays a customer can be associated with credit ratings, shopping habits, travel patterns, entertainment choices, sport habits, etc. Even medical conditions or other private information might be available due to the data integration with a multitude of data sources. Considering all the aforementioned data, each customer becomes very unique and almost equally dissimilar to any other customer. This makes the notion of "similar customers" meaningless. Nevertheless, a meaningful customer segmentation can still be achieved by looking at a subset of dimensions, e.g. travel and sport habits only.

Although similarity between high dimensional objects is not meaningful, similarity according to subsets of attributes is still meaningful. To address this issue, subspace clustering [20] aims at cluster detection in any combination of the given attributes. Even though they enhance clustering quality compared to traditional clustering methods, they come with their own challenges. They suffer from noise sensitivity [1,19], density estimation challenges [4,9,10], many complex parameters [4,10,12,13], or inefficient search through the combinatorial search space [2,9].

In contrast to these methods, we tackle the problem by exploiting local neighborhoods of objects in individual dimensions. These neighborhoods are used as means for cluster detection in combinations of dimensions. We show that clusters can be detected directly from these local neighborhoods. Moreover, exploiting their natural orders, we can avoid common scalability pitfalls in the algorithmic computation of clusters. In particular, we propose a linear-time algorithm for detecting cluster structures in dimensions. Key characteristics of our approach are its scalability w.r.t. both database size and dimensionality, the detection of clusters with variable scales, and its few user-friendly parameters. Further, it allows for individual (dis-)similarity measures for each dimension, which is important for data with different data types, complex distance functions, and the lack of joint (dis-)similarity measures in combinations of different data types and dimensions with complex distance functions.

In summary, our contributions are as follows: - A formal analysis of ordered neighborhoods in the context of cluster detection. - Prove completeness of cluster detection based on ordered neighborhoods. - A scalable algorithm exploiting ordered neighborhoods for cluster detection in data projections. - Exhaustive experiments showing that our method (1) works well with variable densities, (2) scales well with database size and dimensionality, (3) robust w.r.t. noise and irrelevant dimensions, (4) and suitable for exploratory data analysis in real world scenarios.

Please find the supplementary for the paper, in which we cover the topics in detail, along with the implementation and datasets on our website.[1]

---

[1] http://adrem.uantwerpen.be/clon.

2   Formal Properties

We use the ordered local neighborhoods of objects in each individual dimension as indicators for clustered structures. This idea is based on the observation that similarity of objects is captured by the order of objects contained in the local neighborhood [17]. To the best of our knowledge, we are the first ones to exploit this theoretic principle for scalable cluster detection in high dimensional data.

For our solution, objects only need to be sorted once according to the given (dis-)similarity measure. This might even be given (for free) due to the index structures maintained in the relational database system that stores the data. Based on these ordered neighborhoods, we detect clusters by mining object sets that re-occur in a similar order in multiple dimensions. Starting with one dimensional clusters and iteratively refining them allows us to detect clusters in projections (i.e. dimensions which agree in the similarity of objects) of a high dimensional data space.

Although the general idea seems simple to implement, it has several open research questions: Can we guarantee to detect all hidden clusters? How to capture the neighborhood information, and how can it be exploited for clustering high dimensional data? In order to answer all of these questions one by one, we structure our main contributions as follows: We first show that ordered neighborhoods ensure complete cluster detection in Sect. 2.1. Second, we propose a data transformation from relational data to an ordered neighborhood space and investigate its properties in Sect. 2.2. We introduce a scalable algorithm that exploits these properties for cluster detection in Sect. 3.

2.1   Clustering in Neighborhoods

For a *relational database* $\mathscr{D}$ we represent each data object $\mathbf{o}$ in $\mathscr{D}$ as a vector defined over attributes $\mathscr{A} = \{A_1, A_2, \cdots, A_m\}$, with $\mathbf{o}_i$ the value of the object for the attribute A_i. For each attribute A_i, the projected database is denoted as $\mathscr{D}^{A_i}$. Further, we use $\delta(\mathbf{o}, \mathbf{p})$ as a dissimilarity between the objects $\mathbf{o}$ and $\mathbf{p}$. $|S|$ denotes the cardinality of set S.

As basic notion we use local neighborhoods based on k-nearest neighborhood (k-NN), which have already been exploited for lazy classification [7], dimensionality reduction [21], collaborative filtering [16] and many other techniques in various communities [6,11]. For clustering, one has used shared nearest neighborhoods [8], neighborhoods in random projections [18], and frequent co-occurrence of neighborhoods [2]. In our work, we build the basic clustering principle of our method on the k-nearest neighborhood of an object, which is the set of objects that are the most similar to it:

Definition 1 (k-Nearest Neighborhood (k-NN)). *Let* $\mathbf{o} \in \mathscr{D}$, δ *a dissimilarity measure, and the* $NN_k(\mathbf{o})$ *be the* k^{th} *nearest object to* $\mathbf{o}$ *according to* δ, *the* k-*nearest neighborhood* (k-NN) *of* $\mathbf{o}$ *is defined as*

$$k\text{-}NN(\mathbf{o}) = \{\mathbf{p} \in \mathscr{D} \mid \delta(\mathbf{o}, \mathbf{p}) \leq \delta(\mathbf{o}, NN_k(\mathbf{o}))\}.$$

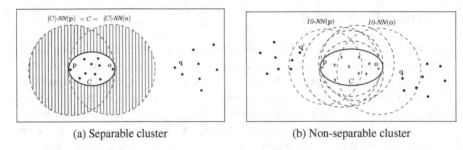

(a) Separable cluster   (b) Non-separable cluster

Fig. 1. Nearest neighborhoods of the objects in a cluster

k-$NN(\mathbf{o})$ captures the similar objects near object $\mathbf{o}$, which we can use for cluster detection. Figure 1 shows the nearest neighborhoods of the objects in clusters. Cluster sized nearest neighborhoods of objects in a *separable cluster* are equal to the cluster itself. In Fig. 1a, $|C|$-NN of all 10 objects in C are equal the C itself. Although the k-NN of some of the objects in a non-separable cluster include objects that are not in the cluster, the k-NN of the majority of objects in the cluster include the whole cluster. For example in Fig. 1b, the neighborhoods of the data points $\mathbf{p}$ and $\mathbf{q}$ are different from each other although they are in the same cluster. On the other hand, they are in the $|C|$-NN of the 8 out of 10 objects in the cluster, shown as ♦. Therefore, they can be identified as being in the same cluster by checking their co-occurrences in neighborhoods of other objects instead of comparing their neighborhoods. This is one of the advantages of using co-occurrences in neighborhoods instead of shared nearest neighborhoods.

2.2   Ordered Neighborhoods

In the original space, computing the co-occurrence of object sets in neighborhoods of objects in $\mathscr{D}$ requires expensive neighborhood computations. Instead, we compute the neighborhood of every object once and conduct co-occurrence search in these neighborhoods.

Definition 2 (Ordered Neighborhood). *The ordered neighborhood of* $\mathbf{o} \in \mathscr{D}^{A_i}$ *is the ordered list of the objects in* k-$NN(\mathbf{o})$, *in which the order reflects the order of objects in* A_i.

$$k\text{-}NN(\mathbf{o}) = (\cdots, \mathbf{p}, \cdots, \mathbf{q}, \cdots) \iff \mathbf{p} < \mathbf{q}$$

Definition 3 (Ordered Neighborhood Database). *An ordered neighborhood database* $\mathscr{N}$ *is the ordered list of ordered neighborhoods. Order of the neighborhoods is the order of objects in* $\mathscr{D}^{A_i}$. *If* $\mathbf{o} < \mathbf{p}$, *then*

$$\mathscr{N}^{A_i} = (\cdots, k\text{-}NN(\mathbf{o}), \cdots, k\text{-}NN(\mathbf{p}), \cdots)$$

For the remaining of the paper, we use *neighborhood* and k-$NN(\mathbf{o})$ to refer to the ordered neighborhood, and *neighborhood database* to refer to the ordered

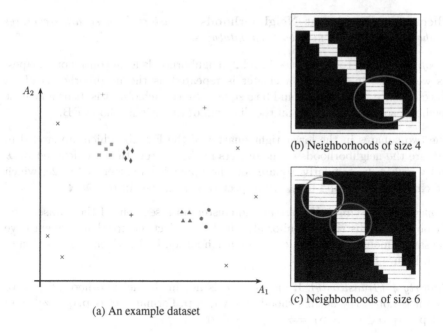

(b) Neighborhoods of size 4

(c) Neighborhoods of size 6

(a) An example dataset

Fig. 2. An example dataset and its neighborhood databases for different sizes

neighborhood database unless it is noted otherwise. Further, if the projection attribute is clear from the context or it is irrelevant, we drop the attribute and use $\mathcal{N}$ instead.

Let us look at the example dataset in Fig. 2a. Figures 2b and 2c show $\mathcal{N}^{A_1}$ of the example dataset with $k = 4$ and $k = 6$. Each column in a neighborhood database represents an object while each row represents the neighborhood of an object. If an object occurs in a neighborhood, the corresponding cell is white. E.g., in Fig. 2b each horizontal white line, which denotes 4-NN, consists of exactly 4 cells.

With a formal analysis, the following properties hold for the neighborhood DBs, their proofs are provided in Sect. S.1 (in Supplementary):

Property 1 (Consecutive Objects). If two objects are in a neighborhood, all the objects in between them are also in that neighborhood. Let $\mathbf{o}, \mathbf{p}, \mathbf{q}, \mathbf{r} \in \mathscr{D}^{A_i}$ and $\mathbf{o} < \mathbf{p} < \mathbf{q}$. $\mathbf{o} \in k\text{-}NN(\mathbf{r}) \wedge \mathbf{q} \in k\text{-}NN(\mathbf{r}) \implies \mathbf{p} \in k\text{-}NN(\mathbf{r})$.

Property 2 (Consecutive Neighborhoods). If an object is in the neighborhood of two objects, then it is in the neighborhood of all the objects in-between them. Let $\mathbf{o}, \mathbf{p}, \mathbf{q}, \mathbf{r} \in \mathscr{D}^{A_i}$ and $\mathbf{o} < \mathbf{p} < \mathbf{q}$. If $\mathbf{r} \in k\text{-}NN(\mathbf{o}) \wedge \mathbf{r} \in k\text{-}NN(\mathbf{q}) \implies \mathbf{r} \in k\text{-}NN(\mathbf{p})$.

We can see Properties 1 and 2 in Fig. 2b. While the objects in a neighborhood are consecutive, an object only appears in consecutive neighborhoods, hence, they respectively form straight horizontal and vertical lines.

Theorem 1 (Recurrent Neighborhoods). *Clusters form square structures on the diagonal of the neighborhood databases.*

Proof. According to Properties 1 and 2, neighborhoods form continuous shapes. As we discuss in Sect. 2.1, a cluster is repeated as the neighborhoods of its objects, hence, the shape should be a square. Since each object is its own nearest neighbor, squares should be on the diagonal of the neighborhood DB.   □

The two squares in the lower right quarter of the Fig. 2b, which are circled in red, are the neighborhoods of the objects in the clusters that are denoted by ▲ and ● in Fig. 2a. Similarly, squares on the upper left quarter of Fig. 2c, which are circled in green and blue, are respectively from the clusters ■, ♦.

Property 3 (Baseline). If there are no clusters in a segment of the dataset, i.e., the objects are distributed uniformly, all of the object sets for that segment have the same amount of recurrency in the neighborhood DB, which is a function of k, cf. Figure S.1.

Property 4 (Transitivity). If an object o is not in the neighborhood of $p > o$, then it is not in the neighborhood of any $q > p$. Formally let $o, p, q \in \mathscr{D}^{A_i}$ and $o < p < q$, $o \notin k\text{-}NN(p) \implies o \notin k\text{-}NN(q)$.

In Fig. 2b, we can see that the 6th object occurs in neighborhoods 4 to 9. It does not occur in any neighborhood before 4 and not in any neighborhood after 9. We use the transitivity to limit the search space of our Algorithm, cf. Sect. 3, while the baseline property helps us to discard the segments without a cluster structure and to eliminate the artifacts that are caused by the transformation.

3   Mining the Neighborhoods

In this section we introduce our proposed algorithm CLON, which efficiently finds all subspace clusters by using ordered neighborhoods. A detailed explanation and pseudo code of the algorithm can be found in Section S.3 (in the supplementary).

If a cluster exists only in a subset of dimensions, the irrelevant dimensions hinder the search and the quality of the cluster. As we mentioned before, this is often the case in high dimensional spaces. To eliminate the negative effects of irrelevant dimensions, CLON follows a bottom-up strategy and exploits the fact that clusters in lower dimensional projections are supersets of clusters in higher dimensional spaces [9]. Hence, CLON finds the one dimensional clusters first and then iteratively refines them by evaluating their higher dimensional subsets.

In its first phase, CLON converts the relational database into a neighborhood database. Values in each dimension have to be sorted only once, after which the neighborhoods can be computed very fast by using a sliding window. A neighborhood DB is created for each of the dimensions. CLON already benefits from the ordered neighborhoods while creating the neighborhood DBs: It exploits the consecutiveness properties to store the neighborhood information so that instead of storing the whole neighborhoods, only the start and end of them are stored. This provides a dramatic memory saving, cf. Algorithm S.1.

In the second phase, the clusters in individual projections are found. As Theorem 1 states, clusters form recurrent object sets in the neighborhood DB. It has been proposed to find these recurrent object sets by using frequent itemset mining methods, however, since these methods are computationally expensive, only approximate methods are feasible for non-trivial datasets [2]. Our method CLON exploits the properties of ordered neighborhoods, so that it can mine these recurrent object sets in linear time instead of exponential time (both in the number of objects). Thanks to this speed improvement, *all* maximally large objects sets that are more recurrent than a certain threshold can be found in a reasonable time, cf. Sect. 4.3. Therefore in this phase, CLON can identify all of the recurrent objects sets in all individual projections, that is all one dimensional clusters, cf. Algorithm S.2.

In the third and the last phase, one dimensional clusters are used to find higher dimensional clusters. For any higher dimensional cluster, there are clusters that have more or equal number of objects, in the subsets of the dimensions of the original cluster [2,9,10]. For example, if a cluster exists in dimensions 1, 2, and 3, then a superset of its objects form clusters in dimension pairs (1,2), (2,3) and (1,3). Therefore, we can find higher dimensional clusters by refining lower dimensional ones, i.e. removing the objects that are not in the cluster. CLON refines each one dimensional cluster by checking whether any of its subsets are recurrent in other dimensions. If there are any, CLON continues with a depth-first search of higher dimensional clusters by traversing all the possible dimensions until none of the subsets are large or recurrent enough. Since CLON uses neighborhood DBs also for higher dimensional search, properties of ordered neighborhoods are exploited during this phase too. Lastly, the recurrent object sets are output as clusters along with the dimensions that they are recurrent in, cf. Algorithm S.3.

CLON requires two user-friendly parameters: (1) Minimum size of a cluster and (2) the neighborhood size. Selecting the minimum size should be straightforward, it depends on the size of the cluster that the user would like to find. The neighborhood size should be larger than the cluster size. Because of the recurrency oriented mining, CLON is robust to these parameters. A detailed discussion about parameter selection can be found in Section S.2.

3.1 Complexity Analysis

For a numeric $\mathscr{D}$ that has n objects and d attributes, the computational complexity of transforming it to the neighborhood DB is $O(d \times n \times \log(n) \times k)$. Thanks to the consecutiveness properties, we do not have to store the entire database but only the start and end of the neighborhoods are enough. Therefore, the space requirement for the neighborhood DBs is $O(n \times d)$ which is equivalent to the storage of the original $\mathscr{D}$. Exploiting the properties of the neighborhood DB allows us to find one dimensional clusters extremely efficiently, in $O(n)$. Therefore, the complexity of finding all one dimensional clusters is $O(d \times n)$. Total time to find all subspace clusters depends on the dimensionality of the clusters in the data. Although the computational complexity to search a cluster in a dimension is linear, the number of dimensions to search is combinatorial. Empirical results in Sect. 4.3 show that the scalability and efficiency of CLON is not hindered by the dimensionality of clusters.

4   Empirical Evaluation

4.1   Experimental Setup

We evaluate our method on heterogeneous datasets, i.e., datasets with different scales, both in terms of dimension scale and cluster distribution. We also conduct a set of experiments to evaluate the scalability of our method according to the number of objects, number of dimensions and the amount of noise. To see whether our method is a good fit for real world scenarios, we conduct experiments on very high dimensional real world datasets. We compare the quality of the clusters that are found by our method with the state of the art subspace clustering methods, such as CartiClus [2], FIRES [10], PROCLUS [1], STATPC [12], and SUBCLU [9], cf. Section S.4.

Object cluster finding capabilities of the methods are evaluated by the supervised F1 Measure. Where appropriate, we evaluate the subspace discovery capabilities by using the established E4SC score [14]. Runtime results are given for the performance versus scale experiments. More information about the experimental setup can be found in Section S.5. For the reproducibility of the experiments, a cross-platform implementation of the method and the information about the datasets are available on our website.

4.2   Heterogeneous Datasets

We measure the cluster discovery capabilities of our method on datasets that (1) have different scaling in different dimensions and (2) have clusters with different spreads. To evaluate these capabilities we generate two sets of datasets: One with different scales of dimensions and another one with clusters of different scales. Details about these datasets can be found in Section S.5.

Figures 3a and 4a show the F1 scores of the methods on a selected subset of these datasets. Different scale factors are indicated with different colors. On both set of experiments, CLON produces very stable and high quality clusters. CartiClus is comparable with CLON, which shows the effectiveness of nearest neighborhood based clustering. Radius based neighborhood evaluation of FIRES and distance sensitive projections of PROCLUS cannot cope with scaling. SUBCLU and STATPC produce high quality, although unstable, results thanks to their thorough search. Considering the execution times of CLON and its closest competitors, cf. Fig. 5, we can see the advantages of cluster refining through dimension search in ordered neighborhoods: CLON consistently produces better or comparable clusters in an order of magnitude less time.

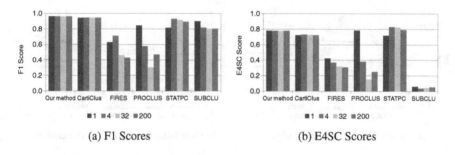

(a) F1 Scores   (b) E4SC Scores

Fig. 3. Quality scores of the methods on datasets with dimension of various scales

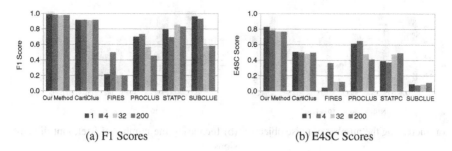

(a) F1 Scores   (b) E4SC Scores

Fig. 4. Quality scores of methods on datasets that have clusters of various scales

Figures 3b and 4b show the E4SC scores for the same sets of datasets. While some of the methods perform poorly on dimension detection, dimension finding capabilities of the most of them are in parallel with the object cluster finding capabilities. CLON performs better than CartiClus in dimension detection although they both mine recurrent object sets, cf. Fig. 4b. This is because CLON is fast enough to mine all subspace clusters instead of a sample of them.

4.3 Scalability Results

We conduct experiments to understand the scalability of our algorithm according to the data size, dimensionality, and the noise ratio. For these experiments we use the datasets that are used in the literature for similar comparative studies [2,14].

Figure 5a shows the run times of the methods on datasets with an increasing number of objects. This set of datasets include 5 different datasets which have approximately 1500, 2500, 3500, 4500, and 5500 objects and 20 dimensions. Our method scales well compared to the algorithms that search the combinations of dimensions.

Figure 5b shows the run times for different methods on datasets with an increasing number of dimensions. We conduct experiments on 6 different datasets that have approximately 1500 objects in 5, 10, 15, 25, 50, and 75 dimensions. In all of these datasets, each cluster exists in approximately 80 % of the dimensions. Therefore, these experiments also evaluate the performance of the algorithms

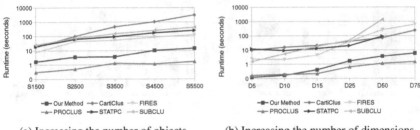

(a) Increasing the number of objects   (b) Increasing the number of dimensions

Fig. 5. Execution times

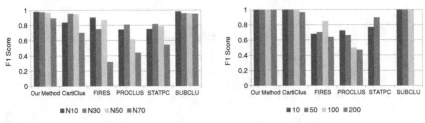

(a) Increasing the number of noise objects   (b) Increasing the number of irrelevant dimensions

Fig. 6. Noise detection capability experiments

regarding dimensionality of clusters. Our method scales well with the increasing number of dimensions while utilizing the information in combinations of dimensions. Note that the missing values indicates that the method did not complete in a practical time or failed because of excessive memory requirements.

Figure 6a shows the quality results for the datasets that contain 10 %, 30 %, 50 %, and 70 % noise. Our method and CartiClus can effectively discard noise thanks to their recurrency and neighborhood based foundation. Only SUBCLU is on par with these methods because of its thorough search which also makes it slower around 100 times than our method, cf. Fig. 5.

Figure 6b shows the capability of irrelevant dimension detection. A dataset which has 10 clusters hidden in subsets of 10 dimensions is generated. Then, 10, 50, 100 and 200 uniformly randomly generated dimensions are added to the dataset. Here again, we see that our method, CartiClus and SUBCLU can effectively discard irrelevant dimensions even though the meaningful structures are hidden in noise that is 20 times of its size. Some of the methods did not complete in meaningful time for some of the datasets, including SUBCLU for more than 100 dimensions.

4.4 Real World Datasets

To evaluate our method, we used two gene expression datasets, *Nutt* and *Alon*, along with a movie rating dataset, *movies*. *Alon* is a dataset of 2000 gene expression across 62 tissues from a colon cancer research. The tissues are grouped into 2 categories: 40 healthy tissues and 22 tissues with tumor [3]. *Nutt* dataset contains expressions of 1377 genes on 50 glioma tissue samples that are grouped into 4 different pathology categories [15]. High dimensional nature of both datasets make the clustering challenging even for the subspace clustering methods. Table 1 shows the F1 scores of the methods. "n/a" indicates that the method did not complete in a practical time or failed because of excessive memory requirements. These results show that, although our method searches the clusters in combinations of dimensions, it keeps being scalable even for very high dimensional datasets.

Table 1. F1 scores for gene expression datasets

	Alon	*Nutt*
Our method	**0.78**	**0.75**
PROCLUS	0.46	0.44
FIRES	0.52	0.55
SUBCLU	0.58	n/a
STATPC	n/a	n/a
CartiClus	n/a	n/a

We conduct an exploratory data analysis on a movie ratings dataset from GroupLens.[2] We use 10M movielens dataset which includes 10000054 ratings applied to 10681 movies by 71567 users from the website http://movielens.org. We use movies as objects and users as attributes. We start with finding the most similar 3 movies according to the user ratings. The result is, not surprisingly, the original Star Wars trilogy. When we lower the similarity threshold, we start to see a cluster of the Lord of the Rings Trilogy along with some other high profile movies in clusters of science fiction movies, action movies and crime movies. Some of the selected clusters are given in Table S.4a.

We continue our analysis by increasing the number of similar movies to 6. Considering the similarity of the original Star Wars trilogy, we search for a cluster of all 6 of the released Star Wars movies, with no luck. Actually, lack of this 6 pack of Star Wars movies cluster does not contradict with the general opinion of the fans of the franchise because the last 3 movies are not as popular as the originals. Table S.4b shows some of the interesting clusters of size 6, which includes a cluster of two most popular trilogies, a cluster of distopian movies, a cluster of popular thrillers, and a cluster of very popular classic movies.

5 Conclusion

In this paper, we tackle the problem of finding clusters in high dimensional databases. We make a formal analysis of ordered neighborhoods and their intrinsic

---

[2] http://grouplens.org/.

properties. We show that co-occurrences in local neighborhoods of objects are indicators of cluster formations, and hence, clusters can be found by mining recurrent neighborhoods. We propose a scalable algorithm that exploits these intrinsic properties to find all of the clusters and their relevant dimensions. Along with its scalability, key properties of our algorithm include its intuitive and user friendly parameters, its noise detection capabilities, its adaptivity to datasets with various scales, and its capability to exploit multiple (dis-)similarity measures.

Besides conducting experiments on scalability, noise detection, and adaptivity; we tested our method on some very high dimensional gene expression datasets. Further, we did an exploratory analysis on a movie rating dataset to show that our method is a good fit for real world scenarios. Finally, we supply a software tool that can be used to interpret the data for further analysis.

Acknowledgements. Emmanuel Müller is supported by Post-Doctoral Fellowships of the Research Foundation – Flanders (FWO). Further, this work is supported by the Young Investigator Group program of KIT as part of the German Excellence Initiative.

References

1. Aggarwal, C.C., Wolf, J.L., Yu, P.S., Procopiuc, C., Park, J.S.: Fast algorithms for projected clustering. SIGMOD Rec. **28**(2), 61–72 (1999)
2. Aksehirli, E., Goethals, B., Müller, E., Vreeken, J.: Cartification: a neighborhood preserving transformation for mining high dimensional data. In: ICDM, pp. 937–942, December 2013
3. Alon, U., Barkai, N., Notterman, D.A., Gish, K., Ybarra, S., Mack, D., Levine, A.J.: Broad patterns of gene expression revealed by clustering analysis of tumor and normal colon tissues probed by oligonucleotide arrays. PNAS **96**(12), 6745–6750 (1999)
4. Assent, I., Krieger, R., Müller, E., Seidl, T.: INSCY: indexing subspace clusters with in-process-removal of redundancy. In: ICDM, pp. 719–724, December 2008
5. Beyer, K., Goldstein, J., Ramakrishnan, R., Shaft, U.: When is nearest neighbor meaningful? In: Beeri, C., Bruneman, P. (eds.) ICDT 1999. LNCS, vol. 1540, pp. 217–235. Springer, Heidelberg (1998)
6. Emrich, T., Kriegel, H.-P., Mamoulis, N., Niedermayer, J., Renz, M., Züfle, A.: Reverse-nearest neighbor queries on uncertain moving object trajectories. In: Bhowmick, S.S., Dyreson, C.E., Jensen, C.S., Lee, M.L., Muliantara, A., Thalheim, B. (eds.) DASFAA 2014, Part II. LNCS, vol. 8422, pp. 92–107. Springer, Heidelberg (2014)
7. Goldstein, M.: k_n-nearest neighbor classification. IEEE TIT **18**(5), 627–630 (1972)
8. Jarvis, R., Patrick, E.: Clustering using a similarity measure based on shared near neighbors. IEEE TC **C–22**(11), 1025–1034 (1973)
9. Kailing, K., Kriegel, H.P., Kröger, P.: Density-connected subspace clustering for high-dimensional data. In: SDM, vol. 4. SIAM (2004)
10. Kriegel, H.P., Kröger, P., Renz, M., Wurst, S.: A generic framework for efficient subspace clustering of high-dimensional data. In: ICDM, pp. 8, November 2005
11. McCann, S., Lowe, D.: Local naive bayes nearest neighbor for image classification. In: CVPR, pp. 3650–3656, June 2012

12. Moise, G., Sander, J.: Finding non-redundant, statistically significant regions in high dimensional data: a novel approach to projected and subspace clustering. In: KDD, KDD 2008, pp. 533–541. ACM, New York (2008)
13. Müller, E., Assent, I., Günnemann, S., Krieger, R., Seidl, T.: Relevant subspace clustering: Mining the most interesting non-redundant concepts in high dimensional data. In: ICDM, pp. 377–386, December 2009
14. Müller, E., Günnemann, S., Assent, I., Seidl, T.: Evaluating clustering in subspace projections of high dimensional data. PVLDB **2**(1), 1270–1281 (2009)
15. Nutt, C.L., Mani, D.R., Betensky, R.A., Tamayo, P., Cairncross, J.G., Ladd, C., Pohl, U., Hartmann, C., McLaughlin, M.E., Batchelor, T.T., Black, P.M., Deimling, A.V., Pomeroy, S.L., Golub, T.R., Louis, D.N.: Gene expression-based classification of malignant gliomas correlates better with survival than histological classification. Cancer Res. **63**(7), 1602–1607 (2003)
16. Park, Y., Park, S., Jung, W., Lee, S.G.: Reversed CF: a fast collaborative filtering algorithm using a k-nearest neighbor graph. Expert Syst. Appl. **42**(8), 4022–4028 (2015)
17. Rodriguez, A., Laio, A.: Clustering by fast search and find of density peaks. Sci. **344**(6191), 1492–1496 (2014)
18. Schneider, J., Vlachos, M.: Fast parameterless density-based clustering via random projections. In: CIKM, CIKM 2013, pp. 861–866. ACM, New York (2013)
19. Sequeira, K., Zaki, M.: SCHISM: A new approach for interesting subspace mining. In: ICDM, vol. 0, pp. 186–193. IEEE Computer Society (2004)
20. Sim, K., Gopalkrishnan, V., Zimek, A., Cong, G.: A survey on enhanced subspace clustering. Data Min. Knowl. Disc. **26**(2), 332–397 (2013)
21. Weinberger, K., Saul, L.: Unsupervised learning of image manifolds by semidefinite programming. Int. J. Comput. Vis. **70**(1), 77–90 (2006)

Unsupervised Semantic and Syntactic Based Classification of Scientific Citations

Mohammad Abdullatif[✉], Yun Sing Koh, and Gillian Dobbie

Department of Computer Science,
The University of Auckland, Auckland, New Zealand
mabd065@aucklanduni.ac.nz,
{ykoh,gill}@cs.auckland.ac.nz

Abstract. In the recent years, the number of scientific publications has increased substantially. A way to measure the impact of a publication is to count the number of citations to the paper. Thus, citations are being used as a proxy for a researcher's contribution and influence in a field. Citation classification can provide context to the citations. To perform citation classification, supervised techniques are normally used. To the best of our knowledge there are no research that performs this task in a unsupervised manner. In this paper we present two techniques to cluster citations automatically without human intervention. This paper presents two novel techniques to cluster citations according to their contents (semantic) and the citation sentence styles (syntactic). The techniques are validated using external test sets from existing supervised citation classification studies.

1 Introduction

Authors cite papers for different reasons including identifying origin, providing background, giving credit, critiquing others' work and addressing the interestingness [8]. Citations are receiving more attention as the shear number of publications and subsequently the number of citations increase. This leads to the introduction of specialised research areas such as Bibliometrics concerned with utilising the existence of citations to evaluate the performance of researchers. One of the main citation assessment measures is the h-index. A scholar with an index of h means that the author has published h papers each of which has been cited at least h times. This measure presents both the productivity and impact of the researcher in his/her field using one number. A major drawback of this evaluation measure, as well as other similar ones, is that they focus on pure citation counts while ignoring the context or motivation of the citations. This leads to issues when comparing the work of researchers using the h-index. For example, consider two highly cited authors, one that has developed a new algorithm used by many in the field and the other published a paper with wrong, controversial claims, resulting in a high number of papers (and thus citations) disproving it. Using only measures that consider pure citation counts, both authors appear to have made the same contribution to their fields.

© Springer International Publishing Switzerland 2015
S. Madria and T. Hara (Eds.): DaWaK 2015, LNCS 9263, pp. 28–39, 2015.
DOI: 10.1007/978-3-319-22729-0_3

There has been some work in the area of citation analysis to address the pure citation count problem. A solution to this problem is to categorize citations based on their function or purpose. Such categorization will enable the creation of sophisticated evaluation models that are more accurate in reflecting the influence of a publication or the value of the work produced by a researcher.

Previous work tried to categorize citations by classifying them into predefined classes that are referred to as classification schemes [17]. Two approaches were used to perform citation classification. The first one uses manually created rules [18], and did not provide accurate classification. Moreover, the manually crafted rules were domain dependent and worked only for examples that have been accounted for by the domain expert crafting the rules. The second approach employed supervised machine learning techniques [4]. The classification accuracy achieved using this approach is reported to be much higher than the manually created rules. However, it was not possible to compare between the accuracies of the supervised machine learning techniques because they used different citation schemes for the classification. A supervised technique requires a manually coded dataset for training and testing. The difficulty lies in the fact that citation schemes proposed in the literature vary and thus a dataset created to train a classifier that uses one citation scheme cannot be used to train another classifier that uses a different citation scheme.

We are taking a different approach to citation classification. Instead of classifying citations to a pre-defined set of categories, we cluster a set of citations based on the reason behind them eliminating the need for a pre-defined classification scheme and the need for a training dataset, used in the other studies. Our unsupervised technique enables different levels of granularity when clustering, and does not require a scheme or a predefined training set.

In this paper we propose two novel techniques that can automatically cluster citations. As a brief overview, it is worth mentioning the most relevant inferences implemented aimed at solving the problem of citation clustering: First, syntactic inferences based on lexical distance measures. For instance, matching of consecutive subsequences, Jaro distance, Cosine distance, ISF specificity based on word frequencies extracted from corpora. Second, semantic inferences focused on semantic distances between concepts. These inferences implement several well-known WordNet-based similarity measures such as verb similarities between verbs.

The rest of the paper is organised as follows. Related work is outlined in Sect. 2. In Sect. 3 we describe our approaches to automatically cluster citation sentences. Section 4 presents our results and evaluation. We conclude and describe future work in Sect. 5.

2   Related Work

Drawing on classical bibliometrics papers, many researchers have focused their research on citation frequency and citation impact and applied it in different domains. Others have studied the association between the citation frequency of articles and various characteristics of journals, articles, and authors. Traditional citation analysis treats all citations equally. In reality, not all citations are equal. Some researchers consider location to be a factor affecting the relative importance of a citation. For example, the work by Herlach [11] found that a publication cited in the introduction or literature review section and re-mentioned in the methodology or discussion sections is more likely to make a greater contribution to the citing publication than others that have been mentioned only once. The stylistic aspects of a citation also matter. Bonzi [2] distinguished between a number of broad categories of citations. Examples of these include citations that are not specifically mentioned in the text and those barely mentioned or those with direct quotations.

Work on developing citation schemes goes back to the mid 1960's and was started by Garfield [9]. Garfield's work involved looking at the intentions of authors when citing and resulted in 15 different classes for citations. Other schemes have been developed since then and the number of classes vary between 3 and 35; some of which have mutually exclusive classes and others allow multiple classes to be attached to a single citation [21]. As an example, Moravcsik and Murugesan formed a scheme comprising four dimensions [17]. A citation can belong to one or more dimensions, which are presented in Table 1.

Table 1. Example of dimensions in a citation scheme

Dimension	Meaning
Conceptual	Concept or theory used
Operational	A tool or physical technique used
Organic	The reference is truly needed for the understanding of the paper
Perfunctory	The reference is just an acknowledgment of previous work
Evolutionary	The paper is a continuation of the cited work or the cited work is a foundation for the paper
Juxtapositional	The paper is an alternative to the work cited
Confirmative	The referenced paper is correct/the current paper agrees with the referenced paper
Negational	The paper contradicts or diminishes the referenced paper

All of the schemes described in the literature have been created by manually analyzing citation sentences and forming a scheme. The manually formed schemes are not compatible making it difficult to compare between the accuracy of classifiers trained using the different schemes.

3 Citation Clustering

In our approach we used a citation clustering approach to classify the citations. Ideally each cluster within the clustering results would represent a dimension/category in citation scheme. We looked at two different approaches to cluster citation sentences: Semantic-based models and Syntactic-based models. In this section we describe the measures used in each of these models.

3.1 Semantic-Based Model

The proposed semantic-based model aims to cluster citations by meaning. The introduced model consists of semantic-based term analysis and a semantic-based similarity measure. This work presents an ongoing project for semantic-based analysis of terms to enhance the quality of the citation clustering. This work introduces deep semantic analysis for concepts extracted from WordNet to enhance the quality of text clustering.

To automatically cluster a citation sentence, we concentrate on the verbs in each of the sentences. A verb is a word (part of speech) that in syntax conveys an action (bring, read, walk, run, learn), an occurrence (happen, become), or a state of being (be, exist, stand). We use verbs as the main indicator of the relationship between the citing and cited work. There are two approaches for selecting a verb from a sentence. The first one is using a supervised learning technique for verb selection such as Support Vector Machines [3]. The benefit of supervised learning is that it is easy to train given a training dataset. However, selecting the appropriate features for training is not always trivial. Also, a training dataset is not available. The second approach is using a systematically formed set of rules for deciding the most relevant verb to select. We opted for the second heuristic approach because it gives us human understandable reasons for why one verb was selected from a sentence as opposed to another. The rules also enable us to incrementally amend them as necessary to improve the performance of selecting the relevant verbs. Each citation sentence is processed through a semantic role labeler followed by our verb analysis process to extract the main verb relevant to the citations in the sentence. For example, passing the following sentence through our verb analysis process yields the verb "expand".

> We **expand** on the work of James et al. [34] validated by the machine learning community.

The relevant verb extraction phase produces a set of relevant verbs $V = \{v_1, v_2, v_3, \ldots, v_k\}$ where k represents the total number of verbs found. The set of relevant verbs may not be unique. Each of the relevant verbs is derived from one citation sentence. Our citation relevant verb selection technique is described in [1].

Verb Similarity Computation. Given the set of relevant verbs V we need to sort the verbs in some methodological manner. To do so we measure the similarity between the verbs in the set. If we have k verbs, we need to compute the similarity between each verb and the rest of the verbs within the set. Thus we will have to make 2^k similarity comparisons. For example, if we have a set of verbs $\{v_1, v_2, v_3\}$, we will calculate the similarity for (v_1, v_2), (v_1, v_3), (v_2, v_3), (v_2, v_1), (v_3, v_1), (v_3, v_2). Each of the pairs will produce a similarity value, $s(v_i, v_j)$ where i and j are the index of two verbs where $i \neq j$. We then store the similarity values in a similarity matrix M of dimension $k \times k$.

We calculate the similarity between each pair of verbs. In our research the similarity measures we use are calculated using the graph structure of WordNet. A number of similarity measures that utilize WordNet for similarity computation have been proposed in the literature [13,19]. WordNet is a hand-crafted lexical database for the English language. The database groups words into synonym sets called synsets [6]. Synsets are linked together with many semantic links including the super-subordinate relation (also called hyperonymy, hyponymy or IS-A relation) and the part-whole relation (also called Meronymy). Synsets are organized in hierarchies where words become more abstract as you go up to the root node (called Entity for Nouns, and a simulated root node is created for verbs).WordNet has been used in a wide variety of applications including Information Retrieval [23] and Natural Language Processing tasks [16].

We used three verb similarity measures: Path Similarity, Wu-Palmer similarity [24], and Leacock Chodorow similarity [13] to measure the similarity between the pairwise verbs. We then present a comparison between them in order to select the most suitable for our application in Sect. 4. We note that all three similarity measures are symmetrical, thus the similarity value for $s(v_1, v_2)$ and $s(v_2, v_1)$ are the same. In this context the computational complexity of our similarity calculations is $\frac{2^k}{2}$.

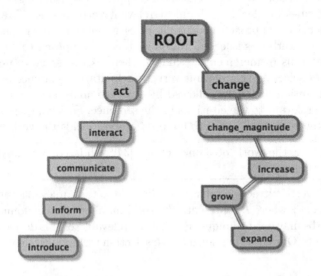

Fig. 1. WordNet hierarchy for verbs "expands" and "introduced"

The Path similarity measure is calculated as shown in Eq. (1) where $L(a, b)$ is the shortest path connecting verbs a and b in the IS-A (hypernym/hypnoym) taxonomy. The shortest path is based on the number of edges. Figure 1 shows the IS-A hierarchy for verbs "introduced" and "expands". In this example, the shortest path between the two verbs is 10. Using Eq. 1, the Path Similarity score is $1/(10 + 1) = 0.0909$.

$$PATH_{(a,b)} = \frac{1}{L_{(a,b)} + 1} \tag{1}$$

The Wu-Palmer similarity measure [24] is calculated using Eq. 2. $D(LCS(a, b))$ is the depth of the lowest common subsumer (deepest common ancestor/parent node of a and b) and $L(a, b)$ is the shortest path between nodes a and b. Using Eq. 2, the Wu-Palmer similarity score between verbs "introduced" and "expands" is as follows: The LCS of both verbs is the root node with a depth of 1. The shortest path is 10.

$$WUP_{(a,b)} = \max \left[\frac{2 \times D(LCS_{(a,b)})}{L_{(a,b)} + 2 \times D(LCS_{(a,b)})} \right] \tag{2}$$

Therefore the score is $(2 \times 1)/(10 + (2 \times 1) = 0.1667$.

The Leacock Chodorow similarity measure [13] is shown in Eq. 3 where $L(a, b)$ is the shortest path connecting a and b and $D_{\max}$ is the maximum depth from the root to the deepest leaf in the hierarchy in which the verbs occur. The distance for cases when a and b are the same verb will result in an infinite similarity score (log of zero). Therefore we always add 1 to the shortest path distance to avoid such a scenario. Using Eq. 3, the Leacock Chodorow similarity score between verbs "introduced" and "expands" is as follows: $L(a, b)$ is 10. $D_{\max}$ is 13. This is based on WordNet version 3.0. Therefore, the Leacock Chodorow similarity score is $-\log((10 + 1)/2 \times 13) = 1.0609$.

$$LCH_{(a,b)} = \max \left[-log(\frac{L_{(a,b)}}{2D_{\max}}) \right] \tag{3}$$

Clustering of Verb Similarity Vectors. Calculating the similarity between all the pair of relevant verbs from the dataset of citation sentences results in a similarity matrix. Each row in the similarity matrix M represents the similarity between one verb and all the other verbs in the dataset. A row in the matrix is known as a similarity vector, SV_{v_k}, for its associated verb, v_k.

We cluster the vectors representing citations using the well-known clustering algorithm k-means [15]. We chose to use k-means because of its intuitive nature and its ability to allow us to specify the number of clusters we want. The ability to set a k value allows us to have control over the number of clusters which enables us to evaluate our technique using test sets from the supervised based citation classification work described in the literature. The end result of the clustering technique is a set of clusters $C = \{c_1, c_2, \ldots, c_k\}$. Each cluster c contains a set of verbs $\{v_j, v_{j+1}, \ldots, v_k\}$. Sentences associated with the verbs are clustered together. Each cluster represents one dimension or reason for citing.

3.2 Syntactic-Based Model

Instead of concentrating only on a particular verb, the entire citation sentence could be used to form the similarity matrix. There are various measures such as Jaro distance, Ratio, and Levenshtein distance that allow for character-based string similarity computation. We adapted these techniques to use words instead of characters as tokens. We then preprocess the citation sentences by performing stemming and stop words removal. Stemming is the process of reducing words to a base or root form. For example, words "advanced", "advancing", "advance" would all be reduced down to "advanc" after stemming is applied. Many stemming algorithms are available in the literature, we used Porter's stemmer [20] to stem words in our experiments. The intuition behind this is authors in the same field tend to write and cite literature in a similar fashion, thus looking at the syntactic nature of the sentences is a viable way of differentiating the types of citations.

Syntactic Similarity Computation. Current distance techniques measure similarity between the sentences based on the words used. However each of these techniques, produces similarity based on a slightly different approximation. The choice of the similarity distance can be dependent on the corpus used and allowable error estimates.

Here we used four different measures. The first measure is Levenshtein edit distance [14]:

$$
Lev_{a,b}\,(i,j) = \begin{cases} \max(i,j) & \text{if } \min(i,j) = 0, \\ \min \begin{cases} lev_{a,b}(i-1,j)+1 \\ lev_{a,b}(i,j-1)+1 \\ lev_{a,b}(i-1,j-1)+1_{(a_i \neq b_j)} \end{cases} & \text{otherwise.} \end{cases} \tag{4}
$$

where $lev_{a,b}(i-1,j)+1$ corresponds to a deletion operation, $lev_{a,b}(i,j-1)+1$ to an insertion and $lev_{a,b}(i-1,j-1)+1_{(a_i \neq b_j)}$ to a match or mismatch. $1_{(a_i \neq b_j)}$ is equal to 0 when $a_i = b_j$ and equal to 1 otherwise, where a, b are citations within the sentences.

The second measure is Jaro Measure [12]:

$$
Jaro(s,t) = \frac{1}{3} \cdot \left(\frac{|s'|}{|s|} + \frac{|t'|}{|t|} + \frac{|s'| - T_{(s',t')}}{|s'|} \right) \tag{5}
$$

where s is a string with $a_1 \ldots a_K$ characters, t is a string with $b_1 \ldots b_L$ words. Here a character a_i in s is in common with t, if there is a $b_j = a_i$ in t such that $i-H \leq j \leq i+H$, where $H = \frac{min(|s|,|t|)}{2}$, s' has $a'_1 \ldots a'_K$ words that are common between s and t in the same order they appear in s, t' has $b'_1 \ldots b'_K$ words that are common between t and s in the same order they appear in t, $T_{(s',t')}$ is half the number of transpositions for s' and t'.

The third measure is Ratio:

$$
Ratio(s,t) = \frac{2 \cdot M}{T} \tag{6}
$$

Table 2. Similarity vectors

SV_{v_1}	$s(v_1, v_1)$	$s(v_1, v_2)$	$\ldots$	$s(v_1, v_k)$
SV_{v_2}	$s(v_2, v_1)$	$s(v_2, v_2)$	$\ldots$	$s(v_2, v_k)$
$\ldots$				
SV_{v_k}	$s(v_k, v_1)$	$s(v_k, v_2)$	$\ldots$	$s(v_k, v_k)$

where M is the number of matches and T is the number of elements in both s and t.

The final measure is Hamming distance, whereby we calculate the distance between two sentences by the number of positions at which the corresponding words are different [10]. On top of that we also used a TF-ISF based method that takes into account the word frequencies adjusted by the factor to account for very frequent words and computes the Cosine similarity between the resulting TF-ISF vectors.

Clustering of Syntactic Similarity Vectors. Similar to the clustering of verb similarity, the similarity between all the pairs of citation sentences in the dataset is stored in a similarity matrix. Each row in the similarity matrix M represents the similarity between one sentence and all the other sentences in the dataset. A row in the matrix is known as a similarity vector, SV_{s_k}, for its associated sentence, s_k. We then built a matrix similar to Table 2 and used k-means for clustering similar to the approach in Sect. 3.1.

4 Experiments, Results and Evaluation

To evaluate the viability of our technique we compared our technique to human annotated datasets. We used the weighted F-measure [7] to evaluate the performance of our techniques. The experiments were conducted on two different datasets used by Dong et al. [5] and Teufel et al. [22].

In the first set of experiments using the dataset from Dong et al. two different numbers of clusters were chosen: 3 and 6. These were chosen because in their schema they divide the citations into the 3 and 6 general categories. The corpus comes from the ACL (Association for Computational Linguistics) Anthology, a comprehensive collection of scientific conference and workshop papers in the area of computational linguistics and language technology. The authors randomly chose papers from proceedings of the ACL conference in 2007 and 2008. We compared the accuracy of our technique using the F-measure against the human annotated datasets. The results of the F-measures are shown in Figs. 2 and 3. Table 3 presents the name of each syntactic based experiment, the variation and the shorthand. We used both syntactic and semantic based approaches. The semantic based results are shown in the last three bars in the figures. In all the experiments we repeated k-means 200 times and averages are reported along with one standard deviation as an error bar with each average.

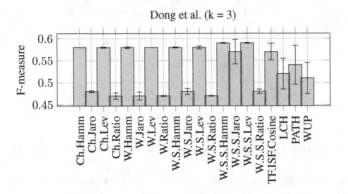

Fig. 2. F-measure on dataset from Dong et al. using k= 3

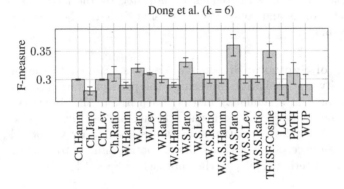

Fig. 3. F-measure on dataset from Dong et al. using k= 6

The number of independent experiments we ran for each test set is 20 where the difference between the experiments is the way the similarity vectors are formed. There are eight similarity measures that are used to form the similarity vectors. All of the measures except TF-ISF Cosine, WUP, PATH and LCH have four variations. The variations for each of the four measures include (1) characters: similarity between strings at the character level, (2) words: similarity between strings at the word level, (3) words stemmed: words are stemmed before the similarity between strings is computed, and (4) words stemmed and stop words removed: stop words are removed and the remaining words stemmed before the similarity is computed.

Table 3. Shorthands used to represent the experiments being run on each test set

Similarity measure	Hamming distance	Levenshtein distance	Jaro distance	Ratio
Characters	Ch. Hamm	Ch. Lev	Ch. Jaro	Ch. Ratio
Words	W. Hamm	W. Lev	W. Jaro	W. Ratio
Stemmed words	W.S. Hamm	W.S. Lev	W.S. Jaro	W.S. Ratio
Stemmed/Stopwords removed Words	W.S.S. Hamm	W.S.S. Lev	W.S.S. Jaro	W.S.S. Ratio

In the second set of experiments using the corpus of Teufel et al. three different numbers of clusters were chosen: 3, 4, and 12. This was chosen as in their schema they divide the citations into 3, 4, and 12 general categories. The dataset comes from a corpus of 360 conference articles in computational linguistics, drawn from the Computation and Language E-Print Archive. The results were compared against on the human annotated datasets and are shown in Figs. 4, 5 and 6. In total, 20000 clustering runs are performed (20 independent experiments × 200 runs × 2 Dong datasets × 3 Teufel datasets).

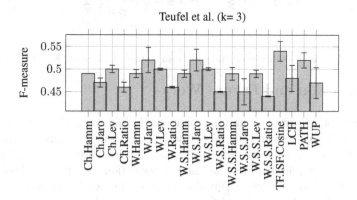

Fig. 4. F-measure on dataset from Teufel et al. using k= 3

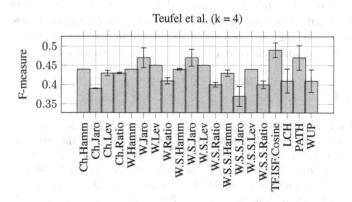

Fig. 5. F-measure on dataset from Teufel et al. using k= 4

From both sets of experiments we can see that the results are very consistent, whereby the top three performing measures were, Jaro, TF-ISF, and Path. Overall all the top performing measures performs well on the dataset across the experiments. This research is definitely a good starting point to how unsupervised automatic citation classification techniques can be built.

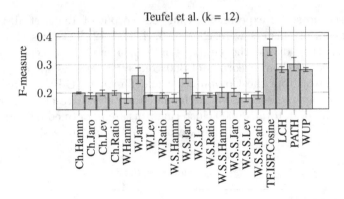

Fig. 6. F-measure on dataset from Teufel et al. using k = 12

5   Conclusion and Future Work

Citation Classification plays an important role in improving the current citation based research evaluation techniques such as the h-index. Most existing citation classification techniques perform the classification based supervised learning algorithms that require training data and the selection of a citation classification scheme. Comparison between the performance of the existing techniques is difficult because the techniques use different schemes and the supervised algorithms are therefore trained using different training sets. In this paper we proposed two novel techniques that can automatically cluster citations in an unsupervised manner based on semantic and syntactic inferences without the need for a citation scheme or a training corpus and we show that we can achieve reasonable results. Future work on the research includes combining both semantic and syntactic inference together and using an ensemble method to provide a better citation clustering and generate a citation scheme automatically.

References

1. Abdullatif, M., Koh, Y.S., Dobbie, G., Alam, S.: Verb selection using semantic role labeling for citation classification. In: Proceedings of the 2013 Workshop on Computational Scientometrics: Theory and Applications CompSci 2013, pp. 25–30. ACM, New York (2013). http://doi.acm.org/10.1145/2508497.2508502
2. Bonzi, S.: Characteristics of a literature as predictors of relatedness between cited and citing works. J. Am. Soc. Inf. Sci. **33**(4), 208–216 (1982). http://dx.doi.org/10.1002/asi.4630330404
3. Cortes, C., Vapnik, V.: Support-vector networks. Mach. Learn. **20**, 273–297 (1995)
4. Dong, C., Schäfer, U.: Ensemble-style self-training on citation classification. In: Proceedings of 5th International Joint Conference on Natural Language Processing, pp. 623–631. Asian Federation of Natural Language Processing, Chiang Mai, November 2011
5. Dong, C., Schäfer, U.: Ensemble-style self-training on citation classification. In: Proceedings of the 5th International Joint Conference on Natural Language Processing, pp. 623–631. Association for Computational Linguistics, November 2011

6. Fellbaum, C.: WordNet: An Electronic Lexical Database. Language, Speech, and Communication, MIT Press, Cambridge (1998)
7. Fung, B.C., Wang, K., Ester, M.: Hierarchical document clustering using frequent itemsets. In: SDM, vol. 3, pp. 59–70. SIAM (2003)
8. Garfield, E.: Journal impact factor: a brief review. Can. Med. Assoc. J. **161**(8), 979–980 (1999)
9. Garfield, E., Stevens, M.E., Giuliano, V.E., Heilprin, L.B.: Can citation indexing be automated? Essay Inf. Sci. **1**, 189–192 (1965)
10. Hamming, R.W.: Error detecting and error correcting codes. Bell Syst. Tech. J. **29**(2), 147–160 (1950)
11. Herlach, G.: Can retrieval of information from citation indexes be simplified? multiple mention of a reference as a characteristic of the link between cited and citing article. J. Am. Soc. Inf. Sci. **29**(6), 308–310 (1978). http://dx.doi.org/10.1002/asi.4630290608
12. Jaro, M.A.: Advances in record-linkage methodology as applied to matching the 1985 census of tampa, florida. J. Am. Stat. Assoc. **84**(406), 414–420 (1989)
13. Leacock, C., Chodorow, M.: Combining local context and WordNet similarity for word sense identification. In: Fellbaum, C. (ed.) WordNet: An Electronic Lexical Database, pp. 305–332. MIT Press (1998)
14. Levenshtein, V.I.: Binary codes capable of correcting deletions, insertions and reversals. Sov. Phys. Dokl. **10**, 707 (1966)
15. MacQueen, J., et al.: Some methods for classification and analysis of multivariate observations. In: Proceedings of the Fifth Berkeley Symposium on Mathematical Statistics and Probability, vol. 1, p. 14, California (1967)
16. Morato, J., Marzal, M.A., Lloréns, J., Moreiro, J.: Wordnet applications. In: Proceedings of the 2nd Global Wordnet Conference, vol. 2004 (2004)
17. Moravcsik, M., Murugesan, P.: Some results on the function and quality of citations. Soc. Stud. Sci. **5**(1), 86 (1975)
18. Nanba, H., Okumura, M.: Towards multi-paper summarization using reference information. In: International Joint Conference on Artificial Intelligence, vol. 16, pp. 926–931. Lawrence Erlbaum Associates Ltd (1999)
19. Patwardhan, S., Pedersen, T.: Using WordNet-based context vectors to estimate the semantic relatedness of concepts. In: Proceedings of the EACL 2006 Workshop on Making Sense of Sense: Bringing Computational Linguistics and Psycholinguistics Together, pp. 1–8, Trento, April 2006
20. Porter, M.F.: An algorithm for suffix stripping. Prog. Electron. Lib. Inf. Syst. **14**(3), 130–137 (1980)
21. Radoulov, R.: Exploring automatic citation classification. University of Waterloo, Waterloo (2008)
22. Teufel, S., Siddharthan, A., Tidhar, D.: Automatic classification of citation function. In: Proceedings of the 2006 Conference on Empirical Methods in Natural Language Processing EMNLP 2006, pp. 103–110. Association for Computational Linguistics, Stroudsburg, PA (2006). http://dl.acm.org/citation.cfm?id=1610075.1610091
23. Varelas, G., Voutsakis, E., Raftopoulou, P., Petrakis, E.G., Milios, E.E.: Semantic similarity methods in wordnet and their application to information retrieval on the web. In: Proceedings of the 7th Annual ACM International Workshop on Web Information and Data Management, pp. 10–16. ACM (2005)
24. Wu, Z., Palmer, M.: Verbs semantics and lexical selection. In: Proceedings of the 32nd Annual Meeting on Association for Computational Linguistics ACL 1994, pp. 133–138. Association for Computational Linguistics, Stroudsburg (1994)

Data Mining

HI-Tree: Mining High Influence Patterns Using External and Internal Utility Values

Yun Sing Koh[1(✉)] and Russel Pears[2]

[1] Department of Computer Science, University of Auckland,
Auckland, New Zealand
ykoh@cs.auckland.ac.nz
[2] School of Computing and Mathematical Sciences,
AUT University, Auckland, New Zealand
rpears@aut.ac.nz

Abstract. We propose an efficient algorithm, called HI-Tree, for mining high influence patterns for an incremental dataset. In traditional pattern mining, one would find the complete set of patterns and then apply a post-pruning step to it. The size of the complete mining results is typically prohibitively large, despite the fact that only a small percentage of high utility patterns are interesting. Thus it is inefficient to wait for the mining algorithm to complete and then apply feature selection to post-process the large number of resulting patterns. Instead of generating the complete set of frequent patterns we are able to directly mine patterns with high utility values in an incremental manner. In this paper we propose a novel utility measure called an influence factor using the concepts of external utility and internal utility of an item. The influence factor for an item takes into consideration its connectivity with its neighborhood as well as its importance within a transaction. The measure is especially useful in problem domains utilizing network or interaction characteristics amongst items such as in a social network or web click-stream data. We compared our technique against state of the art incremental mining techniques and show that our technique has better rule generation and runtime performance.

Keywords: Prefix-tree · High influence patterns · FP-growth

1 Introduction

The accumulation of massive data sets has increased substantially as the ability to record and store more data has increased rapidly. This has led to the development of various techniques that enable us to obtain valuable information from massive data sets. Although frequent pattern mining [4] plays an important role in data mining applications, there is one major limitation: it treats all items with the same importance. However, in reality, each item in a supermarket for example, has a different importance based on its profit margin. In turn, items having high and low selling frequencies may have low and high profit values,

© Springer International Publishing Switzerland 2015
S. Madria and T. Hara (Eds.): DaWaK 2015, LNCS 9263, pp. 43–56, 2015.
DOI: 10.1007/978-3-319-22729-0_4

respectively. For example, some frequently sold items such as bread, milk, and chips may have lower profit values compared to that of infrequently sold higher profit value items such as vodka and caviar. This means that finding only frequent patterns in a dataset cannot fulfill the objective of identifying the most valuable customers that contribute to the major part of the total profits in a retail business. Focusing on high utility items and itemsets has two major benefits: firstly, a smaller set of potentially useful patterns and rules are presented to the user; secondly, the execution time overhead for pattern and rule generation are substantially reduced. Our experimental results clearly illustrate the advantages of the influence factor measure in achieving these two benefits.

A recent approach using a utility mining [5, 7, 11] model was proposed to discover knowledge from a dataset. Using utility mining, several important business objectives such as maximizing revenue or minimizing inventory costs [1] can be considered, and in turn knowledge about itemsets contributing to the majority of the profit can be discovered. Utility mining allows us to represent real world market data. This can be applied in other areas, such as stock tickers, web server logs, network traffic measurements, and data feeds from sensor networks. However, one limitation of these techniques is that we need to have predefined utility values from domain experts for each item within the dataset. This may be impossible if no domain knowledge is available. Moreover, it is difficult for domain experts to constantly update these utility values. Unlike current techniques, we propose a technique that automatically determines the utility value as a dataset evolves incrementally. In situations where there is a lack of domain knowledge, our technique derives an influence factor for each item based on its connections with other items in the dataset. The influence factor measure is based on two factors: the external utility and internal utility of an item. For internal utility, an item that appears multiple times in a single transaction is given a higher internal weighting compared to an item that only appearing once within a transaction. This gives more importance to items that appear frequently within a single transaction over items that appear less frequently. For external utility, we consider the strength of the item based on connections from other items associated with it or co-occurring from it. Such a basis for the derivation of influence factor is intuitive for domains such as social networks and web server logs. In a web server log for an online retailer, we would be interested in identifying how many times a particular product page is viewed in a single session, which will represent internal weighting component. We would also be interested in product pages browsed through together with that particular product, which would account for the external weighting component. Products that are constantly viewed together with co-associated products could be expected to have a major impact on marketing strategies. Most previous work in these areas is based on static datasets and hence do not investigate strategies for constantly changing data. In such dynamic data environments one or more transactions could be deleted, inserted, or modified in the dataset. Using a form of incremental pattern mining, we can use previously mined results, and thereby avoid unnecessary computation when the dataset is updated or the mining threshold used is changed.

Motivated by the limitations existing in previous approaches, we propose an efficient solution to overcome the problems in existing work by proposing a technique to automatically compute the utility of a pattern with the influence factor measure. A further contribution of this research is a new tree structure called the HI-Tree that supports the "build once mine many times" property for high utility pattern mining in an incremental data environment.

The remainder of this paper is organized as follows. In Sect. 2, we describe related studies. In Sect. 3, we describe the high influence item problem. In Sect. 4, we describe our proposed tree structures for high utility pattern mining. In Sect. 5, our experimental results are presented and analyzed. In Sect. 6 we conclude the paper and discuss the future research.

2   Related Work

In utility itemset mining or utility pattern mining, every item in an itemset is associated with an additional value, an external utility is attached to an item, showing its utility (e.g. profit). With such a utility-based dataset, high utility itemsets (or patterns) are mined, including those satisfying the minimum utility. Mining high utility itemsets is much more challenging than discovering frequent itemsets, because the fundamental downward closure property in frequent itemset mining does not hold in utility itemsets. Several algorithms are available. UMining [11] was proposed for mining high utility patterns, but it cannot extract the complete set. IHUP [1] maintains the high utility patterns in an incremental environment, avoiding multiple scans of the dataset. UP-Growth [10] also uses a tree structure, UP-Tree, to mine high utility patterns. Compared to IHUP, UP-Growth is more efficient, since it further reduces the number of promising patterns which cannot be pruned in IHUP. The Two-Phase [8] algorithm was developed to find high utility itemsets using the downward closure property of Apriori. The authors have defined the transaction-weighted utilization (TWU) property. For the first dataset scan, the algorithm finds all the one-element transaction-weighted utilization itemsets, and based on that result, it generates the candidates for two-element transaction-weighted utilization itemsets. This algorithm suffers from the same problem as the level-wise candidate generation-and-test methodology. The expected utility approach usually overestimates the importance of the itemset, especially in the beginning stages, where the number of candidates approaches the number of all the combinations of items. This trait renders the approach impractical whenever the number of distinct items is large and the utility threshold value is low. Furthermore, updating the utility values is a difficult task in an incremental dataset.

Research into developing techniques for incremental and interactive mining has been carried out in the area of traditional frequent pattern mining, and they have shown that incremental prefix-tree structures such as CanTree [6], CP-tree [9] are quite efficient in using currently available memory. Current efficient dynamic dataset updating algorithms are designed for mining datasets when data deletion is performed frequently in any dataset subset. However, these solutions are not applicable to incremental or interactive high utility pattern mining approaches. Therefore, we propose new tree structures for incremental high utility pattern mining. Our algorithm inserts items into a tree based on the influence factor of the item, and uses the FP-growth mining operation.

3   Preliminaries

In this section, we present some background concepts that are used throughout the paper. Moreover, we formalize the definition of influence factor of a particular item.

3.1   Online Frequent Itemsets Mining

A transaction data stream $S = \{T_1, T_2, \ldots, T_L\}$ is a continuous sequence of transactions where L is the tid of the latest incoming transaction T_L. A transaction sliding window, SW, in the data stream is a window that slides forward for every transaction. The window at each slide has a fixed number, w, of transactions where w is the size of the window. In the current window $SW_{L-w+1} = \{T_{L-w+1}, T_{L-w+2}, \ldots, T_L\}$, $L - w + 1$ is the window identifier. The support of an itemset X over SW, denoted as $supp(X)_{SW}$ is the number of transactions in SW that contains X as a subset. An itemset X is called a frequent itemset if $supp(X)_{SW} \geq minsup$. Given a sliding window SW, and a $minsup$, the problem of online mining of frequent itemsets in recent transaction data streams is to mine the set of all frequent itemsets using one scan.

3.2   Influence Factor

In our algorithm we use confidence measure of a discovered itemset to measure its influence. Given a high influence item x, it may lead to a potentially high impact rule $x \rightarrow y$ whereby the purchase of a high influence (high impact) item x triggers the purchase of another high influence item y. We divide the influence factor for an item into two components: internal and external utility values. An item x receives an internal utility value based on the number of times it occurs within a single transaction. Most association rule mining techniques only consider whether an item appears or not within a transaction in a binary sense. The information arising from multiple occurrences of an item within a single transaction is disregarded. For example when generating rules such as a customer who buy bread $\rightarrow$ buy milk we do not consider the quantity of purchase of each of the items, such information may lead to more interesting rules being uncovered. We include this information by including an internal utility component. An item x also receives an external utility value based on the connections between item x and its neighborhood. We measure the external utility of an item x based on the sum of the conditional probability gain from x with the set of items that co-occurred with x.

This measure is applicable to many different types of applications and datasets including web click stream data and network data. To analyze these types of data it is necessary to utilize connections between the items whilst considering the importance of the item by itself. Other types of applications that would benefit are social network analysis applications for Twitter where the importance (weight) of a person is measured on the basis on the number of followers re-tweeting the person's messages (number of connections) and the number of tweets produced by a person (individual importance of a person). Thus it is reasonable for us to combine these two factors to generate an influence factor when no domain knowledge can be found or when the data is constantly evolving. For example the number of followers may constantly change and the number of tweets produced may also fluctuate over time. Given a dynamic environment setting a constant utility value would not suffice. Our influence factor allows us to mimic the importance of an item whilst adapting to the network and data changes.

Definition 1 (Internal Utility). *The internal utility of an item,* x, $I(x)$, *is given by*

$$I(x) = \frac{1}{|t|} \sum_{\exists x \in t} \frac{|x \in t|}{|F|}$$

where t *a transaction and* $t \in SW$, *and* F *is the largest set of an item with multiple occurrences within the same transaction in the neighborhood of* x.

For example given that we have three transactions $\{a, a, b, b, b, c\}$, $\{a, b, c\}$, $\{a, c\}$ in SW. If we are considering the $I(b)$, $|F|$ is 3 whereby item b occurred at most three times within the SW. The $I(b)$ is $\frac{1}{3} \left(\frac{3}{3} + \frac{1}{3} + \frac{0}{3} \right) = \frac{4}{9}$.

Definition 2 (Directed Influence Factor). *The directed influence factor of an item,* x

$$\epsilon_x = I(x) \sum_{y \in N_x} \left(- Pr(xy) \log Pr(y|x) - Pr(x, \neg y) \log Pr(\neg y|x) \right)$$

where $Pr(x|y)$ *is the conditional probability for* x *and* y *and* $\neg x$ *represents the absence of* x. $I(x)$ *is the internal utility for item* x.

Here N_x represents the neighborhood of x, which is the set of items that occur with x. For example, given in the sliding window item x co-occurs with a, b, c, this means that $N_x = \{a, b, c\}$ as shown in Fig. 1. Here item x received an external utility value from its neighborhood:

$$E(x) = \sum_{y = \{a, b, c\}} \left(- \Pr(xy) \log \Pr(x|y) - \Pr(\neg xy) \log \Pr(\neg x|y) \right).$$

This particular measure is analogous to information gain. The main reason information gain is not used by itself is it does not consider the internal utility of an item. The directed influence factor would work in directed datasets, where the interactions within the dataset follows a specific directions. For example in web-clicks incoming and outgoing links represents the directionality of the interactions.

However there are undirected datasets, where the interactions within the dataset does not follow a specific direction, such as market basket and retail datasets. This measure does not take into account the directionality of the items which is more representative of the importance of an item within the structure.

Definition 3 (Undirected Influence Factor). *The undirected influence factor of an item, x:*

$$\epsilon_x = \frac{1}{2} \sum_{y \in N_x} \left(\left(-Pr(xy) \log Pr(y|x) - Pr(x, \neg y) \log Pr(\neg y|x) \right).I(x) + \left(-Pr(xy) \log Pr(x|y) - Pr(\neg x, y) \log Pr(\neg x|y) \right).I(y) \right)$$

where $Pr(\neg x|y)$ is the conditional probability for $\neg x$ and y.

Figure 2 shows the undirected influence. This measure is stricter compared to the directed influence factor. As per the previous measure a higher ϵ_x value for an item x represents a higher influence factor.

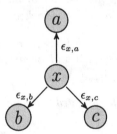

Fig. 1. Directed utility for the neighborhood of x

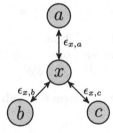

Fig. 2. Undirected utility for the neighborhood of x

A high ϵ value represents a high influence factor, potentially giving rise to more interesting rules. Instead of using high influence factors in place of utility scores, we use them to obtain a projection list of how incoming items for a particular transaction will be inserted into a tree structure for mining.

4   High Influence Tree (HI-Tree)

In our proposed technique, we order incoming transactions based on a projection order which is based on descending ϵ values for the items within a transaction. We then insert the ordered transaction into a prefix tree which will be mined when needed. The main rationale for inserting transactions into the tree based on this order is that items with higher ϵ values will have higher influence factor and will lead to more concise rules. Items which are leaves have less impact on rules produced and should not be considered during the mining phase. In this section we will discuss the construction of our High Influence Tree (HI-tree). For incrementally maintaining the patterns in a sliding window, a summary structure called HI-tree was designed. The HI-tree construction has two phases: a tree construction phase followed by a tree mining phase.

4.1   HI-Tree Structure

The HI-tree has two parts: a prefix subtree which saves the patterns in the sliding window, and an item-index (header) table which indexes the items in the prefix tree. We also have a structure $\prec S$ which stores the projection order of the items based on decreasing value of influence factor, ϵ.

Definition 4 (Projection Order). *Given a current sliding window SW and a set of items I in the sliding window, the projected order of items, $\prec S$, is defined by*

$$\prec S = (s_0, s_1, s_3, \ldots, s_k)$$

where s_k is an item, $s_k \in I$ for $1 \leq k \leq m$, and $\epsilon_k > \epsilon_{k+1}$.

In an HI-Tree prefix subtree, an item can be represented as node(s) within the tree.

4.2   HI-Tree Construction

The tree construction phase is divided into two parts: insertion of transactions and re-sorting transactions. In the insertion step, items in a transaction are inserted into the tree based on the current predetermined order called $\prec S$. Given a transaction T, a data sequence S, called the projection of T is obtained by arranging the items in T according to a predefined order $\prec S$. In order to ensure the completeness of patterns for the data stream, it is necessary to store not only the information related to high Influence items, but also that of low Influence items since they may transit to high Influence status later on. Therefore, when a new transaction T arrives, the completeness information of T is incrementally updated in the HI-tree.

Suppose that the size of the sliding window SW is $N = 9$, and that the window is equally divided into three blocks ($B = 3$), each containing three

Table 1. Example of transactions within a sliding window

Table 2. Example of projection list for t_1 to t_3

TID	Transactions	TID	Transactions	TID	Transactions
t_1	{a,b,c,d,e}	t_4	{d,a,b}	t_7	{a,b,d}
t_2	{a,b,e}	t_5	{a,c,b}	t_8	{b,a,d}
t_3	{b}	t_6	{c,b,a,e}	t_9	{c,e,d, f}

TID	Transactions	Projection S
t_1	{a,b,c,d,e}	{a,b,c,d,e}
t_2	{a,b,e}	{a,b,e}
t_3	{b}	{b}

transactions. A block is the unit at which re-sorting of transactions take place. For reasons of efficiency, the re-sorting operation does not take place with every new transaction as the change will be minimal and unnecessary overhead will b expended. The items of the transactions in the sliding window at a certain moment in time are shown in Table 1. Assume that all items in the HI-tree are arranged in descending order of the high Influence projection list for the sliding window SW. Given a projection list of $\prec S = (a, b, c, d, e)$ where the influence factor of $\epsilon_a > \epsilon_b > \epsilon_c > \epsilon_d > \epsilon_e$, the projection of each transaction is shown in Table 2.

Figure 3 shows the resulting tree after transactions t_1 to t_3 are added. The main difference between our technique and previous techniques is that each transaction is sorted based on the order of the projected order S before insertion into the tree.

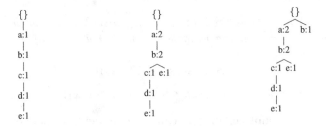

Fig. 3. Insertion of t_1 to t_3

In the re-sorting step, we update the projection order list and re-sort the tree based on the latest projection order list. Subsequent transactions are inserted according to the new projection order. Given a projection list of $\prec S = a, b, e, c, d$ where the influence factor of $\epsilon_a > \epsilon_b > \epsilon_e > \epsilon_c > \epsilon_d$; the projection of each transaction is also shown in the Table 3.

In the re-sorting step for transactions t_7 to t_9 we update the projection order list and re-sort the tree based on the latest projection order list. Subsequent transactions are inserted according to the new projection order. Given a projection list of $\prec S = a, b, d, e, c$ where the Influence factor of $\epsilon_a > \epsilon_b > \epsilon_d > \epsilon_e > \epsilon_c$. The projection of each transaction is shown in the Table 4.

Table 3. Example of projection list for t_4 to t_6

TID	Transactions	Projection S
t_4	{d,a,b}	{a,b,d}
t_5	{a,c,b}	{a,c,b}
t_6	{c,b,a,e}	{,a,b,e,c}

Table 4. Example of projection list for t_7 to t_9

TID	Transactions	Projection S
t_7	{a,b,d}	{a,b,d}
t_8	{b,a,d}	{a,b,d}
t_9	{c,e,d,f}	{d,e,c,f}

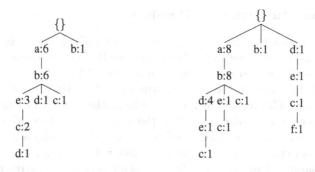

Fig. 4. Insertion of t_4 to t_6 **Fig. 5.** Insertion of t_7 to t_9

4.3   HI-Tree Mining

The tree mining phase follows the FP-Growth mining technique. Once the HI-tree is constructed we use FP-growth to mine patterns. FP-Growth is used in the mining phase in CP-Tree. Instead of using minimum support threshold as seen in previous research, we push two different constraints into our algorithm.

In the HI-Tree, the influence factor of any item is no smaller than those of its descendants. When we incrementally maintain the transactions, the less influential are also stored in the tree to ensure completeness of the patterns within a stream. Additionally as the sliding window moves, new transactions enter as the oldest transactions are removed. The transaction that represents the old transactions may cause errors in the mining results. Therefore after the current data segment has been processed and before a new data segment arrives, the insignificant patterns in the HI-tree should be deleted by performing a pruning operation. Keeping obsolete and low influence not only degrades performance, it also causes errors in the mining results.

If the influence factor, ϵ, of an item in the projection order list is less than a minimum ϵ_{min} threshold, then it is redundant and its items are not considered in the prefix tree. The rationale for this is that a low influence would not contribute to a strong rule, thus excluding items with low influence would provide us with a set of strong influential patterns.

5   Experimental Evaluation

All experiments were performed on a Intel Core i7 PC with 1.9 GB of memory, running Windows 7. The experiments were performed on real world datasets. We compared the performance of HI-Tree using undirected Influence factor on the 4 datasets while varying ϵ_{min}. In the initial experiments we compare the performance of our approach with the CP-tree approach [9] on runtime and size of rule base produced. Due to space constraints we show the overall end result to process the entire dataset as opposed to each individual block. The datasets used in our experiments were from both the UCI [2] and FIMI [3] repository.

5.1   Varying Minimum ϵ_{min} Threshold

Figures 6 and 7 show the execution time and number of rules produced for each of the two approaches, HI-tree and CP-tree. For the BMS-POS dataset the window size was set to 50,000 and ϵ_{min} ranged from 1.2 to 2.0.

For the Retail dataset the window size was set to 10,000 and ϵ_{min} ranged from 0.2 to 0.8. The dataset contained information on 5,133 customers who have purchased at least one product in the supermarket. Over the entire data collection period, the supermarket store carries 16,470 unique items. For the Kosarak dataset the window size was set to 100,000 and ϵ_{min} ranged from 1.0 to 2.2. The Kosarak dataset contains 990,000 sequences of click-stream data from an Hungarian news portal. For the Mushroom dataset the window size was set to 1,600 and ϵ_{min} ranged from 10.0 to 15.0. This data set includes descriptions of hypothetical samples corresponding to 23 species of gilled mushrooms with 8124 instances. Overall the runtime performance using our undirected Influence measure was far superior to that of CP-Tree using similar minimum support and confidence thresholds. In the next section we compute reduction ratios on the

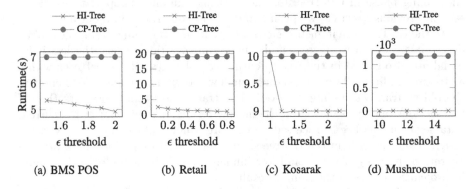

Fig. 6. Runtime performance

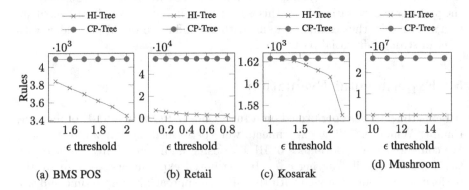

Fig. 7. Rules produced

number of rules generated of the HI-tree approach in relation to the CP-tree approach.

5.2 Reduction Ratio

We show the implication of ϵ_{min} threshold using the reduction ratio of the rules produced as compared to the traditional incremental mining (CP-Tree) technique. This experiment is set to investigate the number of rules produced when undirected Influence measure is used on the datasets above. The reduction ratio in this case describes the ratio between the number of rules produced by CP-Tree versus the number of rules produced by HI-tree. Table 5 shows the number of rules generated by our algorithm as compared to CP-Tree technique. From this table, it can be seen that when ϵ_{min} increases the reduction ratio also increases and vice versa. We notice that the ratio of rules produced varies between 1:1 to 1:79807 rules in the four datasets. A large reduction ratio means that the amount of rules produced by our technique is proportionally fewer as compared to CP-Tree. We do not always assume that a larger reduction ratio would be better for all cases; it only signifies that there were redundant rules produced by a traditional technique in such cases.

Table 5. Reduction ratio

ϵ_{min}	1.5	1.6	1.7	1.8	1.9	2.0
BMS-POS	1.1	1.1	1.1	1.1	1.2	1.2
ϵ_{min}	0.1	0.2	0.3	0.4	0.6	0.8
Retail	7.5	9.5	11.9	14.4	19.2	23.7
ϵ_{min}	10.0	11.0	12.0	13.0	14.0	15.0
Mushroom	280.3	312.0	793.8	7620.1	9253.0	79807.4
ϵ_{min}	1.0	1.0	1.4	1.6	1.8	2.0
Kosarak	1.0	1.0	1.0	1.0	1.0	1.0

5.3 Scalability Test and Rule Validation

To test the scalability of HI-Tree by varying the number of transactions, we used the Kosarak dataset, since it is a large sparse dataset with a large number of distinct items and transactions. We divided this dataset into blocks, each containing 100,000 transactions. Then we tested the performance on Kosarak by mining incrementally with the current block accumulated with previous blocks and setting $\epsilon_{min} = 1.0$. The time in the y-axis of Fig. 8 specifies the total required time with increasing dataset size. Clearly, as the size of the dataset increases, the overall time increases. However, as shown in the figure, HI-Tree showed a linear scalability over the dataset size.

To assess the validity of the rules produced we examined the average confidence, Lift, Information Gain and Kappa value of rules produced on the four

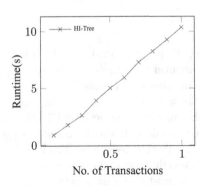

Fig. 8. Scalability

Table 6. Rule validation

Datasets	ϵ_{min}	Conf	Lift	IGain	Kappa
BMS-POS	1.5	0.95	1.59	0.45	0.02
	1.6	0.95	1.59	0.45	0.02
	1.7	0.95	1.58	0.45	0.02
	1.8	0.95	1.58	0.45	0.02
	1.9	0.95	1.58	0.45	0.02
	2.0	0.95	1.58	0.44	0.02
Kosarak	1.0	0.96	3.13	0.81	0.06
	1.2	0.96	3.13	0.81	0.06
	1.5	0.96	3.13	0.81	0.06
	1.8	0.96	3.13	0.81	0.06
	2.0	0.96	3.09	0.80	0.06
Mushroom	10	0.97	1.22	0.19	0.71
	11	0.97	1.21	0.19	0.69
	12	0.97	1.13	0.12	0.59
	13	0.97	1.07	0.07	0.43
	14	0.97	1.06	0.06	0.40
	15	0.97	1.08	0.07	0.40
Retail	0.1	0.95	13.97	1.10	0.03
	0.2	0.96	9.41	1.06	0.02
	0.4	0.96	3.26	0.98	0.01
	0.6	0.96	3.30	1.01	0.01
	0.8	0.96	3.52	1.07	0.02

Table 7. Sample of top 10 rules from CP-tree and HI-tree

CP-Tree (Top 10 rules from block 1)
gill-spacing=c stalk-color-above-ring=w stalk-color-below-ring=w → stalk-shape=e
stalk-shape=e → gill-spacing=c stalk-color-above-ring=w stalk-color-below-ring=w
bruises=t stalk-shape=e → gill-spacing=c stalk-color-above-ring=w stalk-color-below-ring=w
gill-spacing=c ring-type=p stalk-color-above-ring=w stalk-color-below-ring=w → stalk-shape=e
gill-spacing=c stalk-color-above-ring=w stalk-color-below-ring=w → ring-type=p stalk-shape=e
ring-type=p stalk-shape=e → gill-spacing=c stalk-color-above-ring=w stalk-color-below-ring=w
stalk-shape=e → gill-spacing=c ring-type=p stalk-color-above-ring=w stalk-color-below-ring=w
bruises=t ring-type=p stalk-shape=e → gill-spacing=c stalk-color-above-ring=w stalk-color-below-ring=w
bruises=t stalk-shape=e → gill-spacing=c ring-type=p stalk-color-above-ring=w stalk-color-below-ring=w
gill-spacing=c stalk-color-above-ring=w stalk-color-below-ring=w stalk-surface-above-ring=s → stalk-shape=e

HI-Tree (Top 10 rules from block 1)
ring-type=p → stalk-surface-above-ring=s
gill-attachment=f ring-type=p → stalk-surface-above-ring=s
ring-type=p → gill-attachment=f stalk-surface-above-ring=s
ring-type=p veil-type=p → stalk-surface-above-ring=s
ring-type=p → stalk-surface-above-ring=s veil-type=p
ring-type=p veil-color=w → stalk-surface-above-ring=s
ring-type=p → stalk-surface-above-ring=s veil-color=w
ring-number=o ring-type=p → stalk-surface-above-ring=s
ring-type=p → ring-number=o stalk-surface-above-ring=s

datasets, as shown in Table 6. Based on the results from these objective measures, all of the measures agree that the set of rules generated by our technique is interesting. A sample of the top ten rules produced by HI-Tree and CP-Tree in Table 7. Overall, we observe that the average size of the precedent terms of rules produced by HI-Tree is shorter, thus making the rules more actionable, in general.

6   Conclusion and Future Work

In this research we proposed the HI-Tree approach, an incremental approach for pattern mining of influential rules. We developed two measures that automatically estimated a utility value for items using the both the internal and external characteristics of an item. Our experimental results show that potentially high influence rules can be efficiently generated from the HI-Tree with a single scan of the dataset. Our experimentation on real world data revealed that mining performance is enhanced significantly since both the search space and the number of rules generated are effectively reduced by the proposed strategies.

In the future we plan to incorporate an adaptive sliding window size in our approach. By adapting the sliding window size we have greater control over changes in the computation of influence factor measure as well as the ability to react to changes faster when the changes have occurred within the underlying data stream. We will also consider the effects of drift, which will allow us to analyze items that have had rapid fluctuations in their influence factor values.

References

1. Ahmed, C., Tanbeer, S., Jeong, B.S., Lee, Y.K.: Efficient tree structures for high utility pattern mining in incremental databases. IEEE Trans. Knowl. Data Eng. **21**(12), 1708–1721 (2009)
2. Bache, K., Lichman, M.: UCI machine learning repository (2013). http://archive. ics.uci.edu/ml
3. Goethals, B.: Frequent itemset mining dataset repository (2013). http://fimi.ua. ac.be/data/
4. Han, J., Pei, J., Yin, Y., Mao, R.: Mining frequent patterns without candidate generation: a frequent-pattern tree approach. Data Min. Knowl. Discov. **8**(1), 53–87 (2004)
5. Lan, G.C., Hong, T.P., Tseng, V.S., Wang, S.L.: Applying the maximum utility measure in high utility sequential pattern mining. Expert Syst. Appl. **41**(11), 5071–5081 (2014)
6. Leung, C.K.S., Khan, Q.I., Li, Z., Hoque, T.: Cantree: a canonical-order tree for incremental frequent-pattern mining. Knowl. Inf. Syst. **11**(3), 287–311 (2007)
7. Lin, C.W., Hong, T.P., Lan, G.C., Wong, J.W., Lin, W.Y.: Incrementally mining high utility patterns based on pre-large concept. Appl. Intell. **40**(2), 343–357 (2014)
8. Liu, Y., Liao, W., Choudhary, A.K.: A two-phase algorithm for fast discovery of high utility itemsets. In: Ho, T.-B., Cheung, D., Liu, H. (eds.) PAKDD 2005. LNCS (LNAI), vol. 3518, pp. 689–695. Springer, Heidelberg (2005)

9. Tanbeer, S.K., Ahmed, C.F., Jeong, B.-S., Lee, Y.-K.: CP-Tree: a tree structure for single-pass frequent pattern mining. In: Washio, T., Suzuki, E., Ting, K.M., Inokuchi, A. (eds.) PAKDD 2008. LNCS (LNAI), vol. 5012, pp. 1022–1027. Springer, Heidelberg (2008)

10. Tseng, V.S., Wu, C.W., Shie, B.E., Yu, P.S.: Up-growth: an efficient algorithm for high utility itemset mining. In: Proceedings of the 16th ACM SIGKDD International Conference on Knowledge Discovery and Data Mining, pp. 253–262. ACM, New York (2010)

11. Yao, H., Hamilton, H.J.: Mining itemset utilities from transaction databases. Data Knowl. Eng. **59**(3), 603–626 (2006)

Balancing Tree Size and Accuracy in Fast Mining of Uncertain Frequent Patterns

Carson Kai-Sang Leung[✉] and Richard Kyle MacKinnon

University of Manitoba, Winnipeg, MB, Canada
kleung@cs.umanitoba.ca

Abstract. To mine frequent patterns from uncertain data, many existing algorithms (e.g., UF-growth) directly calculate the expected support of a pattern. Consequently, they require a significant amount of storage space to capture all existential probability values among the items in the data. To reduce the amount of required storage space, some existing algorithms (e.g., PUF-growth) combine nodes with the same item by storing an upper bound on expected support. Consequently, they lead to many false positives in the intermediate mining step. There is trade-off between storage space and accuracy. In this paper, we introduce a new algorithm called MUF-growth for achieving a tighter upper bound on expected support than PUF-growth while balancing the storage space requirement. We evaluate the trade-off between storing more information to further tighten the bound and its effect on the performance of the algorithm. Our experimental results reveal a diminishing return on performance as the bound is increasingly tightened, allowing us to make a recommendation on the most effective use of extra storage towards increasing the efficiency of the algorithm.

1 Introduction and Related Works

Frequent pattern mining from precise data has become popular since the introduction of the *Apriori algorithm* [1]. Users definitely know whether an item is present in, or absent from, a transaction in databases of precise data. Common pattern mining techniques applied to precise data can be extended to the mining of interaction patterns [6], sequential patterns [7], social patterns [10], and popular patterns [17]. However, there are also situations in which users are uncertain about the presence or absence of items [3,4,9,20]. For example, a physician may highly suspect (but cannot guarantee) that a coughing patient suffers from the Middle East respiratory syndrome (MERS). The uncertainty of such suspicion can be expressed in terms of existential probability (e.g., a 70% likelihood of suffering from the MERS). With this notion, each item in a transaction t_j in databases containing precise data can be viewed as an item with a 100% likelihood of being present in t_j.

To handle uncertain data, the tree-based *UF-growth algorithm* [15] was proposed. In order to compute the *exact* expected support of each pattern, paths in the corresponding UF-tree are shared only if tree nodes on the paths have the

© Springer International Publishing Switzerland 2015
S. Madria and T. Hara (Eds.): DaWaK 2015, LNCS 9263, pp. 57–69, 2015.
DOI: 10.1007/978-3-319-22729-0_5

same item and same existential probability. Hence, due to this more restrictive path sharing requirement, the UF-tree may be quite large when compared to the FP-tree [8] for precise data. This issue has been addressed [13,14,16]. For instance, Calders et al. [5] mined uncertain data with sampling. Expected support of a pattern X is then estimated based on the average actual support of X in several random samples (or instantiations). We [18] previously proposed the *PUF-growth algorithm* to utilize a concept of *item caps* (which provide upper bounds to expected support) together with aggressive path sharing (in which paths are shared if nodes have the same item in common regardless of existential probability) to yield a more compact tree. Here, the expected support of X is estimated based on relevant existential probability values.

In this paper, we study the following questions: Can we further tighten the upper bound on expected support in PUF-growth? Can the resulting tree still be as compact as the FP-tree in terms of number of nodes? At what point does continuing to tighten the upper bound no longer provide an acceptable decrease in runtime? Our *key contributions* of this paper are as follows:

1. a metal value uncertain frequent pattern **tree (MUF-tree)**, which can be as compact as the original FP-tree;
2. a mining algorithm — called **MUF-growth** — that is guaranteed to find *all and only those* frequent patterns with *no* false negatives (frequent patterns mistakenly categorized as infrequent) and *no* false positives (infrequent patterns mistakenly categorized as frequent) from uncertain data at the end of the mining process; and
3. an empirical analysis of the upper bound in MUF-growth.

The remainder of this paper is organized as follows. The next section presents background. Then, we present our MUF-tree structure and MUF-growth algorithm in Sects. 3 and 4, respectively. Experimental results are shown in Sect. 5, and conclusions are given in Sect. 6.

2   Background

Let (i) Item be a set of m domain items and (ii) $X = \{x_1, x_2, \ldots, x_k\}$ be a k-itemset (i.e., a pattern consisting of k items), where $X \subseteq$ Item and $1 \leq k \leq m$. Then, a transactional database $= \{t_1, t_2, \ldots, t_n\}$ is the set of n transactions, where each transaction $t_j \subseteq$ Item. The projected database of X is the set of all transactions containing X. Each item x_i in a transaction $t_j = \{y_1, y_2, \ldots, y_h\}$ in an uncertain database is associated with an *existential probability value* $P(y_i, t_j)$, which represents the likelihood of the presence of y_i in t_j. Note that $0 < P(y_i, t_j) \leq 1$. The *existential probability* $P(X, t_j)$ of a pattern X in t_j is then the product of the corresponding existential probability values of items within X when these items are independent [11,12]: $P(X, t_j) = \prod_{x \in X} P(x, t_j)$. The *expected support expSup(X)* of X in the database is the sum of $P(X, t_j)$ over all n transactions in the database: $expSup(X) = \sum_{j=1}^{n} P(X, t_j)$. A pattern X

is *frequent* in an uncertain database if $expSup(X) \geq$ a user-specified minimum support threshold *minsup*.

Given a database and *minsup*, the research problem of *frequent pattern mining from uncertain data* is to discover from the database a complete collection of frequent patterns having expected support $\geq$ *minsup*.

3   Our MUF-tree Structure

Recall that PUF-growth utilizes an upper bound to expected support to first find all potentially frequent patterns, which include (i) true positives (i.e., patterns with expected support $\geq$ *minsup*) and (ii) false positives (i.e., patterns with expected support $<$ *minsup* but with upper bound $\geq$ *minsup*). Then, PUF-growth verifies each of the patterns and returns only those truly frequent ones. To further tighten the upper bound to expected support, we propose (i) the metal value uncertain frequent pattern **tree (MUF-tree)** structure in this section and (ii) the corresponding **MUF-growth** algorithm in the next section.

The key idea behind the MUF-tree is to keep track of the i-th highest probability values in the prefix of t_j and use them every time a frequent extension $(k > 2)$ is added to the suffix item during the mining process. Hence, each node in an MUF-tree keeps (i) an item y_r, (ii) an *item cap* $IC(y_r, t_j)$, and (iii) a list of "*metal*" *values*, which are the i-th highest existential probabilities in the prefix of y_r in t_j. Figure 1(c) shows the contents of an MUF-tree for the database in Table 1, in which each node maintains (i) an item, (ii) its item cap, and (iii) its first three metal values (i.e., "silver value" $M_2(y_r, t_j)$, "bronze value" $M_3(y_r, t_j)$

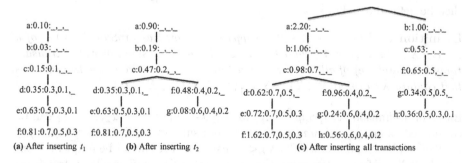

(a) After inserting t_1 (b) After inserting t_2 (c) After inserting all transactions

Fig. 1. Our MUF-tree for the database in Table 1 with *minsup* $= 0.5$ and 3 metal values

Table 1. A transactional database of uncertain data

Transactions	Transactions
$t_1 = \{a{:}0.1,\ b{:}0.3,\ c{:}0.5,\ d{:}0.7,\ e{:}0.9,\ f{:}0.9\}$	$t_4 = \{b{:}0.3,\ c{:}0.6,\ f{:}0.5,\ g{:}0.1,\ h{:}0.6\}$
$t_2 = \{a{:}0.8,\ b{:}0.2,\ c{:}0.4,\ f{:}0.6,\ g{:}0.1\}$	$t_5 = \{a{:}0.4,\ b{:}0.6,\ c{:}0.1,\ f{:}0.8,\ g{:}0.2,\ h{:}0.7\}$
$t_3 = \{b{:}0.7,\ c{:}0.5,\ f{:}0.5,\ g{:}0.4\}$	$t_6 = \{a{:}0.9,\ b{:}0.7,\ c{:}0.5,\ d{:}0.3,\ e{:}0.1,\ f{:}0.9\}$

and "copper value" $M_4(y_r, t_j)$). With this new data structure, a new *tighter* upper bound can be given by the product of $IC(y_r, t_j)$ and the various "metal" values. The *compounded (prefixed) item cap*—denoted as $CIC(X, t_j)$—of any k-itemset $X = \{x_1, x_2, \ldots, x_k\}$ in $t_j = \{y_1, y_2, \ldots, y_r, \ldots, y_h\}$ (where $x_k = y_r$) can then be defined as in Definition 1.

Definition 1. *Let* $t_j = \{y_1, y_2, \ldots, y_{r-1}, y_r, \ldots, y_h\}$ *where* $h = |t_j|$. *Also, let* $M_i(x_r, t_j)$ *be the i-th metal value, which is the i-th highest existential probability value among all $(r - 1)$ items in the proper prefix $\{y_1, y_2, \ldots, y_{r-1}\} \subset t_j$. If $X = \{x_1, x_2, \ldots, x_k\}$ is a k-itemset in t_j such that $x_k = y_r$, then*

$$CIC(X, t_j) = \begin{cases} IC(X, t_j) & \text{if } k = 2, \\ IC(X, t_j) \times \prod_{i=2}^{k-1} M_i(y_r, t_j) & \text{if } k \geq 3, \end{cases} \tag{1}$$

where $IC(X, t_j) = P(y_r, t_j) \times M_1(y_r, t_j)$. □

It is interesting to observe that the item cap $IC(X, t_j)$ provided by PUF-growth is a special case of the compounded item cap $CIC(X, t_j)$ provided by MUF-growth when $k = 2$ (i.e., where no metal values are stored). The generalization we present here benefits us by obtaining an increasingly tighter bound when multiplying successive metal values for larger itemsets $k \geq 3$. For example, when $k = 4$, $CIC(X, t_j)$ involves $IC(X, t_j)$, as well M_2 ("silver value") and M_3 ("bronze value"). Moreover, in theory, the maximum number of metal values we can use is ultimately limited above by the length of the largest transaction in the database. In practice, we will usually use a smaller number. See Sect. 5.

Since the expected support of X is the sum of all existential probabilities of X over all the transactions containing X, the *cap* of expected support of X can then be defined as follows.

Definition 2. *The cap of expected support*—*denoted as* $expSup^{Cap}(X)$—*of a pattern $X = \{x_1, \ldots, x_k\}$ (where $k > 1$) is defined as the sum (over all n transactions in a database) of all the compounded item caps of x_k in all the transactions that contain X: $expSup^{Cap}(X) = \sum_{j=1}^{n} \{CIC(X, t_j) \mid X \subseteq t_j\}$.* □

Based on Definition 2, $expSup^{Cap}(X)$ for any k-itemset $X = \{x_1, \ldots, x_k\}$ can be considered as an *upper bound* to the expected support of X, i.e., $expSup(X) \leq expSup^{Cap}(X)$. So, if $expSup^{Cap}(X) < minsup$, then X cannot be frequent. Conversely, if X is a frequent pattern, then $expSup^{Cap}(X) \geq minsup$. We take advantage of this property to safely mine all frequent patterns using $expSup^{Cap}(X)$ in place of $expSup(X)$.

Lemma 1. *The cap of expected support of a pattern X satisfies the partial downward closure property: All non-empty subsets Y of a frequent pattern X (such that X and Y share the same suffix item y_r) are also frequent.* □

An MUF-tree can be constructed as follows. With the first scan of the uncertain database, we find all distinct *frequent* items. Then, the MUF-tree is constructed with the second database scan in a fashion similar to that of the FP-tree.

A key difference is that, when inserting a transaction item, we compute its item cap and its metal values. Only frequent items are inserted into the MUF-tree, and they are inserted according to some canonical order. If a node containing that item already exists in a tree path, in addition to updating its item cap, we also update its metal values by taking the *maximum* of each computed metal value and the corresponding existing value. Otherwise, we create a new node with this item cap and metal values. Similar to the FP-tree, our MUF-tree maintains horizontal node traversal pointers whose links are adjusted whenever a new node is created. For a better understanding of the MUF-tree construction, see Example 1.

Example 1. Consider the uncertain database in Table 1. Let the user-specified support threshold *minsup* be set to 0.5 and the number of stored metal values be equal to three. For simplicity, we arrange items in alphabetical order. After the first database scan, the expected supports of all items (after removing infrequent single items) are a:2.2, b:2.3, c:2.6, d:1.0, e:1.0, f:4.2, g:0.8 and h:1.3.

With the second database scan, we insert only the frequent items of each transaction (with their respective item cap and metal values). For instance, as shown in Fig. 1(a), when inserting transaction $t_1 = \{a$:0.1, b:0.3, c:0.5, d:0.7, e:0.9, f:0.9$\}$, we store:

1. a:0.10:_:_:_ because $IC(a, t_1) = P(a, t_1) = 0.10$;
2. b:0.03:_:_:_ because $IC(b, t_1) = P(b, t_1) \times M_1(b, t_1) = 0.3 \times 0.1 = 0.03$;
3. c:0.15:0.1:_ because $IC(c, t_1) = P(c, t_1) \times M_1(c, t_1) = 0.5 \times 0.3 = 0.15$ and $M_2(c, t_1) = 0.1$;
4. d:0.35:0.3:0.1:_ because $IC(d, t_1) = P(d, t_1) \times M_1(d, t_1) = 0.7 \times 0.5 = 0.35$, $M_2(d, t_1) = 0.3$ and $M_3(b, t_1) = 0.1$;
5. e:0.63:0.5:0.3:0.1 because $IC(e, t_1) = P(e, t_1) \times M_1(e, t_1) = 0.9 \times 0.7 = 0.63$, $M_2(e, t_1) = 0.5$, $M_3(e, t_1) = 0.3$ and $M_4(e, t_1) = 0.1$; as well as
6. f:0.81:0.7:0.5:0.3 because $IC(f, t_1) = P(f, t_1) \times M_1(f, t_1) = 0.9 \times 0.9 = 0.81$, $M_2(f, t_1) = 0.7$, $M_3(f, t_1) = 0.5$ and $M_4(f, t_1) = 0.3$.

As t_2 shares a common prefix $\langle a, b, c \rangle$ with an existing path in the MUF-tree created when t_1 was inserted, (i) the item cap values of those items in the common prefix are *added* to their corresponding nodes, and (ii) the metal values of those items are checked against the metal values for their corresponding nodes, with only the *maximum* saved for each node. In this case, $IC(a, t_2) = 0.8$ is added to the previous value of 0.1 to get 0.9. Similarly, $IC(b, t_2) = 0.16$ is added to the previous value of 0.03 to get 0.19. For c, $IC(c, t_2) = 0.32$ is added to the previous value of 0.15 to get 0.47. As $M_2(c, t_2) = 0.2$ is higher than the previous M_2 value of 0.1, the new value of 0.2 is stored. As f and g in t_2 not sharing any common prefix with t_1, new nodes are created as shown in Fig. 1(b). Figure 1(c) shows the status of the MUF-tree after inserting all of the remaining transactions. □

4   Our MUF-growth Algorithm

Here, we propose a pattern-growth mining algorithm called **MUF-growth**, which mines frequent patterns from our MUF-tree structure. Recall that the construction of an MUF-tree is similar to that of a FP-tree, except that metal values are additionally stored. Thus, the basic operation in MUF-growth for mining frequent patterns is to construct a projected database for each potential frequent pattern and recursively mine its potentially frequent extensions.

Based on Lemma 1, we apply the MUF-growth algorithm to our MUF-tree for generating only those k-itemsets (where $k > 1$) with caps of expected support $\geq minsup$. Similar to other algorithms (e.g., UFP-growth [2]), the mining process of our MUF-growth may also lead to some false positives in the intermediate step (e.g., at the end of the second database scan), but all these false positives will be filtered out with a quick third scan of the database. Hence, our MUF-growth is guaranteed to return to the user *all* and *only those* truly frequent patterns with *neither* false positives *nor* false negatives at the end of the mining process.

Example 2. The MUF-growth algorithm mines extensions of every frequet item. Consider the MUF-tree in Fig. 1(c). The $\{f\}$-conditional tree is created by accumulating all the prefix paths from each f node up to the root. In the very first recursion, each node in the prefix of each f node takes on the current value of $expSup^{Cap}(\{f\})$ since the cap of expected support already contains the frequency and existential probability information of f together with the highest existential probability value in the prefix of f. In addition, each node in the prefix of f in the global tree inherits the metal values of f when they appear in the $\{f\}$-conditional tree. Thus, the $\{f\}$-conditional tree, pictured in Fig. 2(a) consists of two branches, one for each prefix path of the f-nodes up to the root that is not shared. The left branch contains $\langle a, b, c, d, e \rangle$ with (i) d and e having an $expSup^{Cap}$ value of 1.62 and (ii) a, b and c have an $expSup^{Cap}$ value of $1.62 + 0.96 = 2.58$ due to the shared prefix between the two f-nodes. All the nodes in the left branch contain the metal values $\langle 0.7, 0.5, 0.3 \rangle$ as they are (i) the default values for d and e by default and (ii) higher than its corresponding value in $\{0.4, 0.2, \_\}$ for a, b and c. The right branch contains $\langle b, c \rangle$, each having the $expSup^{Cap}$ value of 0.65 and the M_1 value of 0.5.

For the $\{f, c\}$-conditional tree in Fig. 2(b), the compounding effect of the metal values is observed because the $expSup^{Cap}$ value for each node in the prefix of c is further multiplied by the first metal value in each list for c. The left branch in the $\{f, c\}$- conditional tree contains $\langle a, b \rangle$ with an $expSup^{Cap}$ value of

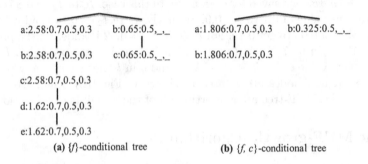

(a) $\{f\}$-conditional tree   (b) $\{f, c\}$-conditional tree

Fig. 2. Conditional trees in the MUF-growth algorithm

$1.806 = 2.58 \times 0.7$ and the metal values $\langle 0.7, 0.5, 0.3 \rangle$. The right branch contains just b with an $expSup^{Cap}$ value of $0.325 = 0.65 \times 0.5$ and the metal value 0.5 (which is just repeated as no further values were previously stored). At this point, if $minsup = 2.2$, then using just a single metal value M_2 has saved us having to further generate any more candidates which are extensions of $\{f, c\}$.

The $\{f, c, b\}$-conditional tree consists of a single node a:0.903:0.7,0.5,0.3. If we take advantage of the second metal value in b's list, each node in the prefix of b will take on the $expSup^{Cap}$ value of $0.903 = 1.806 \times 0.5$. If instead we only decided to store a single metal value in this implementation of MUF-growth, then we would reuse the first metal value in b's list. Consequently, each node in the prefix of b would take on the $expSup^{Cap}$ value of $1.2642 = 1.806 \times 0.7$. In this case, using extra storage space to accommodate more metal values in each node allows us to bring the $expSup^{Cap}$ value closer to the expected support with fewer recursive calls to generate candidate extensions. Every time an extra metal value is used in this manner, the chance goes up that the sub-tree generation stops earlier and fewer false positives are generated.   □

Since each metal value used during the candidate generation process is a maximum among all values in the prefix of an item, MUF-growth finds a complete set of patterns from an MUF-tree without any false negatives. In addition, with each additional unique metal value that is stored in the MUF-tree nodes, MUF-growth does so while generating fewer false positives than existing algorithms. In much larger databases, the runtime savings caused by this effect are significant.

5 Evaluation Results

This section presents our results on evaluating the following aspects of our MUF-tree structure and its corresponding MUF-growth algorithm in mining frequent patterns from uncertain data.

5.1 Analytical Evaluation

Tree Compactness. For any uncertain databases, *our MUF-tree has the same number of nodes as in the existing PUF-tree* [18]. Thus, the node count of the MUF-tree (i) can be equal to that of the FP-tree [8] (when the MUF-tree is constructed using the frequency-descending order of items) and (ii) is bounded above by total number of frequent items summed over every transaction in the database.

Tree Completeness. The complete set of mining results (with no false positives and no false negatives) can be generated because the *MUF-tree contains the set of all of the frequent items from every transaction in the database along with the compounded item cap (CIC) in each node. Mining based on this compounded item cap ensures that no frequent k-itemset (k > 1) will be missed.*

Tightness of CIC. *Recall that the PUF-tree utilizes an item cap (IC), which is based on the existential probability value $P(y_r, t_j)$ of y_r and the single highest existential probability value $M_1(y_r, t_j)$ in its prefix. In contrast, our MUF-tree* utilizes a CIC, which is based on various metal values $M_i(y_r, t_j)$ in addition to $P(y_r, t_j)$ and $M_1(y_r, t_j)$.Note that the CIC used in our MUF-tree provides a tighter upper bound than the IC used in the PUF-tree because candidates are generated during the mining process with increasing cardinality of X in the CIC (cf. the IC has no such compounding effect).

5.2   Empirical Evaluation

We also compared the performance of our MUF-growth algorithm (with varying numbers of metal values stored per node) with the existing PUF-growth algorithm. Recall that PUF-growth [18] was shown to outperform UF-growth [15], UFP-growth [2] and UH-Mine [2]. So, we compare our MUF-growth solely with PUF-growth in order to clearly evaluate the effect of the tightened upper bound. Similar to other papers [2] in the sub-community of uncertain frequent pattern mining, we also used both real life and synthetic databases for our tests. The synthetic databases, which are generally sparse, are generated within a domain of 1000 items by the data generator developed at IBM Almaden Research Center [1]. We also considered several real life databases such as mushroom, retail and kosarak. We assigned a (randomly generated) existential probability value from the range (0 %,100 %] to each item in every transaction in the databases. The name of each database indicates some of its characteristics. For example, the database u10k5L10100_2 contains 10 K transactions with average transaction length of 5, each item in a transaction is associated with an existential probability value that lies within a range of [10 %, 100 %] and the probability values are spaced out with minimum increment of 5 % (i.e., every 10 % increment is separated into 2 different values). Due to space constraints, we present here the results on some of the above databases. All programs were written in C++ and ran in a Linux environment on an Intel Core i5-661 CPU with 3.33 GHz and 8 GB of RAM. Unless otherwise specified, runtime includes CPU and I/Os for tree construction, mining, and false-positive removal. While the number of false positives generated at the end of the second database scan may vary, all algorithms (ours and others) produce the same set of truly frequent patters at the end of the mining process. The results shown in this section are based on the average of multiple runs for each case. In all experiments, *minsup* was expressed in terms of the absolute support value, and all trees were constructed using the descending order of expected support.

Number of False Positives. Although PUF-trees and MUF-trees are compact (in fact, the number of nodes in the global tree can be equal to the FP-tree for both of them), their corresponding algorithms generate some false positives. Hence, their overall performances depend on the number of false positives generated. In this experiment, we measured the number of false positives generated by both algorithms for fixed values of *minsup* with different databases. We present some results in Fig. 3 using one *minsup* value over several probability distributions for the synthetic database u10k5L10100 and the real-life database retail10100. In general, *MUF-growth was observed to remarkably reduce the number of false positives when compared with PUF-growth.* The primary reason for

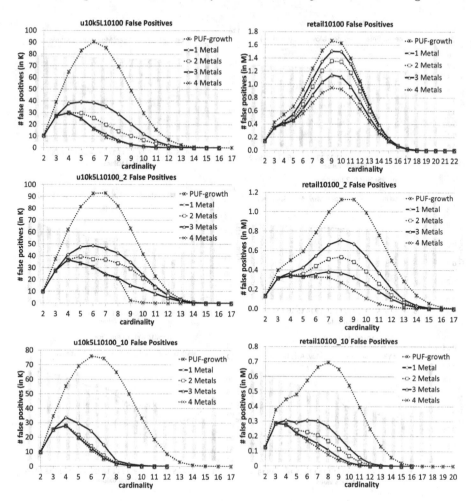

Fig. 3. Number of false positives

this improvement is that the upper bound in this algorithm is much tighter than that in PUF-growth for higher cardinality itemsets ($k > 2$), therefore less total candidates are generated and subsequently less false positives.

On the IBM synthetic database u10k5L10100, MUF-growth generated anywhere from 47 % to as low as 23 % of the false positives generated by the corresponding PUF-growth algorithm as the number of metal values was increased. On the other hand, with the UC Irvine real-life database retail10100, this effect is not quite as pronounced. When the number of distinct probability values is increased, as in the retail10100_10 database, we again see similar numbers as in the synthetic databases. The wider number of distinct probability values leads to a higher chance that additional metal values are significantly lower than the previous values. In synthetic data, the decrease in false positives is more pronounced even with fewer distinct values as the distribution of items in

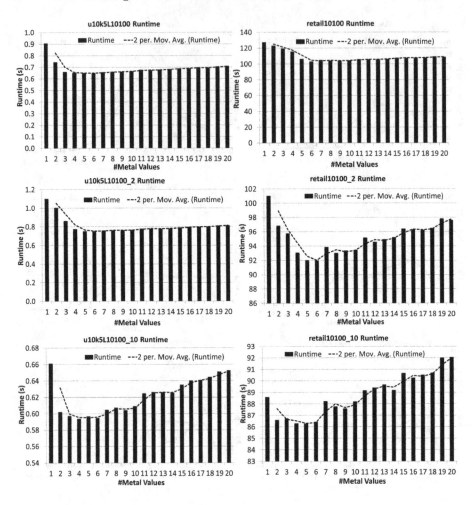

Fig. 4. Runtime

transactions becomes completely independent, whereas items in the retail data-base tend to appear or be absent in predetermined groups.

When changing the range of observed probability values, as in retail_5060 (not shown for brevity), we found that *MUF-growth generated fewer than 1.5 %* *of the total false positives of the corresponding PUF-growth algorithm.* Adding subsequent metal values beyond the first in retail_5060 did not make a significant difference in the number of false positives generated. The reason is that with only two closely spaced existential probability values to choose from, each additional metal value is either the exact same or very close to the previous one, giving little to no benefit over just reusing the previous value.

Runtime. Recall that PUF-growth was shown to outperform UH-Mine [18] and subsequently UFP-growth [2,19]. Hence, we compared our MUF-growth algorithm with different metal values to PUF-growth. *The addition of just a single metal value in MUF-growth makes a remarkable improvement in the runtime when compared to PUF-growth.* The primary reason is that, even though PUF-growth finds the exact set of frequent patterns when mining an extension of X, it may suffer from the high computation cost of generating unnecessarily large numbers of infrequent candidates and their extensions. *The use of the metal values in MUF-growth ensure that those high cardinality candidates are never generated due to their expected support caps being much closer to the expected support.*

When comparing our MUF-growth algorithm to it with different metal values the runtimes are much closer together. We show, in Fig. 4, the runtime of MUF-growth on the same databases we used earlier in this section. With these databases *we notice a diminishing return on runtime after five or six metal values.* For instance, on the retail10100 database, MUF-growth showed a decrease in runtime from 127 seconds to 103 seconds as the number of metal values increased from one to six, followed by subsequent increases in runtime as additional metal values were used. With the u10k5L10100 databases, similar numbers of metals values provided a decrease in run time. On the other hand, with the retail_5060 database we notice a diminishing return on runtime after a smaller number of metal values. For all databases, at the point where the runtime starts increasing again, the upper bound given by the increased number of metal values is so close as to be indistinguishable from the bound obtained with fewer metal values, thus the extra computation is wasted.

Scalability. To test the scalability of MUF-growth, we mined frequent patterns from increasing sizes of input databases. The experimental results indicate that our algorithm (i) is scalable with respect to the number of transactions and (ii) can mine large volumes of uncertain data within a reasonable amount of time.

The experimental results show that our algorithm effectively mines frequent patterns from uncertain data irrespective of distribution of existential probability values (whether they are distributed into a narrow or wide range of values) and whether the data is dense or sparse.

6 Conclusions

In this paper, we proposed (i) the **MUF-tree structure** for capturing important information from uncertain data and (ii) the **MUF-growth algorithm** for mining frequent patterns from the MUF-tree structure. The algorithm obtains upper bounds on the expected supports of frequent patterns by accumulating item caps in its tree structure. These item caps are *compounded* with various metal values (computed based on the i-th highest existential probabilities of any item in the prefix) during the mining of potentially frequent patterns. Hence, they further tighten the upper bound on expected supports of frequent patterns when compared to the existing PUF-growth algorithm. Our proposed MUF-growth algorithm has been shown to generate significantly fewer false positives than

PUF-growth (e.g., up to 1 % of the total value). In all cases, MUF-growth significantly outperformed PUF-growth. Using extra metal values in MUF-growth to further tighten the upper bound was shown to be most effective on sparse data, providing further decreases in both false positives and runtime. Our algorithm is guaranteed to find *all* frequent patterns (with *no* false negatives). Experimental results show the effectiveness of our MUF-growth algorithm in fast mining of uncertain frequent patterns when balancing tree size and accuracy.

Acknowledgement. This project is partially supported by NSERC (Canada) and University of Manitoba.

References

1. Agrawal, R., Srikant, R.: Fast algorithms for mining association rules. In: Bocca, J.B., Jarke, M., Zaniolo, C. (eds.) VLDB 1994, pp. 487–499. Morgan Kaufmann, San Francisco (1994)
2. Aggarwal, C.C., Li, Y., Wang, J., Wang, J.: Frequent pattern mining with uncertain data. In: Elder, J.F., Fogelman-Soulié, F., Flach, P.A., Zaki, M.J. (eds.) ACM KDD 2009, pp. 29–37. ACM, New York (2009)
3. Bernecker, T., Kriegel, H.-P., Renz, M., Verhein, F., Zuefle, A.: Probabilistic frequent itemset mining in uncertain databases. In: Elder, J.F., Fogelman-Soulié, F., Flach, P.A., Zaki, M.J. (eds.) ACM KDD 2009, pp. 119–127. ACM, New York (2009)
4. Calders, T., Garboni, C., Goethals, B.: Approximation of frequentness probability of itemsets in uncertain data. In: Webb, G.I., Liu, B., Zhang, C., Gunopulos, D., Wu, X. (eds.) IEEE ICDM 2010, pp. 749–754. IEEE, Los Alamitos (2010)
5. Calders, T., Garboni, C., Goethals, B.: Efficient pattern mining of uncertain data with sampling. In: Zaki, M.J., Yu, J.X., Ravindran, B., Pudi, V. (eds.) PAKDD 2010, Part I. LNCS (LNAI), vol. 6118, pp. 480–487. Springer, Heidelberg (2010)
6. Fariha, A., Ahmed, C.F., Leung, C.K.-S., Abdullah, S.M., Cao, L.: Mining frequent patterns from human interactions in meetings using directed acyclic graphs. In: Pei, J., Tseng, V.S., Cao, L., Motoda, H., Xu, G. (eds.) PAKDD 2013, Part I. LNCS (LNAI), vol. 7818, pp. 38–49. Springer, Heidelberg (2013)
7. Fournier-Viger, P., Gomariz, A., Šebek, M., Hlosta, M.: VGEN: fast vertical mining of sequential generator patterns. In: Bellatreche, L., Mohania, M.K. (eds.) DaWaK 2014. LNCS, vol. 8646, pp. 476–488. Springer, Heidelberg (2014)
8. Han, J., Pei, J., Yin, Y.: Mining frequent patterns without candidate generation. In: Chen, W., Naughton, J.F., Bernstein, P.A. (eds.) ACM SIGMOD 2000, pp. 1–12. ACM, New York (2000)
9. Jiang, F., Leung, C.K.-S.: Stream mining of frequent patterns from delayed batches of uncertain data. In: Bellatreche, L., Mohania, M.K. (eds.) DaWaK 2013. LNCS, vol. 8057, pp. 209–221. Springer, Heidelberg (2013)
10. Jiang, F., Leung, C.K.-S., Liu, D., Peddle, A.M.: Discovery of really popular friends from social networks. In: IEEE BDCloud 2014, pp. 342–349. IEEE, Los Alamitos (2014)
11. Leung, C.K.-S.: Uncertain frequent pattern mining. In: Aggarwal, C.C., Han, J. (eds.) Frequent Pattern Mining, pp. 417–453. Springer, Switzerland (2014)

12. Leung, C.K.-S., Jiang, F.: A data science solution for mining interesting patterns from uncertain big data. In: IEEE BDCloud 2014, pp. 235–242. IEEE, Los Alamitos (2014)
13. Leung, C.K.-S., MacKinnon, R.K.: BLIMP: a compact tree structure for uncertain frequent pattern mining. In: Bellatreche, L., Mohania, M.K. (eds.) DaWaK 2014. LNCS, vol. 8646, pp. 115–123. Springer, Heidelberg (2014)
14. Leung, C.K.-S., MacKinnon, R.K., Tanbeer, S.K.: Fast algorithms for frequent itemset mining from uncertain data. In: Kumar, R., Toivonen, H., Pei, J., Huang, J.Z., Wu, X. (eds.) IEEE ICDM 2014, pp. 893–898. IEEE, Los Alamitos (2014)
15. Leung, C.K.-S., Mateo, M.A.F., Brajczuk, D.A.: A tree-based approach for frequent pattern mining from uncertain data. In: Washio, T., Suzuki, E., Ting, K.M., Inokuchi, A. (eds.) PAKDD 2008. LNCS (LNAI), vol. 5012, pp. 653–661. Springer, Heidelberg (2008)
16. Leung, C.K.-S., Tanbeer, S.K.: Fast tree-based mining of frequent itemsets from uncertain data. In: Lee, S., Peng, Z., Zhou, X., Moon, Y.-S., Unland, R., Yoo, J. (eds.) DASFAA 2012, Part I. LNCS, vol. 7238, pp. 272–287. Springer, Heidelberg (2012)
17. Leung, C.K.-S., Tanbeer, S.K.: Mining popular patterns from transactional databases. In: Cuzzocrea, A., Dayal, U. (eds.) DaWaK 2012. LNCS, vol. 7448, pp. 291–302. Springer, Heidelberg (2012)
18. Leung, C.K.-S., Tanbeer, S.K.: PUF-tree: a compact tree structure for frequent pattern mining of uncertain data. In: Pei, J., Tseng, V.S., Cao, L., Motoda, H., Xu, G. (eds.) PAKDD 2013, Part I. LNCS (LNAI), vol. 7818, pp. 13–25. Springer, Heidelberg (2013)
19. Tong, Y., Chen, L., Cheng, Y., Yu, P.S.: Mining frequent itemsets over uncertain databases. PVLDB 5(11), 1650–1661 (2012)
20. Zhang, Q., Li, F., Yi, K.: Finding frequent items in probabilistic data. In: Wang, J.T.-L. (ed.) ACM SIGMOD 2008, pp. 819–832. ACM, New York (2008)

Secure Outsourced Frequent Pattern Mining by Fully Homomorphic Encryption

Junqiang Liu[1]([✉]), Jiuyong Li[2], Shijian Xu[1], and Benjamin C.M. Fung[3]

[1] School of Information and Electronic Engineering,
Zhejiang Gongshang University, Hangzhou 310018, China
`jjliu@alumni.sfu.ca`
[2] School of Information Technology and Mathematical Sciences,
University of South Australia, Adelaide, SA 5001, Australia
[3] School of Information Studies, McGill University,
Montreal, QC H3A 1X1, Canada

Abstract. With the advent of the big data era, outsourcing data storage together with data mining tasks to cloud service providers is becoming a trend, which however incurs security and privacy issues. To address the issues, this paper proposes two protocols for mining frequent patterns securely on the cloud by employing fully homomorphic encryption. One protocol requires little communication between the client and the cloud service provider, the other incurs less computation cost. Moreover, a new privacy notion, namely α-pattern uncertainty, is proposed to reinforce the second protocol. Our scenario has two advantages: one is stronger privacy protection, and the other is that the outsourced data can be used in different mining tasks. Experimental evaluation demonstrates that the proposed protocols provide a feasible solution to the issues.

Keywords: Privacy and security · Big data · Data mining · Frequent patterns · Cloud computing

1 Introduction

The big data era is coming with exponentially growing data and increasingly complicated technologies. On one hand, big data are becoming important assets, and people are more and more interested in applying data mining technology to better utilize such assets. On the other hand, the huge volume of big data and the great complexity of technologies make it hard for an average user or business to manage and analyze their data. Therefore, outsourcing both data storage and data mining to cloud service providers is becoming a trend.

While outsourcing data storage and data mining benefits from the scale of economy and greatly reduces the complexity of deploying information technology, it comes with the privacy and security issues [10], one of which is the risk of disclosing sensitive information in the mining process.

In the literature, there are many works on privacy preserving data mining. Some works [8,15,17,21] retain the mining models at the aggregate level and

© Springer International Publishing Switzerland 2015
S. Madria and T. Hara (Eds.): DaWaK 2015, LNCS 9263, pp. 70–81, 2015.
DOI: 10.1007/978-3-319-22729-0_6

preserve the privacy at the data record level. Some [5,20] preserve the privacy of data records when publishing the mining results. Others [12,19,22] retain the precise supports of patterns while observing k-support anonymity. However, no prior work preserves both the privacy of data and the privacy of mining results while retaining the exactness of the mining results in outsourcing frequent pattern mining. Moreover, the randomization and group based approaches [2] employed by the prior works cannot deal with such a scenario.

Fortunately, fully homomorphic encryption emerged as the most promising method to address security and privacy issues in cloud computing. The concept of full homomorphism was first introduced by Rivest et al. [16] in 1978, and not realized until Gentry [9] proposed an ideal lattice based scheme in 2009. Since then, there were efforts applying homomorphic encryption in data mining, for example, in answering private queries [13] and in similarity ranking [6], but to the best of our knowledge no work in frequent pattern mining.

In this paper, we consider a secured outsourcing scenario, that is, to keep encrypted data in the storage provided by a cloud service provider and to outsource data mining tasks to the service provider by employing algorithms that can work on the encrypted data directly. Such a scenario has two advantages: One is that the encrypted data can be used in various mining tasks while the randomized data by the priori works can only be used in a specific task. The other advantage is that by employing fully homomorphic encryption, it releases much less information in the mining process and thus provides more protection.

Our scenario is nontrivial as the fully homomorphic encryption supports no comparison operation, i.e., no predicate for a total order. To deal with such a challenge, we propose two secured protocols. The main contributions are as follows: First, we propose one protocol, for outsourcing both the encrypted data and the frequent pattern mining task to a cloud service provider, that requires little communication between the cloud service provider and the client. Second, we propose a new privacy notion, namely α-pattern uncertainty, and implement this privacy notion by shadow mappings. By employing both α-pattern uncertainty and fully homomorphic encryption, we propose another secured protocol for mining frequent patterns, which incurs less computation cost. Experimental evaluation shows that our protocols provide a feasible solution to addressing security and privacy issues when outsourcing frequent pattern mining.

The rest of this paper is organized as follows: Sect. 2 reviews the related works. Section 3 presents our scenario for securing outsourced frequent pattern mining. Section 4 proposes our first secured protocol based on fully homomorphic encryption. Section 5 proposes the notion of α-pattern uncertainty and our second protocol. Section 6 evaluates our protocols. Section 7 concludes the paper.

2 Related Works

Our work relates to prior works on privacy preserving frequent pattern mining and association rule mining. We can only review the most relevant and representative works due to space limit.

The works [8,15,17,21] retain the data mining models at the aggregate level and preserve the privacy at the level of individual data records by random perturbation or approximation [2]. Concretely, [8,17] proposed to randomize the data for limiting privacy breaches and to send the randomized data to the server for data mining. [15] presented a Bloom filter based solution for outsourcing the task of mining association rules. [21] proposed to generate synthetic data in such a way that the frequent patterns discovered in the original data can be mined in the synthetic data and the privacy leakage can be limited.

The works [5,20] preserve the privacy of individual records when publishing the mining results. [20] hides a subset of sensitive association rules by adding items to the original data or removing items to tune the supports and confidences of sensitive rules. [5] deals with privacy threats embedded in mining results by distorting the supports of patterns to eliminate inference channels.

The works [12,19,22] retain precise patterns with precise supports and observe the k-support anonymity notion, i.e., each (transformed) item is indistinguishable from at least $k - 1$ other items w.r.t. their supports, under the assumption that the attacker knows the exact supports of certain items. Their approach is to map items to meaningless symbols and to add noise to prevent re-identification. In [12], the noise is a set of fake transactions consisting of the original items. In [22], the noise is introduced by adding fake items into the original transactions. In [19], a pseudo taxonomy tree is employed to facilitate hiding of the original items and to limit the fake items.

Our work is the first that employs fully homomorphic encryption in privacy preserving frequent pattern mining although there are works employing fully homomorphic encryption in other tasks. For example, Hu et al. [13] answered private queries over a data cloud, and Chu et al. [6] proposed a privacy preserving similarity ranking protocol based on fully homomorphic encryption.

3   Preliminaries

We consider a secured scenario for outsourcing frequent pattern mining to a cloud service provider with a guarantee of the privacy and the exactness of both the data and the mining results, faciliatated by fully homomorphic encryption.

3.1   Frequent Pattern Mining Problem

Frequent pattern mining is one of the fundamental data mining problems [3,11] with a variety of applications, for example, consumer behavior analysis, design of goods shelves, inventory control, product promotion, and so on. Given a database consisting of a group of transactions, frequent patterns are sets of items that are present in more than a given number of transactions as defined as follows.

Definition 1 (Frequent Patterns). *Let $I = \{i_1, i_2, \cdots, i_m\}$ be a set of items, a database $D = \{(tid_1, t_1), (tid_2, t_2), \cdots, (tid_n, t_n)\}$ is a set of transactions where tid_k is the identifier of transaction $t_k \subseteq I$. A pattern $p \subseteq I$ is a set of items*

Table 1. Transaction database D in different representations

(a) D as sets

tid	items
1	{a, b, c, d, f}
2	{a, b, c, e}
3	{b, f}
4	{a, b, c, d }
5	{a, c, e}

(b) D as bit vectors

tid	a	b	c	d	e	f
1	1	1	1	1	0	1
2	1	1	1	0	1	0
3	0	1	0	0	0	1
4	1	1	1	1	0	0
5	1	0	1	0	1	0

(c) $E(D)$: the encrypted

tid	$\tilde{a}$	$\tilde{b}$	$\tilde{c}$	$\tilde{d}$	$\tilde{e}$	$\tilde{f}$
1	$1'$	$1'$	$1'$	$1'$	$0'$	$1'$
2	$1'$	$1'$	$1'$	$0'$	$1'$	$0'$
3	$0'$	$1'$	$0'$	$0'$	$0'$	$1'$
4	$1'$	$1'$	$1'$	$1'$	$0'$	$0'$
5	$1'$	$0'$	$1'$	$0'$	$1'$	$0'$

from I, and p is contained by a transaction (tid, t) if $p \subseteq t$. The number of transactions in D containing p is the absolute support of p, which is denoted as $p.supp$. Any pattern is a frequent pattern if its support is equal to or greater than a user-defined minimum threshold, $minSup$.

Running Example: Given the database D in Table 1(a) and $minSup = 3$, all frequent patterns are {a}, {b}, {c}, {a,b}, {a,c}, {b,c}, and {a,b,c}.

While some mining algorithms [4,14] represent transactions as sets of items as in Table 1(a), some other algorithms [1,18] employ the bit vector representation where there is a bit vector for each transaction with each bit indicating the corresponding item's presence in or absence from the transaction. For example, Table 1(b) is the bit vector representation of D in Table 1(a).

3.2 Secure Outsourced Mining by Fully Homomorphic Encryption

With our secured scenario, the database is represented by bit vectors [1,18], and encrypted and kept in the cloud in advance, from which various types of knowledge can be discovered by outsourcing the corresponding mining tasks.

The encrypted database, denoted as $E(D) := \mathbf{Encrypt}(pk, \mathbf{Mask}(D))$, is derived as follows. First, the **Mask** function is called to map the items in I to meaningless symbols in a set $\hat{I}$ and to rearrange rows and columns in the bit-matrix, i.e., the bit vector representation. Second, the bits in $\mathbf{Mask}(D)$ are encrypted by calling the **Encrypt** function provided by the fully homomorphic encryption scheme [7,9].

The scheme [7,9] supports any computation that can be expressed by a boolean circuit $\mathbf{C}$ that consists of AND-gates and XOR-gates and takes as input the plaintext bits. Concretely, this scheme is defined by four functions: a key generation function, **KeyGen**, an encryption function, **Encrypt**, a decryption function, **Decrypt**, and an evaluation function, **Evaluate**.

– **KeyGen**(λ) generates a secret key sk and a public key pk given a security parameter λ. The sk is an odd η-bit integer, and the pk is defined as (pk^*, y) where pk^* is a set of $\tau + 1$ random γ-bit integers with $pk_i^* = sk \cdot q_i + 2r_i$ and q_i and r_i randomly selected for $i \in \{0, \cdots, \tau\}$, and y is a vector of Θ-components with κ bits of precision after the binary point that will be used with the encrypted secret key to evaluate a circuit homomorphically.

- **Encrypt**$(pk, m_1, \cdots, m_s)$ encrypts any s plaintext bits by the public key pk. The ciphertext c_i of $m_i \in \{0, 1\}$ for $i \in \{1, \cdots, s\}$ is the sum of a subset of pk^* plus m_i and random noise $2r$.
- **Decrypt**$(sk, c_1, \cdots, c_t)$ decrypts any t ciphertexts by the secret key sk. Each plaintext bit m_i of c_i for $i \in \{1, \cdots, t\}$ is recovered as $(c_i \bmod sk) \bmod 2$.
- **Evaluate**$(pk, \mathbf{C}, c_1, \cdots, c_t)$ takes as input a public key pk, a boolean circuit $\mathbf{C}$, and t ciphertexts c_i, and correctly evaluates the circuit $\mathbf{C}$ on ciphertexts in that if $c_i = \mathbf{Encrypt}(pk, m_i)$ for $i \in \{1, \cdots, t\}$, then $\mathbf{Decrypt}(sk, \mathbf{Evaluate}(pk, \mathbf{C}, c_1, \cdots, c_t)) = \mathbf{C}(m_1, \cdots, m_t)$.

For the running example, Table 1(c) shows $E(D)$ where $I = \{$a, b, c, d, e, f$\}$, $\hat{I} = \{\hat{a}, \hat{b}, \hat{c}, \hat{d}, \hat{e}, \hat{f}\}$, **Mask**$(a) = \hat{a}$, **Mask**$^{-1}(\hat{a}) = a$, and so on. Each entry in Table 1(c) is the encryption of the corresponding bit in Table 1(b).

An important observation is that the encryption of any bit b contains random noise, and thus two invocation of **Encrypt**(pk, b) will yield different results. Therefore, no comparison is supported on encrypted bits, which has profound impact on our design of protocols.

3.3 A Variant of the Apriori Algorithm

The Apriori algorithm [4, 14] is the most influential algorithm for mining frequent patterns, which represents a transaction as a set of items as in Table 1(a). The **BinaryApriori** algorithm in Algorithm 1 is a variant of Apriori for mining data in the bit vector representation [1, 18] as in Table 1(b), which may be converted into a homomorphic counterpart that can mine the encrypted data.

Algorithm 1.   BinaryApriori$(D, minSup)$

1: **for each** candidate $c \in I$ **do** $c.supp :=$ **countSupport**(c, D);
2: $L_1 := \{c \in I | c.supp \geq minSup\}$;
3: **for** ($i := 1$; $L_i \neq \emptyset$; $i{+}{+}$) **do**
4:   $C_{i+1} :=$ **genCandidates**(L_i);
5:   **for each** candidate $c \in C_{i+1}$ **do** $c.supp :=$ **countSupport**(c, D);
6:   $L_{i+1} := \{c \in C_{i+1} | c.supp \geq minSup\}$;
7: **end for**

BinaryApriori computes the set L_1 of frequent patterns of length 1 by counting every item's support (at lines 1 to 2), and then finds frequent patterns of length $i + 1$ for $i \geq 1$ iteratively (at lines 3 to 7) as follows.

First, **BinaryApriori** generates the set C_{i+1} of candidates of length $i + 1$ by calling the **genCandidates** function (at line 4). This function employs the monotone property: the subsets of a frequent pattern must be frequent.

Second, **BinaryApriori** counts the support of every candidate $c \in C_{i+1}$ (at line 5), and identifies those candidates whose supports are no less than $minSup$ to get L_{i+1} (at line 6). The **countSupport** function computes the support of c in D by first performing bitwise AND-operation on the columns corresponding to the constituent items of c and then summing up the resulting bits.

For the running example, to count the support of {a, b}, **countSupport** first performs the bitwise AND-operation on columns a and b in Table 1(b), i.e. $(1, 1, 0, 1, 1)^T$ and $(1, 1, 1, 1, 0)^T$, to produce $(1, 1, 0, 1, 0)^T$, and then sums up the bits to get the result 3. In short, **countSupport** can be expressed by a boolean circuit of AND-gates and OR-gates.

4   Secured Protocol Based on FHE for Pattern Mining

It is nontrivial to protect the security and privacy of outsourced data and frequent pattern mining based on fully homomorphic encryption (FHE), since no comparison can be performed on encrypted data and hence the server, i.e., the cloud service provider, cannot run a frequent pattern mining algorithm on encrypted data without the help of the client. Therefore, we propose a protocol, namely **Sphene - S**ecured protocol by **h**omomorphic **en**umeration, that avoids the comparison operation on the encrypted data at the server side.

In a preparation step of the Sphene protocol (Step 0 in Fig. 1), the client's encrypted database, $E(D)$, is outsourced to the server in advance, and the client keeps a secret key sk to himself and shares a public key pk with the server.

In interaction steps (Steps 1 - 4 in Fig. 1), the server enumerates patterns of a given length and counts the number of frequent patterns in the encrypted data homomorphically, and the client controls the iterations over the pattern length, which is detailed in the following.

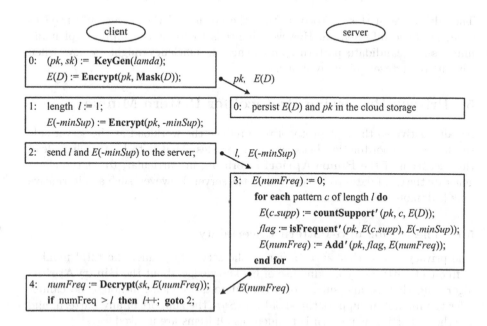

Fig. 1. Sphene - our first protocol for mining frequent patterns based on FHE

Step 1: The client encrypts the complement of $minSup$ in $\lceil log_2|D| \rceil + 2$ bits, denoted as $E(-minSup)$, and starts the iterations with length l of 1.

Step 2: The client sends a pattern length l to the server together with the encrypted complement of $minSup$, $E(-minSup)$, which will be used by the server to evaluate if a pattern is frequent homomorphically.

Step 3: The server enumerates every pattern c of length l, computes the encrypted support of c and an encrypted bit $flag$ that is an encryption of 1 if c is frequent or that of 0, and adds up such bits to get $E(numFreq)$, i.e., the encrypted number of frequent patterns of length l, by employing the following.

– **countSupport′**$(pk, c, E(D))) = $ **Evaluate**$(pk, $ **countSupport**$, c, E(D))$ to compute the encrypted support of c homomorphically as in Sect. 3.
– **isFrequent′**$(pk, E(c.supp), E(-minSup))$ to evaluate 1 XOR-ed with the most significant "bit" of **Add′**$(pk, E(c.supp), E(-minSup))$.
– **Add′**$(pk, \cdot, \cdot)$ to evaluate the sum of encrypted integers homomorphically.

Step 4: The client gets $numFreq$ by decrypting the encrypted number of frequent patterns sent back from the server. If $numFreq$ is greater than l, the client starts the next iteration, otherwise it stops since a frequent pattern of length $l + 1$ should have $l + 1$ frequent subsets of length l.

Theorem 1. *The Sphene protocol is correct, and is secured in the sense that no one except the client can know the minimum support threshold and the support of a pattern, and can infer if a pattern is frequent. An attacker can only infer that the length of a frequent pattern is no more the number of iterations.*

The advantage of this protocol is that the client and the server only need to exchange little information. However, it needs to count the supports of many unnecessary candidate patterns, and hence its efficiency might be low when running on datasets of many items.

5   Privacy Preserving Protocol for Pattern Mining

An alternative to the first protocol is to reduce the workload at the server side through close coordination between the two sides. That is to let the client lead the execution of the **Binary Apriori** algorithm, in particular, to let the client generate the candidates to be counted by the server. However, such an alternative may leak more information.

5.1   The Notion of α-Pattern Uncertainty

The privacy risk is that an attacker or the server may infer the total number of frequent patterns by making use of his knowledge about the **Binary Apriori** algorithm, that is, any subset of a candidate is frequent, although he cannot infer the minimum support threshold $minSup$, the support of any pattern, and whether a particular pattern is frequent as all items are masked.

We propose a new privacy notion, namely α-pattern uncertainty, to limit the server's certainty in inferring the total number of frequent patterns.

Definition 2 (α-pattern Uncertainty). *A set M of (masked) patterns is α-uncertain, if an attacker's certainty in inferring any pattern in M to be a true candidate pattern is no more than a privacy parameter α set by the client.*

For example, if the client sends the masked version of the true candidate pattern set, $\hat{C}_2 = \{\{\hat{a}, \hat{b}\}, \{\hat{a}, \hat{c}\}, \{\hat{b}, \hat{c}\}\}$ to the server, the server will infer that there are 3 frequent items although he does not know what $\hat{a}$, $\hat{b}$, and $\hat{c}$ stand for. But, if the client sends an expanded set $E\hat{C}_2 = \hat{C}_2 \cup \{\{\hat{d}, \hat{e}\}, \{\hat{d}, \hat{f}\}, \{\hat{e}, \hat{f}\}\}$, the attacker's certainty of inference may drop to 50 %.

Definition 3 (shadow mappings and shadow patterns). *The shadow mappings, denoted as $sMaps$, are $(1/\alpha - 1)$ one-to-one and onto functions, that have the same domain and disjoint ranges, from frequent items to infrequent items. For each pattern c, shadow patterns of c are generated by applying on c the functions from $sMaps$, denoted as* **shadows**$(c, sMaps)$.

To investigate the feasibility of applying FHE in outsourcing frequent patten mining, a simple technique, shadow mappings (patterns), is proposed, which however may be subject to further attacks. We leave as a future work an improved technique that prevents any further inference.

5.2 Privacy Preserving Protocol for Counting Candidates

Our second protocol, namely **P3CC - P**rivacy **P**reserving **P**rotocol for **C**ounting **C**andidates homomorphically in outsourcing frequent pattern mining as shown in Fig. 2, employs both α-pattern uncertainty and fully homomorphic encryption as privacy and security measures.

The preparation step of the P3CC protocol (Step 0 in Fig. 2) is the same as the Sphene protocol. In the interaction steps, the client generates the set of candidate patterns and the set of shadow patterns, and sends the masked version EC of their union to the server. The server homomorphically computes the support of each pattern in EC, and sends the result to the client. The client recovers and decrypts the result and gets the frequent patterns. The interaction is further explained with the running example in the following.

Step 1: The client sends the masked candidate set $EC = \mathbf{Mask}(C_1)) = \mathbf{Mask}(I)) = \{\{\hat{a}\}, \{\hat{b}\}, \{\hat{c}\}, \{\hat{d}\}, \{\hat{e}\}, \{\hat{f}\}\}$ to the server.

Step 2: The server counts the support of each masked candidate pattern homomorphically, and sends the result back to client in the encrypted form.

Step 3: The client recovers the candidate set from the masked one, decrypts the support of each candidate in $C_1 = EC'$ and determines the set of frequent patterns of length 1, $L_1 = \{\{a\}, \{b\}, \{c\}\}$. Suppose $\alpha = 50\%$, then only one function is in the shadow mappings, e.g., $sMaps = \{\{(a, d), (b, e), (c, f)\}\}$.

Step 4: The client gets the candidate set $C_2 = \{\{a, b\}, \{a, c\}, \{b, c\}\}$, and sends the masked and expanded candidate set $EC_2 = \{\{\hat{a}, \hat{b}\}, \{\hat{a}, \hat{c}\}, \{\hat{b}, \hat{c}\}, \{\hat{d}, \hat{e}\}, \{\hat{d}, \hat{f}\}, \{\hat{e}, \hat{f}\}\}$ to the server.

Step 5: The server counts the supports and sends the encrypted result back.

Step 6: The client recovers and decrypts the support of each pattern in EC_2 and gets $L_2 = \{\{a, b\}, \{a, c\}, \{b, c\}\}$. As the size of L_2 is greater than 2, the protocol enters a new round by jumping to Step 4, and so on.

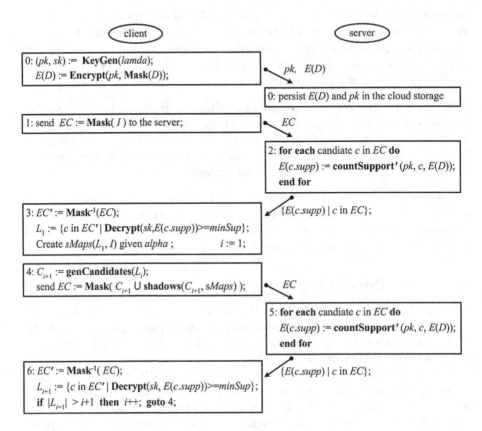

Fig. 2. P3CC - our second protocol for mining frequent patterns

Theorem 2. *The P3CC protocol is correct, and is secured in the sense that no one except the client can know the minimum support threshold and the support of a pattern, and can infer if a pattern is frequent. An attacker can only infer that the length of a frequent pattern is no more the number of iterations, and infer the total number of frequent patterns with a certainty no more than α.*

6 Experimental Evaluation

This section evaluates our protocols experimentally. To the best of our knowledge, there is no prior work on FHE-based frequent pattern mining, and the related works [5,6,8,12,13,15,17,19–22] consider scenarios that are not comparable to ours. We can only evaluate our protocols, Sphene and P3CC.

Two datasets are used in experiments. The first is the artificial dataset, T10I6N50D5kL1k, generated by the generator from the IBM Almaden Quest research group[1]. The second is the Chess dataset from FIMI[2].

The protocols were implemented upon a general-purposed FHE library that supports the scheme [7] and uses the GNU multiple precision arithmetic library[3]. The implementation renders parallelism by multi-threading. The experiments were performed on an HP Pavilion dm4 laptop running Ubuntu 12.04.4.

We first evaluate the Sphene and P3CC protocols on the T10I6N50D5kL1k and chess datasets with varying the minimum support threshold, $minSup$. For T10I6N50D5kL1k as shown in Fig. 3(a), P3CC is up to 3 orders of magnitude more efficient than Sphene. For Chess as shown in Fig. 3(b), P3CC is up to 5 orders of magnitude more efficient than Sphene.

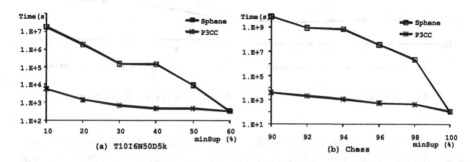

Fig. 3. Running time of two protocols with varying $minSup$ on two datasets

We then evaluate the protocols with varying data features and privacy requirements. First, Fig. 4(a) shows the result for the artificial dataset with the number (D) of transactions ranging from 1,000 (D1k) to 10,000 (D10k), which depicts that both P3CC and Sphene scale almost linearly in database size.

Second, Fig. 4(b) shows the result with the number (N) of distinct items ranging from 10 (N10) to 100 (N100). While the performance of P3CC is not highly correlated to this feature, the performance of Sphene fluctuates depending on which of the two opposite effects with the increase of the number of items will dominate: one is that the number of candidates in a protocol iteration increases, the other effect is the the number of iterations decreases (as data get sparser).

Third, Fig. 4(c) shows the result with the number (L) of possible frequent patterns ranging from 1,000 (L1k) to 6,000 (L6k), which indicates this feature has little impact on the performance of P3CC and Sphene.

Finally, Fig. 4(d) shows the result with the varying privacy parameter α with $k = 1/\alpha$ ranging from 1 to 6. While Sphene does not employ the privacy preserving measure, P3CC performs linearly in the privacy parameter.

---

[1] http://miles.cnuce.cnr.it/~palmeri/datam/DCI/datasets.php/.

[2] http://fimi.cs.helsinki.fi/data/.

[3] https://gmplib.org/.

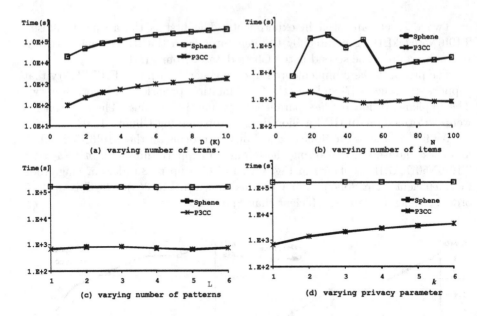

Fig. 4. Evaluation with varying data features on the artificial dataset

In summary, the experimental evaluation demonstrates that while the Sphene protocol may be not efficient enough for real applications, the P3CC protocol provides a solution for outsourcing both data storage and frequent pattern mining tasks, although P3CC has higher communication cost than Sphene.

7   Conclusions and Future Work

This paper proposes two secured protocols for mining frequent patterns on the cloud, which secures the data and the mining process by employing fully homomorphic encryption (FHE), and addresses privacy issues by introducing the α-pattern uncertainty and reinforcing the second protocol by shadow mappings. Preliminary experimental evaluation shows the feasibility of the protocols.

The future work is to address the limitations of this preliminary work, which includes: (1) new techniques other than shadow mappings that prevent further inference attacks when observing the α-pattern uncertainty; (2) handling possible attacks on the α-pattern uncertainty and on FHE by formal analysis; (3) distributed mining approaches based on the MapReduce and Spark framework to improve the efficiency; (4) new homomorphic encryption schemes customized for frequent pattern mining with lower time complexity.

Acknowledgements. This work was supported in part by the National Natural Science Foundation of China (61272306), and the Zhejiang Provincial Natural Science Foundation of China (LY12F02024).

References

1. Agarwal, R., Aggarwal, C., Prasad, V.V.V.: Depth first generation of long patterns. In: 6th SIGKDD, pp. 108–118 (2000)
2. Aggarwal, C.C., Yu, P.S.: Privacy-preserving Data Mining: Models and Algorithms. Springer-Verlag, Boston (2008)
3. Agrawal, R., Imielinski, T., Swami, A.: Mining association rules between sets of items in large databases. In: 1993 SIGMOD, pp. 207–216 (1993)
4. Agrawal, R., Srikant, R.: Fast algorithms for mining association rules. In: Research Report RJ 9839. IBM Almaden Research Center, San Jose, CA (1994)
5. Atzori, M., Bonchi, F., Giannotti, F., Pedreschi, D.: Anonymity preserving pattern discovery. VLDB J. **17**(4), 703–727 (2008)
6. Chu, Y.-W., Tai, C.-H., Chen, M.-S., Yu, P.S.: Privacy-preserving simrank over distributed information network. In: 12th ICDM, pp. 840–845 (2012)
7. van Dijk, M., Gentry, C., Halevi, S., Vaikuntanathan, V.: Fully homomorphic encryption over the integers. In: Gilbert, H. (ed.) EUROCRYPT 2010. LNCS, vol. 6110, pp. 24–43. Springer, Heidelberg (2010)
8. Evfimievski, A., Gehrke, J., Srikant, R.: Limiting privacy breaches in privacy preserving data mining. In: 22nd PODS, pp. 211–222 (2003)
9. Gentry, C.: Fully Homomorphic encryption using ideal lattices. In: 41st ACM Symposium on Theory of Computing, pp. 169–178 (2009)
10. Gellman, R.: Privacy in the clouds: risks to privacy and confidentiality from cloud computing. In: World Privacy Forum, 23 February (2009)
11. Goethals, B.: Survey on frequent pattern mining. Technical report, University of Helsinki (2003)
12. Giannotti, F., Lakshmanan, L.V.S., Monreale, A., Pedreschi, D., Wang, H.: Privacy-preserving mining of association rules from outsourced transaction databases. IEEE Syst. J. **7**(3), 385–395 (2012)
13. Hu, H., Xu, J., Ren, C., Choi, B.: Processing private queries over untrusted data cloud through privacy homomorphism. In: 27th ICDE, pp. 601–612 (2011)
14. Mannila, H., Toivonen, H., Verkamo, A.I.: Efficient algorithms for discovering association rules. In: AAAI Workshop Knowledge Discovery in Databases, KDD 1994, pp. 181–192 (1994)
15. Qiu, L., Li, Y., Wu, X.: Protecting business intelligence and customer privacy while outsourcing data mining tasks. Knowl. Inf. Syst. **17**(1), 99–120 (2008)
16. Rivest, R.L., Adleman, L., Dertouzos, M.L.: On data banks and privacy homomorphisms. In: DeMillo, R.A., et al. (eds.) Foundations of Secure Computation, pp. 169–179. Academic Press, New York (1978)
17. Rizvi, S., Haritsa, J.R.: Maintaining data privacy in association rule mining. In: 28th VLDB, pp. 682–693 (2002)
18. Shenoy, P., Haritsa, J.R., Sudarshan, S., Bhalotia, G., Bawa, M., Shah, D.: Turbocharging vertical mining of large databases. In: 2000 SIGMOD, pp. 22–33 (2000)
19. Tai, C. H., Yu, P.S., Chen, M.S.: k-support anonymity based on pseudo taxonomy for outsourcing of frequent itemset mining. In: 16th SIGKDD, pp. 473–482 (2010)
20. Verykios, V.S., Elmagarmid, A.K., Bertino, E., Saygin, Y., Dasseni, E.: Association rule hiding. IEEE Trans. Knowl. Data Eng. **16**(4), 434–447 (2004)
21. Wang, Y., Wu, X.: Approximate inverse frequent itemset mining: privacy, complexity, approximation. In: 5th ICDM, pp. 482–489 (2005)
22. Wong, W.K., Cheung, D.W., Hung, E., Kao, B., Mamoulis, N.: Security in outsourcing of association rule mining. In: 33rd VLDB, pp. 111–122 (2007)

Supervised Evaluation of Top-k Itemset Mining Algorithms

Claudio Lucchese[1], Salvatore Orlando[2(✉)], and Raffaele Perego[1]

[1] ISTI-CNR, Pisa, Italy
[2] DAIS - Università Ca' Foscari, Venice, Italy
orlando@unive.it

Abstract. A major mining task for binary matrixes is the extraction of approximate top-k patterns that are able to concisely describe the input data. The top-k pattern discovery problem is commonly stated as an optimization one, where the goal is to minimize a given cost function, e.g., the accuracy of the data description.

In this work, we review several greedy state-of-the-art algorithms, namely Asso, Hyper+, and PANDA$^+$, and propose a methodology to compare the patterns extracted. In evaluating the set of mined patterns, we aim at overcoming the usual assessment methodology, which only measures the given cost function to minimize. Thus, we evaluate how good are the models/patterns extracted in unveiling supervised knowledge on the data. To this end, we test algorithms and diverse cost functions on several datasets from the UCI repository. As contribution, we show that PANDA$^+$ performs best in the majority of the cases, since the classifiers built over the mined patterns used as dataset features are the most accurate.

1 Introduction

Binary matrixes can be derived from several typologies of datasets, collected in diverse and popular application domains. Without loss of generality, we can think of a binary matrix as a representation of a transactional database, composed of a multi-set of transactions (matrix rows), each including a set of items (matrix columns). An *approximate pattern* extracted from a binary matrix thus corresponds to a pair of sets, items and transactions, where the items of the former set are mostly included in all the transactions of the latter set. The cardinality of the transaction set (matrix rows) can also be seen as the approximate support of the corresponding set of items.

Top-k pattern mining is an alternative approach to pattern enumeration. It aims at discovering the (small) set of k patterns that best *describes*, or *models*, the input dataset. State-of-the-art algorithms differ in the formalization of the above concept of *dataset description*. For instance, in [1] the goodness of the description is given by the number of occurrences in the dataset incorrectly modeled by the extracted patterns, while shorter and concise patterns are promoted in [2,3]. The goodness of a description is measured with some cost function,

© Springer International Publishing Switzerland 2015
S. Madria and T. Hara (Eds.): DaWaK 2015, LNCS 9263, pp. 82–94, 2015.
DOI: 10.1007/978-3-319-22729-0_7

and the top-k mining task is casted into an optimization of such cost. In most of such formulations, the problem is demonstrated to be NP-hard, and therefore greedy strategies are adopted. At each iteration, the pattern that best optimizes the given cost function is added to the solution. This is repeated until k patterns have been found or until it is not possible to improve the cost function.

In this paper we analyze in depth three state-of-the-art algorithms for mining approximate top-k pattern from binary data: Asso [1], Hyper+ [2] and PANDA+ [4]. Indeed, the cost functions adopted by Asso [1] and Hyper+ [2] share important aspects that can be generalized into a unique formulation. The PANDA+ framework can be plugged with such generalized formulation, which makes it possible to greedily mine approximate patterns according to several cost functions, including the ones proposed in [3,6]. PANDA+ also permits expressing noise constraints [5].

Concerning the evaluation methodology, we observe that state-of-the-art algorithms for approximate top-k mining measure the goodness of the discovered patterns by their capability of minimizing the same cost function optimized by the greedy algorithm. This simple assessment methodology captures the effectiveness of the greedy strategy rather than the quality of the extracted patterns. In this paper, we want to go beyond this common evaluation approach by adopting an assessment methodology aimed at measuring also how good are the concise models extracted in unveiling some hidden supervised knowledge, in particular the class labels associated with transactions.

We test this capability by using the algorithms for approximate top-k mining as a sort of feature extractors. The accuracy of the classifiers built on top of the extracted features is then considered a proxy for the quality of the mined patterns. In this way, we are able to complement internal indices of quality (cost function), with external ones (classification accuracy). PANDA+, which is able to optimize several cost function, permits to extract those patterns/features that when used to train a classifier, produces models that are in the majority of the cases the most accurate.

2   Problem Statement and Algorithms

In this sections we first introduce some notation and the problem statement, and then we briefly discuss the main features and peculiarities of three state-of-the-art algorithms solving our approximate top-k pattern mining problem.

2.1   Notation and Problem Statement

A *transactional dataset* of N transactions and M items can be represented by a *binary* matrix $\mathcal{D} \in \{0,1\}^{N \times M}$ where $\mathcal{D}(i,j) = 1$ if the i–th item occurs in the j–th transaction, and $\mathcal{D}(i,j) = 0$ otherwise. An *approximate pattern* P is identified by the set of items it contains and the set of transactions where these items (partially) occur. We represent these two sets as binary vectors $P = \langle P_I, P_T \rangle$, where $P_I \in \{0,1\}^M$ and $P_T \in \{0,1\}^N$. The outer product $P_T \cdot$

$P_I^\mathsf{T} \in \{0,1\}^{N \times M}$ identifies a sub-matrix of $\mathcal{D}$. Being each pattern approximate, the sub-matrix should mainly cover 1-bits in $\mathcal{D}$ (*true positives*), but it may also cover a few 0-bits too (*false positives*).

Finally, let $\Pi = \{P_1, \ldots, P_{|\Pi|}\}$ be a set of *overlapping patterns*, which approximately cover dataset $\mathcal{D}$, except for some noisy item occurrences, identified by matrix $\mathcal{N} \in \{0,1\}^{N \times M}$:

$$\mathcal{N} = \bigvee_{P \in \Pi} (P_T \cdot P_I^\mathsf{T}) \;\veebar\; \mathcal{D}. \tag{1}$$

where $\vee$ and $\veebar$ are respectively the element-wise *logical or* and *xor* operators. Note that some 1-bits in $\mathcal{D}$ may not be covered by any pattern in Π (*false negatives*). Indeed, our formulation of noise (matrix $\mathcal{N}$) models both *false positives* and *false negatives*. If an occurrence $\mathcal{D}(i,j)$ corresponds to either a false positive or a false negative, we have that $\mathcal{N}(i,j) = 1$.

On the basis of this notation, we can state the top-k pattern discovery problem as an optimization one:

Problem 1 (Approximate Top-k Pattern Discovery Problem). Given a binary dataset $\mathcal{D} \in \{0,1\}^{N \times M}$ and an integer k, find the pattern set $\overline{\Pi}_k$, $|\overline{\Pi}_k| \leq k$, that minimizes the given cost function $J(\Pi_k, \mathcal{D})$:

$$\overline{\Pi}_k = \underset{\Pi_k}{argmin}\, J(\Pi_k, \mathcal{D}) \tag{2}$$

In the following we review some cost functions and the algorithms that adopt them. These algorithms try to optimize specific functions J with some greedy strategy, since the problem belongs to the NP class. In addition, they exploit some specific *parameters*, whose purpose is to make the pattern set Π_k subject to particular *constraints*, with the aim of (1) reducing the algorithm search space, or (2) possibly avoiding that the greedy generation of patterns brings to local minima. As an example of the former type of parameters, we mention the frequency of the pattern. Whereas, for the latter type of parameters, an example is the amount of false positives we can tolerate in each pattern.

2.2   Minimizing Noise (ASSO)

ASSO [1] is a greedy algorithm aimed at finding the pattern set Π_k that minimizes the amount of noise in describing the input data matrix $\mathcal{D}$. This is measured as the L^1-norm $\|\mathcal{N}\|$ (or Hamming norm), which simply counts the number of 1 bits in matrix $\mathcal{N}$ as defined in Eq. (1). ASSO is thus a greedy algorithm minimizing the following function:

$$J_A(\Pi_k, \mathcal{D}) = \|\mathcal{N}\|. \tag{3}$$

Indeed, ASSO aims at finding a solution for the *Boolean matrix decomposition problem*, thus identifying two low-dimensional factor binary matrices of rank k, such that their *Boolean product* approximates $\mathcal{D}$. The authors of ASSO called this matrix decomposition problem the Discrete Basis Problem (DBP). It can be shown that the DBP problem is equivalent to the approximate top-k pattern mining problem when optimizing J_A. The authors prove that the decision version of the problem is NP-complete by reduction to the set basis problem, and that J_A cannot be approximated within any factor in polynomial time, unless P = NP.

ASSO works as follows. First it creates a set of candidate item sets, by measuring the correlation between every pair of items. The minimum confidence parameter τ is used to determine whether two items belong to the same item set. Then ASSO iteratively selects a pattern from the candidate set by greedily minimizing the J_A.

2.3   Minimizing the Pattern Set Complexity (HYPER+)

The HYPER+ [2] algorithm works in two phases. In the first phase (corresponding to the covering algorithm HYPER + [7]), given a collection of frequent item sets, the algorithm greedily selects a set of patterns Π^* by minimizing the following cost function that models the pattern set complexity:

$$J_H(\Pi^*, \mathcal{D}) = \sum_{P \in \Pi_*} (\|P_I\| + \|P_T\|). \tag{4}$$

During this first phase, the algorithm aims to cover in the best way all the items occurring in $\mathcal{D}$, with neither false negatives nor positives, and thus without any noise. The rationale is to promote the simplest description of the input data $\mathcal{D}$. Note that the size of Π^* is unknown, and depends on the amount of patterns that suffice to cover the 1-bits in $\mathcal{D}$. The minimum support parameter σ is used by HYPER+ to select an initial set of frequent item sets for starting the greedy selection phase, thus reducing the search space of the greedy optimization strategy.

Concerning the second phase of the algorithm, pairs of patterns in Π^* are recursively merged as long as a new collection Π', with a reduced number of patterns, can be obtained without generating an amount of *false positive* occurrences larger than a given budget β. Finally, since the pattern set Π' is ordered (from most to least important), we can simply select Π_k as the top-listed k patterns in Π', as done by the algorithm authors in Sect. 7.4 of [2]. Note that this also introduces false negatives, corresponding to all the occurrences $\mathcal{D}(i, j) = 1$ in the dataset that remain uncovered after selecting the top-k patterns Π_k only.

2.4   Minimizing Multiple Cost Functions (PANDA+ Framework)

PANDA+ is a framework recently proposed [4] that inherits the optimization engine of PANDA [3]. Thus it adopts a greedy strategy by exploiting a two-stage heuristics to iteratively select each pattern: (a) discover a noise-less pattern that covers the yet uncovered 1-bits of $\mathcal{D}$, and (b) extend it to form a good approximate pattern, thus allowing some false positives to occur within the pattern. Rather than considering all the possible exponential combinations of items, these are sorted to maximize the probability of generating large cores, and processed one at the time without backtracking.

Table 1. Objective functions for Top-k pattern discovery problem.

Cost Function	Description
$J_A(\Pi_k, \mathcal{D}) = J^+(\Pi_k, \mathcal{D}, \gamma_\mathcal{N}(\mathcal{N}) = \|\mathcal{N}\|, \gamma_P(P) = 0, \rho = 0)$	Minimize noise [1]
$J_H(\Pi_k, \mathcal{D}) = J^+(\Pi_k, \mathcal{D}, \gamma_\mathcal{N}(\mathcal{N}) = 0, \gamma_P(P) = \|P_T\| + \|P_I\|, \rho = 1)$	Minimize pattern set complexity [7]
$J_P(\Pi_k, \mathcal{D}) = J^+(\Pi_k, \mathcal{D}, \gamma_\mathcal{N}(\mathcal{N}) = \|\mathcal{N}\|, \gamma_P = \|P_T\| + \|P_I\|, \rho = 1)$	Minimize noise and pattern set complexity [3,9]
$J_P^{\overline{\rho}}(\Pi_k, \mathcal{D}) = J^+(\Pi_k, \mathcal{D}, \gamma_\mathcal{N}(\mathcal{N}) = \|\mathcal{N}\|, \gamma_P(P) = \|P_T\| + \|P_I\|, \rho = \overline{\rho})$	Extend J_P to leverage the trade-off between noise and pattern set complexity
$J_E(\Pi, \mathcal{D}) = J^+(\Pi, \mathcal{D}, \gamma_\mathcal{N}(\mathcal{N}) = \mathsf{enc}(\mathcal{N}), \gamma_P(P) = \mathsf{enc}(P), \rho = 1)$	Minimize the encoding length [8] of the pattern model according to [6]

The PANDA$^+$ [3] original cost function J_P can be replaced without harming its greedy heuristics. Indeed, J_P, which is simply the sum of $J_A(\cdot)$ and $J_H(\cdot)$ – see Eqs. (3) and (4) – is generalized as follows:

$$J^+(\Pi_k, \mathcal{D}, \gamma_\mathcal{N}, \gamma_P, \rho) = \gamma_\mathcal{N}(\mathcal{N}) + \rho \cdot \sum_{P \in \Pi_k} \gamma_P(P) \tag{5}$$

where $\mathcal{N}$ is the noise matrix defined by Eq. (1), $\gamma_\mathcal{N}$ and γ_P are user defined functions measuring the cost of the noise and patterns descriptions respectively, and $\rho \geq 0$ works as a regularization factor weighting the relative importance of the patterns cost.

Table 1 shows how the cost function defined by Eq. (5) can be instantiated to obtain all the functions discussed above, and allows for new functions to be introduced, by fully leveraging the trade-off between patterns description cost and noise cost (thanks to parameter ρ). Note the $J_P^{\overline{\rho}}$ is a generalization of the function J_P already proposed for PANDA, with parameter $\overline{\rho}$ that determines a different trade-off between patterns description cost and noise cost. In addition, one instance of the generalized cost function is J_E, originally proposed in [6] to evaluate a pattern set Π_k. According to the MDL principle [8], the regularities in $\mathcal{D}$, corresponding to the discovered patterns Π_k, can be used to *lossless compress* $\mathcal{D}$: thus the best pattern set Π_k is the one that induces the smallest encoding of $\mathcal{D}$. Finally, in order to avoid the greedy search strategy accepting too noisy patterns, we introduced into PANDA$^+$ two maximum noise thresholds $\epsilon_r, \epsilon_c \in [0, 1]$, inspired by [5], aimed at bounding the maximum amount of noise generated by adding a new item or a new transaction to a pattern.

3   Evaluation Methodology and Experiments

We observe the lack of common real-world benchmarks, e.g., datasets for which the most important embedded patterns are known, and that can be used as a

Table 2. Characteristics of the datasets used for the experiments

Dataset	# Classes	# Items	# Transactions	Avg. trans. length
Abalone	3	40	4177	8.0
Anneal	5	108	898	38.0
Audiology	24	154	226	67.6
Auto	6	129	205	24.7
Congres	2	32	434	15.1
Credita	2	70	690	14.9
CylBands	2	120	540	33.2
Dermatology	6	43	366	12.0
Diabetes	2	40	768	8.0
Ecoli	8	26	336	7.0
Flare	8	30	1389	10.0
Glass	6	40	214	9.0
Heart	5	45	303	13.0
Hepatitis	2	50	155	17.9
HorseColic.D85	2	81	368	16.8
Ionosphere	2	155	351	34.0
Iris	3	16	150	4.0
Mushroom	2	88	8124	21.7
Pima	2	36	768	8.0
Sick	2	75	3772	27.4
Soybean-large	19	99	683	31.6
Vehicle	4	90	846	18.0
Wine	3	65	178	13.0
Zoo	7	35	101	16.0

ground truth to evaluate and compare algorithms. We thus propose a *supervised evaluation methodology* of the extracted top-k patterns that measures and compares the ability of each algorithm in discovering interesting transaction features that we can use to train accurate classifier.

To this end we focus on 24 datasets from the UCI repository[1] which have been discretized for pattern mining[2]. The main characteristics of the datasets are reported in Table 2. In all the experiments, we thus remove class labels before feeding any top-k approximate item sets mining algorithm. Such patterns are used as features to train and test an SVM classifier.

3.1   Parameter Setting of Pattern Mining Algorithms

We apply HYPER+, ASSO, and PANDA$^+$ over the 24 UCI datasets, by requiring the mining algorithms to extract a number of patterns equal to *twice* the number

---

[1] http://archive.ics.uci.edu/ml/.

[2] http://www.csc.liv.ac.uk/~frans/KDD/Software/LUCS_KDD_DN/.

of classes present in each distinct dataset. Altough the number of classes is a good proxy for the actual number of interesting patterns occurring in the data, we doubled this number to deal with the presence of multiple dense subgroups that characterize a single class, as found in [6]. Also, note that all the three algorithms may produce less patterns than required if none is found to improve the objective function.

Although HYPER+ allows for setting the maximum number of patterns k to extract, unfortunately we found this option to perform poorly, resulting in excessive noise. This is due to the covering constraint of the algorithm: false negatives are not allowed, and the extracted k patterns must cover all the occurrences $\mathcal{D}(i, j) = 1$ in the dataset. It is possible to achieve much better results by tuning the algorithm noise budget β, and then accepting only the k top-listed patterns. We fine-tuned the β parameter on every single dataset by choosing β in the set $\{1\%, 10\%\}$. Moreover, we used frequent closed item sets in order to take advantage of a lower minimum support thresholds, which is the most sensitive parameter of HYPER+. We swept the minimum support threshold in the interval $[10\%, 90\%]$ by increments of 10%.

The ASSO algorithm has a minimum correlation parameter τ which determines the initial patterns candidate set. Indeed, ASSO is very sensitive to this parameter. We fine-tuned the algorithm independently on every single dataset, by tuning τ in the range $[0.5, 1]$ with steps of 0.05. In our experiments we always tested the best performing variant of the algorithm which is named ASSO + *iter* in the original paper.

We evaluated four variants of PANDA$^+$ optimizing different cost functions: J_A, J_P, J_E, $J_P^{1.2}$. Recall that J_A measures the noise cost only, J_P adds the cost of each pattern, J_E measures the cost of an optimal MDL encoding, and $J_P^{1.2}$ mimics J_E by using a larger weight for the cost of the patters cost. In addition, PANDA$^+$ uses two maximum noise thresholds ϵ_r and ϵ_c, to control the maximum amount of noise on each pattern row or column. Also in this case, on each dataset we swept parameters ϵ_r and ϵ_c in the range $[0.0, 0.5]$, with steps of 0.05, and also considering the value 1.0 which is equivalent to ignoring the thresholds.

3.2   Supervised Evaluation of Pattern Set

Each binary UCI dataset, from which the class labels is removed in advance, is first subdivided into training and test set. Indeed, we use 10-fold cross validation, so a training set includes $9/10$ of all transactions, and the corresponding test set the remaining $1/10$ ones. We employ the three algorithms, instantiated with a given parameter set, to extract a distinct a pattern set Π_k from the training set. Then, every transaction in the training and test sets is mapped into the pattern space by considering a binary feature for each approximate pattern in Π_k indicating its presence/absence in the transaction. Note that the algorithms we evaluated produce, for each pattern $\langle P_I, P_T \rangle \in \Pi_k$, the set of transactions P_T in the training where P_I it occurs, and this set is used for the mapping. For what regard the test set, we say that an approximate pattern $P \in \Pi_k$ occurs in an unseen test transaction t *iff* $|P_I \cap t|/|P_I| \geq \eta$, where η is the minimum intersection ratio $|P_I \cap t^*|/|P_I|$ for every training transaction $t^* \in P_T$. The rationale is to accept a pattern P for a test transaction t if it does not generate more noise than what it has been observed in the training set.

After mapping the transactions into such pattern space, the class labels are restored in the transformed training and test sets. The training set is thus used to train an SVM classifier. Finally, the classifier is evaluated on the mapped test set. Specifically we adopt the implementation provided by [10], in which we use a radial basis function as SVM kernel, and estimate its crucial regularization parameter as in [11].

It is worth noting that, unlike [12], the features extracted from each transaction only correspond to the patterns discovered by the given algorithm: we do not consider the singletons as a transaction feature in classifier training/test data, unless such patterns composed of single items are actually extracted by the mining algorithm. This is because we want to evaluate exclusively the predictive power of the mined collection of patterns Π_k.

Since the various algorithms take as input several parameters summarized in Fig. 1(a), we exploit a parameter sweeping technique, like in [1] and [6], aimed at understanding the full potential of the algorithms taken into consideration.

Specifically, we adopt two distinct approaches to derive the best classifier for each dataset, and thus testing its accuracy. In the former approach – see Fig. 1(b) – we sweep over the input parameters of the given algorithm, and select 'ex ante' the best parameters on the basis of the specific cost function J used to evaluate the quality of the mined pattern sets. We thus conduct 10-fold cross validation to evaluate the goodness of the algorithm as follows: at each fold the top-k patterns are extracted from the training set using the best parameters previously found, then they are exploited as features to map the data in a new feature space, and finally, the mapped data is used to train and test an SVM classifier. In the latter approach – see Fig. 1(c) – we use parameter sweeping to explore the maximum accuracy achievable by a given algorithm. The classifier resulting from every combination of the algorithm parameters is evaluated with

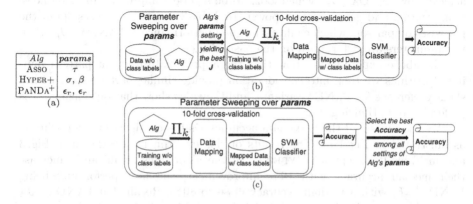

Fig. 1. Supervised evaluation: (a) algorithms' parameters, (b) unsupervised and (c) supervised selection of the algorithms' parameters.

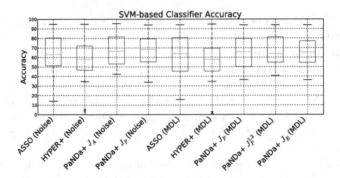

Fig. 2. Box-plots of the accuracy of the SVM-based classifier with pre-optimized algorithm parameters across the various UCI datasets (HYPER+ did not complete on audiology). Leftmost data correspond to noise-based parameter tuning and rightmost data to MDL-based parameter tuning.

same 10-fold cross validation process, and the best accuracy is used to measure the goodness of the algorithm.

3.3 Experimental Results

The first of our experiments compares ASSO, HYPER+ and PANDA$^+$ when their parameters are chosen according to their ability to optimize J_A or J_E, as illustrated in Fig. 1(b). Results reported in Fig. 2 show that the performance of HYPER+ are better than expected. Even if HYPER+ is not able to optimize well neither J_A nor J_E, the resulting patterns provide a reasonably good accuracy. We observe that the J_E cost function leads to poorer patterns on average. This is quite surprising, since we expected that a purely noise-based cost function would lead to over-fitting patterns and that the MDL-based cost would promote patterns with a larger generalization power. The best median is achieved by PANDA$^+$ J_P, which looks to be a good compromise between noise minimization and generalization power. In Table 3 we report the results of the pairwise comparison, which confirm the good performance of PANDA$^+$ J_P when its parameters are determined on the basis of J_A.

The above experiments, show that the approximate frequent pattern mining algorithms we took into consideration provide similar performance, with a slight preference for PANDA variants, and they also show that optimizing J_A is preferable to optimizing J_E.

We also measured the maximum accuracy that each algorithm can achieve, as illustrated in Fig. 1(c). The results of this experiment are shown in Fig. 3 and in Table 4. Except for HYPER+, all the algorithms significantly increase their median accuracy with 10 % improvement, the best performing being PANDA$^+$ J_A with a median accuracy close to 80 %. Recall that PANDA$^+$ J_A was not the best algorithm at optimizing J_A. Again, we can state that neither J_A nor J_E are good hints for the discovery of predictive patterns. Moreover, PANDA$^+$ J_A performs almost always better the HYPER+, and beats ASSO in

Table 3. Number of times an algorithm (column) generated better/equal/worse patterns than the baseline (row) based on the corresponding SVM accuracy.

		Noise-based parameters			MDL-based parameters				
		Asso	Hyper+	PANDA+ J_P	PANDA+ J_A	Asso	Hyper+	PANDA+ J_P	PANDA+ $J_P^{1.2}$	PANDA+ J_E
Noise-based	Asso	(24)	10/2/11	**16/1/7**	11/1/12	6/9/9	10/1/12	11/1/12	11/1/12	13/2/9
	Hyper+	11/2/10	(23)	**15/1/7**	14/0/9	11/2/10	7/6/10	13/1/9	13/1/9	12/2/9
	PANDA+ J_P	7/1/16	7/1/15	**(24)**	9/2/13	7/2/15	5/1/17	5/2/17	8/2/14	7/2/15
	PANDA+ J_A	12/1/11	9/0/14	**13/2/9**	(24)	11/1/12	9/0/14	7/1/16	9/1/14	10/1/13
MDL-based	Asso	9/9/6	10/2/11	**15/2/7**	12/1/11	(24)	10/1/12	12/2/10	10/2/12	12/3/9
	Hyper+	12/1/10	10/6/7	**17/1/5**	14/0/9	12/1/10	(23)	13/1/9	15/1/7	13/1/9
	PANDA+ J_P	12/1/11	9/1/13	**17/2/5**	16/1/7	10/2/12	9/1/13	(24)	12/2/10	11/2/11
	PANDA+ $J_P^{1.2}$	12/1/11	9/1/13	**14/2/8**	14/1/9	12/2/10	7/1/15	10/2/12	(24)	12/2/10
	PANDA+ J_E	9/2/13	9/2/12	**15/2/7**	13/1/10	9/3/12	9/1/13	11/2/11	10/2/12	(24)

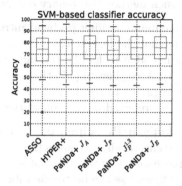

Fig. 3. Box-plots of the accuracy of the SVM-based classifier with parameter sweeping across the various UCI datasets (Hyper+ did not complete on audiology).

Table 4. Number of times an algorithm (column) achieves better/equal/worse MDL score than the baseline (row) on the test datasets. Between parentheses the number of datasets where the algorithm succeeded in extracting the top-k patterns. In boldface the best results.

	Asso	Hyper+	PANDA+ J_P	PANDA+ $J_P^{1.2}$	PANDA+ J_E
Asso	(24)	0/0/21	**13/0/11**	9/0/15	12/0/12
Hyper+	**21/0/0**	(21)	**21/0/0**	**21/0/0**	**21/0/0**
PANDA+ J_P	11/0/13	0/0/21	(24)	8/0/16	**15/0/9**
PANDA+ $J_P^{1.2}$	15/0/9	0/0/21	16/0/8	(24)	**20/0/4**
PANDA+ J_E	**12/0/12**	0/0/21	9/0/15	4/0/20	(24)

17 datasets out of 24. The difference between PANDA$^+$ J_A and ASSO is statistically significant according to the two-tailed sign test with $p = 0.023$, and according to the two-tailed Wilkoxon signed-rank test with $p = 0.008$.

4 Related Work

We classify related works in three large categories: matrix decomposition based, database tiling and Minimum Description Length based.

Matrix Decomposition Based. The methods in this class aim at finding a product of matrices that describes the input data with a smallest possible amount of error. These methods include Probabilistic latent semantic indexing (PLSI) [13], Latent Dirichlet allocation (LDA) [14], Independent Component Analysis (ICA) [15], Non-negative Matrix Factorization, etc. However, ASSO was shown to outperform such methods on binary datasets.

Database Tiling. The maximum k-tiling problem introduced in [16] requires to find the set of k tiles having the largest coverage of the given database $\mathcal{D}$. However, this approach is not able to handle the false positives present in the data, similarly to HYPER [7]. According to [17], tiles can be hierarchical. Unlike our approach, low-density regions are considered as important as high density ones, and inclusion of tiles is preferred instead of overlapping.

Minimum Description Length Principle. In [18] a set of item sets, called *cover* or *code table*, is used to encode all the transactions in the database, meaning that every transaction is represented by the union of some item sets in the cover. The MDL principle is used to choose the best code table. The proposed KRIMP algorithm selects the item sets of the cover from a pool of candidates. In their experiments, the authors exploited the collection of all the item sets occurring in the dataset to achieve good results. However, patterns that cover a given transaction must be disjoint, thus increasing the size and redundancy of the model, and noisy occurrences are not allowed.

A similar MDL-based approach is adopted in [19], but in this case, knowledge on the data marginal distributions is assumed to be known, thus generating a different kind of patterns. In this work, assume no knowledge on the data in evaluating the extracted patterns. In [20] a novel framework is proposed for the comparison of different pattern sets. From a given pattern set, a probability distribution over $\mathcal{D}$ is derived according to the maximum entropy principle, and then the Kullback-Leibler distance between these two distributions is used to measure the dissimilarity of two pattern sets. After comparing several algorithms including ASSO, HYPER+, KRIMP, and others, the authors conclude that HYPER+ and ASSO are the two best performing algorithms, i.e. producing the set of patterns whose probability distribution is closer to the underlying class label distribution. Note that HYPER+ and ASSO are the algorithms analyzed in this work.

5   Conclusions

In this paper we have given a deep insight into the problem of mining approximate top-k patterns from binary matrixes. Our analysis has explored ASSO, HYPER+, and PANDA$^+$. The main contribution is a methodology to assess the various solutions at hand. In fact, the lack of standard ground truth datasets, which we can use for evaluating and comparing the quality of pattern sets extracted by the various algorithms – namely their greedy strategies, cost functions to optimize, and parameter setting – makes very hard an objective judgement. In this paper we have thus considered the accuracy of SVM classifiers, built on top of the top-k patterns used as feature sets, as a strong signal for the quality of the pattern set extracted. The experiments conducted on 24 datasets from the UCI repository show that the patterns extracted by only optimizing J_A (noise) or J_E (pattern set complexity) exhibit limited performance. The best results achieved by full parameter sweeping show that PANDA$^+$ outperforms all the other algorithms with a statistically significant improvement.

References

1. Miettinen, P., Mielikainen, T., Gionis, A., Das, G., Mannila, H.: The discrete basis problem. IEEE TKDE **20**(10), 1348–1362 (2008)
2. Xiang, Y., Jin, R., Fuhry, D., Dragan, F.F.: Summarizing transactional databases with overlapped hyperrectangles. Data Min. Knowl. Discov. **23**(2), 215–251 (2011)
3. Lucchese, C., Orlando, S., Perego, R.: Mining top-k patterns from binary datasets in presence of noise. In: SDM, pp. 165–176. SIAM (2010)
4. Lucchese, C., Orlando, S., Perego, R.: A unifying framework for mining approximate top-k binary patterns. IEEE TKDE **26**, 2900–2913 (2014)
5. Cheng, H., Yu, P.S., Han, J.: AC-Close: efficiently mining approximate closed itemsets by core pattern recovery. In: Proceedings of ICDM, pp. 839–844. IEEE Computer Society (2006)
6. Miettinen, P., Vreeken, J.: Model order selection for boolean matrix factorization. In: Proceedings of KDD, pp. 51–59. ACM (2011)
7. Xiang, Y., Jin, R., Fuhry, D., Dragan, F.F.: Succinct summarization of transactional databases: an overlapped hyperrectangle scheme. In: Proceedings of KDD, pp. 758–766. ACM (2008)
8. Rissanen, J.: Modeling by shortest data description. Automatica **14**(5), 465–471 (1978)
9. Lucchese, C., Orlando, S., Perego, R.: A generative pattern model for mining binary datasets. In: SAC, pp. 1109–1110. ACM (2010)
10. Joachims, T.: Learning to Classify Text Using Support Vector Machines: Methods, Theory and Algorithms. Kluwer Academic Publishers, Norwell (2002)
11. Cherkassky, V., Ma, Y.: Practical selection of svm parameters and noise estimation for SVM regression. Neural Netw. **17**(1), 113–126 (2004)
12. Cheng, H., Yan, X., Han, J., wei Hsu, C.: Discriminative frequent pattern analysis for effective classification. In: Proceedings of ICDE, pp. 716–725 (2007)
13. Hofmann, T.: Probabilistic latent semantic indexing. In: Proceedings of SIGIR, pp. 50–57. ACM (1999)

14. Blei, D.M., Ng, A.Y., Jordan, M.I.: Latent dirichlet allocation. J. Mach. Learn. Res. **3**, 993–1022 (2003)
15. Hyvärinen, A., Karhunen, J., Oja, E.: Independent Component Analysis. John and Wiley, Chichester (2001)
16. Geerts, F., Goethals, B., Mielikäinen, T.: Tiling databases. In: Suzuki, E., Arikawa, S. (eds.) DS 2004. LNCS (LNAI), vol. 3245, pp. 278–289. Springer, Heidelberg (2004)
17. Gionis, A., Mannila, H., Seppänen, J.K.: Geometric and combinatorial tiles in 0–1 data. In: Boulicaut, J.-F., Esposito, F., Giannotti, F., Pedreschi, D. (eds.) PKDD 2004. LNCS (LNAI), vol. 3202, pp. 173–184. Springer, Heidelberg (2004)
18. Vreeken, J., van Leeuwen, M., Siebes, A.: Krimp: mining itemsets that compress. Data Min. Knowl. Discov. **23**(1), 169–214 (2011)
19. Kontonasios, K.N., Bie, T.D.: An information-theoretic approach to finding informative noisy tiles in binary databases. In: SDM, pp. 153–164. SIAM (2010)
20. Tatti, N., Vreeken, J.: Comparing apples and oranges. In: Gunopulos, D., Hofmann, T., Malerba, D., Vazirgiannis, M. (eds.) ECML PKDD 2011, Part III. LNCS, vol. 6913, pp. 398–413. Springer, Heidelberg (2011)

Finding Banded Patterns in Data: The Banded Pattern Mining Algorithm

Fatimah B. Abdullahi[✉], Frans Coenen, and Russell Martin

The Department of Computer Science, The University of Liverpool,
Ashton Street, Liverpool L69 3BX, UK
{f.b.abdullahi,coenen,russell.martin}@liverpool.ac.uk

Abstract. The concept of Banded pattern mining is concerned with the identification of "bandings" within zero-one data. A zero-one data set is said to be fully banded if all the "ones" can be arranged along the leading diagonal. The discovery of a banded pattern is of interest in its own right, at least in a data analysis context, because it tells us something about the data. Banding has also been shown to enhances the efficiency of matrix manipulation algorithms. In this paper the exact N dimensional Banded Pattern Mining (BPM) algorithm is presented together with a full evaluation of its operation. To illustrate the utility of the banded pattern concept a case study using the Great Britain (GB) Cattle movement database is also presented.

Keywords: Banded patterns · Zero-one data · Banded pattern mining

1 Introduction

Banded pattern mining is concerned with the identification of "bandings" within zero-one data [7–10]. The idea is, given a zero-one data set, to rearrange the elements in each dimensions so that a banding is revealed. A zero-one data set is said to be fully banded if all the "ones" can be arranged along the leading diagonal. Typically a perfect banding cannot be discovered, however it is still possible to rearrange the data to reveal a nearest possible banding. The discovery of a banded pattern is of interest in its own right, at least in a data mining context, because it tells us something about the data. Banding has also been shown to improve the operation of various n dimensional matrix manipulation algorithms in that only the portion of the matrix located near the leading diagonal needs to be considered.

Previous work on banding identification [7,10] has been mostly directed at two dimensional data and has focussed on using heuristics to identify permutations in the data. Two examples are the Minimum Banded Augmentation (MBA) algorithm [8] and the Barycentric (BC) algorithm [9]. These approaches worked well but are difficult to scale up to encompass n dimensional data because of the exponential increase in the number of permutations that need to be considered.

© Springer International Publishing Switzerland 2015
S. Madria and T. Hara (Eds.): DaWaK 2015, LNCS 9263, pp. 95–107, 2015.
DOI: 10.1007/978-3-319-22729-0_8

The idea proposed in this paper, instead of considering large numbers of permutations, is to use a "banding score" mechanism to iteratively rearrange the elements in each dimension.

The rest of this paper is organised as follows. Section 2 presents some relevant previous work, concentrating on the MBA and BC algorithms. Section 3 presents some formalism concerning the banded pattern problem, while Sect. 4 presents the proposed BPM algorithm. To aid in the understanding of the BPM algorithm Sect. 5 presents a worked example. A full evaluation, using both artificial and real data sets, is presented in Sect. 6. Some conclusion are given in Sect. 7.

2   Previous Work

The concept of banded matrices, from the data analysis perspective, occurs in many applications domains; examples can be found in paleontology [2], Network data analysis [3] and linguistics [8]. More recent work can be found in [7] and [10]. The current state of the art algorithm, the Minimum Banded Augmentation (MBA) proposed by [8], focuses on minimizing the distance of non-zero entries from the main diagonal of a matrix by reordering the original matrix. The MBA algorithm operates by "flipping" zero entries (0s) to one entries (1s) and vice versa to identify a banding. Gemma et al. fixed the column permutations of the data matrix before executing their algorithm [7]. Here the basic idea is to solve optimally the consecutive one property on the permuted matrix M and then resolve "Sperner conflicts" to eliminate all the overlapping row intervals between each row of the permuted matrix, by going through all the extra rows and making them consecutive. While it can be argued that the fixed column permutation assumption is not a very realistic assumption with respect to many real world situations, heuristical methods were proposed in [7] to determine a suitable fixed column permutation. The MBA algorithm use the Accuracy (Acc) measure to evaluate the performance of the banding produced. Another frequently referenced banding strategy is the Barycentric (BC) algorithm that transposes a matrix. It was originally designed for graph drawing, but more recently used to reorder binary matrices [9]. The BC algorithm uses the Mean Row Moment (MRM) measure to evaluate the performance of the banding produced. The distinction between these previous algorithms and that presented in this paper is that the previous algorithms were all directed at 2D data, while the proposed algorithm operates in ND. The above MBA and BC algorithms are the two examplar banding algorithms with which the operation of the proposed BPM algorithm was compared and evaluated as discussed later in this paper. Note that Bandwidth minimization of binary matrices is known to be NP-Complete [5].

3   Formalism

The data space of interest comprises a set of n dimensions, with $DIM = \{dim_1, dim_2, \ldots, dim_n\}$. Each dimension dim_i comprises a set of sequential coordinates (indexes) commencing with the coordinate 0, $dim_i = \{0, 1, 2 \ldots\}$. Each dimension dim_i also has a set of index positions associated with it

$pos_i = \{p_1, p_2, p_3, \ldots\}$. The set of positions is indicated by $POS = \{pos_1, pos_2, \ldots, pos_n\}$. Note that dimensions are not necessarily all of the same size. However there is a correspondence between each dim_i and pos_i pairing ($|dim_i| \equiv |pos_i|$). Initially the content of each set dim_i will be the same as the content of each set pos_i, however it is the elements in each set pos_i in POS that we wish to rearrange so as to identify a banding.

Each "cell" in the n dimensional data space is thus identifiable by a coordinate tuple of length n, $\langle c_1, c_2 \ldots c_n \rangle$; where $c_1 \in dim_i$, $c_2 \in dim_2$ and so on. If $n = 2$ we might think of these as x-y coordinates, and if $n = 3$ we might think of these as x-y-z coordinates. Each cell contains either a 1 or 0. For illustrative purposes, in the remainder of this paper, we say that a cell with a 1 value contains a "dot", and a cell with a 0 is empty.

The data sets $D = \{d_1, d_2, \ldots\}$ we are interested in thus comprise a sequence of k coordinate tuples each of size n and each representing a "dot". Note that a specific coordinate tuple can appear only once in D. Each tuple thus describes the location of a dot in the data space. It is these dots that we wish to arrange in such a manner that they are as close to the leading diagonal as possible by rearranging the position of the indexes (coordinates).

4   The Banded Pattern Mining Algorithm

A high level view of the proposed Banded Pattern Mining (BPM) algorithm is presented in Algorithm 1. The algorithm describes an iterative approach whereby the elements in each dimension are repeatedly rearranged until a best banding score is arrived at or until some maximum number of iterations is reached. The algorithm is founded on ideas presented in [1] where an approximate BPM algorithm (BPM_A) was presented. Comparison are made later in Sect. 6 using both the proposed exact BPM algorithm and the previous BPM_A algorithm. The inputs to the algorithm are: (i) a zero-one data set D which is to be banded, (ii) a desired maximum number of iterations max and (iii) the set of dimensions DIM describing the n dimensional data space to which D subscribes. The Output is a rearranged data set D'.

On each iteration each dimension dim_i in DIM is considered in turn. For each element j in dim_i a normalised banding score bs_{ij} is calculated (line 12) using Eq. 2 (the derivation of the equation is described in more detail below). The elements in dimension dim_i are then rearranged (line 14) in descending order according to the banding scores calculated earlier. If two or more elements have the same banding score (an unlikely event given a large data set) then the number of dots featured in each element is taken into consideration together with the relative position of the new location with respect to the centroid of the data space (essentially we wish to place elements with larger numbers of dots closer to the centroid of the current data sub-space than elements with smaller numbers of dots). The rearranging of elements in dimensions is repeated for each dimension in turn.

At the end of each complete iteration the global banding score gbs for the new configuration is determined (line 16) by summing the banding score values derived earlier (Eq. 2). If the newly calculated banding score is equal or less (worse) than the previously calculated score we break (line18), otherwise we continue onto the next iteration. On completion (line 25) we derive the new dataset D' using the positions of the indexes contained in DIM. The situation where a worse gbs than that obtained on the previous iteration may be arrived at is where we have a "poling" effect where we are rearranging that data one way, and then another way, without improving the banding.

Algorithm 1. The BPM Algorithm

1: **Input**
2: D = Binary valued input data set
3: max = The maximum number of iterations
4: **Output**
5: D' = Data set D rearranged so as to display as near a banding as possible
6: $gbs = 0$ (The global banding score so far)
7: $counter = 0$
8: **while** $counter < max$ **do**
9:      **for all** $dim_i \in DIM$ **do**
10:         **for** $j = 1$ to $j = |dim_i|$ **do**
11:             bs_{ij} = Banding score for element j in $|dim_i|$ calculated using
12:                 Eq. 2
13:         **end for**
14:         dim_i' = The set dim_i rearranged according to the calculated bs
15:             (in descending order)
16:     **end for**
17:     gbs' = Global banding score for current configuration described by
18:         $DIM' = \{dim_1', dim_2', \ldots, dim_n'\}$ calculated using Equation 3
19:     **if** $gbs' \leq gbs$ **then**
20:         **break**
21:     **else**
22:         $gbs = gbs'$
23:         $DIM = DIM'$
24:     **end if**
25:     $counter = counter + 1$
26: **end while**
27: D' = Input dataset D rearranged according to index positions in POS

The banding score bs_{ij} for a particular element j in dimension dim_i is determined according to the location of the subset of dots S in D whose c_i coordinate is equal to j (recall that each dot in D is define by a coordinate tuple of the form $\langle c_1, c_2 \ldots c_n \rangle$). For each dot in S we calculate the distance to the origin of the data sub-space that does not include dim_i. We exclude the current dimension because this is the dimension we want to rearrange. Thus a banding score bs is calculated as follows:

$$bs = \sum_{i=1}^{i=|S|} distToOrigin(dot_i) \tag{1}$$

However, to allow for comparison of bs we need to normalise the score. Given a dot that is at the origin of the sub-space of interest the normalised bs should be 0. Given a dot that is as far away from the the origin of the sub-space of interest as is geometrically possible the normalised bs should be 1. Thus to normalise bs we need to divide by the sum of the set of $|S|$ maximum distances that can be attained in the given sub space:

$$bs = \frac{\sum_{i=1}^{i=|S|} distToOrigin(dot_i)}{\sum_{i=1}^{i=|M|} max_i} \tag{2}$$

where M is a set of maximum distances corresponding to the number of dots in S. Note that the content of M will vary according to the nature of the set DIM for a given data set D. Given that we can identify the coordinate value that features most frequently in the coordinate tuples in D we know the maximum required size of max. Given our knowledge of DIM we can therefore calculate the values to be included in M at the start of the process (not shown in Algorithm 1) and thus we can calculate these values in advance and store them in a table to be used as necessary. In the evaluation section presented later in this paper a comparison is presented between using a pre calculated BMP "M-table" and calculating maximum distances as required (not using a BPM "M-table"). Using an BPM M-table means that values are only calculated once, although it may be the case that some values are calculated that are never used.

The global banding score, gbs for a configuration is then given by:

$$gbs = \frac{\sum_{i=1}^{i=|DIM|} \sum_{j=1}^{j=|dim_i|} bs_{ij}}{\sum_{k=1}^{k=|DIM|} |dim_k|} \tag{3}$$

the sum of all the identified banding scores for each element in each dimension. Thus if every element within a given data space is filled with a dot the global banding score will be 1. Distances can be calculated in a variety of ways. Two obvious mechanisms are Manhattan distance and Euclidean distance (both are considered in the evaluation presented later in this paper).

5   Worked Example

The operation of the BPM algorithm can best be illustrated using a worked example. Consider the 2 dimensional 4×4 configuration given in Fig. 1(a) (the origin is in the top left hand corner). It has dimensions $Dim = \{x, y\}$, and:

$$D = \{\langle 0, 1\rangle, \langle 0, 2\rangle, \langle 0, 3\rangle, \langle 1, 0\rangle, \langle 1, 1\rangle, \langle 2, 0\rangle, \langle 2, 1\rangle, \langle 2, 3\rangle, \langle 3, 2\rangle, \langle 3, 3\rangle\}.$$

Considering dimension x first, we calculate banding scores as shown in Table 1. This produces the banding scores 1.00, 0.20, 0.67 and 1.00. We thus rearrange the elements in x in ascending order of their banding scores. Note that in the case where two elements have the same score we arrange things so

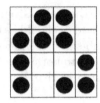
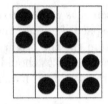
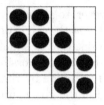

| (a) Raw Data | (b) After rearrangement of dim_1 | (c) After rearrangement of dim_2 |

Fig. 1. Example of operation of BPM algorithm

Table 1. Calculation of banding scores for dimension x

# Element	Dist from origin	Max. dist. from origin	bs
0	$1+2+3=6$	$1+2+3=6$	1.00
1	$0+1=1$	$2+3=5$	0.20
2	$0+1+3=4$	$1+2+3=6$	0.67
3	$2+3=5$	$2+3=5$	1.00
		Total	2.87

Table 2. Calculation of banding scores for dimension y

# Element	Dist from origin	Max. dist. from origin	bs
0	$0+1=1$	$2+3=5$	0.20
1	$0+1+3=4$	$1+2+3=6$	0.67
2	$2+3=5$	$2+3=5$	1.00
3	$1+2+3=6$	$1+2+3=6$	1.00
		Total	2.87

that the element with the largest number of dots associated with it is nearest to the centre of the data space. The result is as shown in Fig. 1(b).

Considering dimension y next, we calculate banding scores as shown in Table 2. This produces the banding scores 0.20, 0.67, 1.00 and 1.00. The elements in y are more or less already in ascending order of bs. We only need to swap the last two elements so that the element with the greater number of dots is nearer the centre of the data space. The result is as shown in Fig. 1(c).

The global banding score is then the sum of the individual banding scores divided by the total number of elements in the configuration:

$$gbs = \frac{2.87 + 2.87}{8} = \frac{5.73}{8} = 0.72$$

The process is repeated on the next iteration (not shown here) and the same gbs value produced because we already have a best banding. The rearranged data set D' arrived at is:

$$D' = \{\langle 0,0 \rangle, \langle 0,1 \rangle, \langle 1,0 \rangle, \langle 1,1 \rangle, \langle 1,2 \rangle, \langle 2,1 \rangle, \langle 2,2 \rangle, \langle 2,3 \rangle, \langle 3,2 \rangle, \langle 3,3 \rangle\}.$$

6   Evaluation

This section presents an evaluation and discussion of the proposed exact BPM algorithm. The objectives of the evaluation were as follows:

- To determine the effect of data set size and density on the BPM algorithm using artificial data.
- To compare the operation of the BPM algorithm with existing algorithms (MBA and BC) using real data sets.
- To consider the application of the BPM algorithms with respect to a large scale application (the GB cattle movement database).

Each is discussed in further detail in the following three sub-sections.

6.1   Effect of Data Set Size

To determine the efficiency of the proposed BPM algorithm in comparison with the MBA and BC all three algorithms were run using artificial data sets of varying size generated using the LUCS-KDD data generator [6][1]. Using this generator ten data sets measuring 100×100, 141×141, 173×173, 200×200, 224×224, 245×245, 265×265, 283×283, 300×300 and 316×316 were generated, corresponding to numbers of elements approximately increasing from $10,000$ to $100,000$ in steps of $10,000$. A density of $10\,\%$ was used (in other words $10\,\%$ of the cells in each row contained a dot). A further five 100×100 data sets were generated using densities from $10\,\%$ to $50\,\%$ increasing in steps of 10.

The recorded runtime results obtained by applying the proposed BPM algorithm and the MBA and BC algorithms to data sets of increasing size are presented in graph form in Fig. 2. From the graph it can be seen that there is a clear correlation between the dataset and the run-time, as the dataset size increases the processing time also increases (this is to be expected).

The recorded runtime results obtained by applying the proposed BPM algorithm and the MBA and BC algorithms to data sets of increasing density are presented in Fig. 3. From the graph it can be seen that there is a correlation between the density of the datasets and the run-time as the density of the datasets increases the processing time also increases.

6.2   Comparison with BPM, MBA and BC

To compare the nature of the bandings produced using BPM, MBA and BC the average width of the banding produced was used as an independent measure (as oppose to global banding score gbs, accuracy Acc and Mean Row Moment MRM). Average Banding Width (ABW) was calculated as shown in Eq. 4. Similarly, Acc and Mean Row Moment (MRM) are calculated as shown in Eqs. 5 and 6:

---

[1] Available at http://cgi.csc.liv.ac.uk/frans/KDD/Software/LUCS-KDD-DataGen/ generator.html.

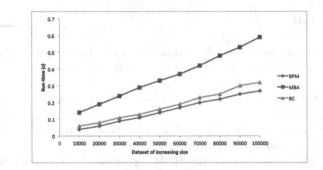

Fig. 2. Recorded run time (sec.) when the BPM, MBA and BC algorithms are applied to data sets of increasing size ($10,000$ to $100,000$ elements in steps of $10,000$)

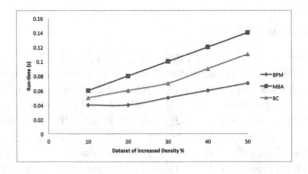

Fig. 3. Recorded run time (sec.) when the BPM, MBA and BC algorithms are applied to data sets of increasing density (10% to 50% elements in steps of 10)

$$ABW = \frac{\sum_{i=1}^{i=|D|} distance \; d_i \; from \; leading \; diagonal}{|D|} \qquad (4)$$

$$Acc = \frac{TP + TN}{TP + TN + FP + FN} \qquad (5)$$

$$MRM = \frac{\sum_{j=1}^{n} j a_{ij}}{\sum_{j=1}^{n} a_{ij}} \qquad (6)$$

where TP is the number of "true positives" corresponding to original 1 entries, TN is the number of "true negatives" corresponding to original 0 entries, FP is the number of "false positives" corresponding to transformed 0 entries, FN is the number of "false negatives" corresponding to transformed 1 entries, a_{ij} is the jth entry in row (resp. column) i and n is the number of columns (resp. rows). For the experiments to compare the nature of the bandings produced we applied the algorithms to a range of data sets taken from the UCI machine learning data repository [4]. As in the case of the artificial data sets the UCI data sets are all two dimensional with one dimension representing records and the other data attributes. The data sets were discretised using the LUCS-KDD Algorithm [4].

Table 3. *ABW* and *gbs* results presented using: *BPM*, *MBA* and *BC* algorithms

Dataset	(column	BPM	MBA	BC	BPM	MBA	BC
	× rows)	*ABW*	*ABW*	*ABW*	*gbs*	*gbs*	*gbs*
Auto	205	**0.6125**	0.8182	0.7802	**0.7604**	0.6545	0.6109
Breast	699	**0.9852**	0.9982	0.9957	**0.8890**	0.7281	0.7423
Car	1728	**0.9916**	0.9983	0.9973	**0.8053**	0.7783	0.7697
Congress	435	**0.9588**	0.9918	0.9881	**0.8807**	0.8086	0.8018
Cylband	540	**0.8913**	0.9659	0.9496	**0.8405**	0.7854	0.7417
Dematology	366	**0.9205**	0.9729	0.9709	**0.8189**	0.7742	0.7553
Ecoli	336	**0.9149**	0.9908	0.9717	**0.7767**	0.7544	0.7697
Flare	1389	**0.9794**	0.9981	0.9924	**0.8014**	0.7379	0.7807
Glass	214	**0.8392**	0.9468	0.9391	**0.7744**	0.7503	0.6963
Ionosphere	351	**0.7152**	0.8295	0.8696	**0.7906**	0.7393	0.6882

Table 4. *Acc* (%) and *MRM* results presented using: *BPM*, *MBA* and *BC* algorithms

Dataset	(column	BPM	MBA	BC	BPM	MBA	BC
	× rows)	*Acc*	*Acc*	*Acc*	*MRM*	*MRM*	*MRM*
Auto	205	71.47	**73.37**	73.35	**135.64**	129.78	130.81
Breast	699	**52.62**	51.25	51.25	**437.72**	358.75	379.13
Car	1728	61.97	**62.01**	61.13	1168.23	1146.58	**1171.21**
Congress	435	55.52	**55.68**	**56.34**	**352.21**	348.66	322.72
Cylband	540	62.86	**63.42**	63.32	**352.21**	348.66	322.72
Dematology	366	**65.70**	65.05	61.79	242.99	238.17	**245.77**
Ecoli	336	**61.94**	60.36	60.32	**254.70**	239.50	249.17
Flare	1389	**63.91**	63.27	63.21	**1051.57**	1031.53	1015.01
Glass	214	**74.88**	72.68	72.86	149.84	141.35	**150.84**
Ionosphere	351	**66.17**	65.94	65.90	**244.55**	243.76	230.42

For each dataset we applied the three algorithms, for each algorithm we recorded four scores (i) *ABW*, (ii) *gbs*, (iii) *Acc* and (iv) *MRM*. Tables 3 and 4 show the results obtained (best scores highlighted using bold font). In terms of the independent *ABW* metric (and the author's *gbs* metric) the tables clearly demonstrate that the proposed BPM algorithm outperformed the MBA and BC algorithms. Similarly, in terms of *Acc* and *MRM* Table 4 shows that in 6 out of the 10 cases, the proposed BPM algorithm still outperformed MBA and BC. Figures 4, 5 and 6 show the bandings obtained using the glass data set and the BPM, MBA and BC algorithms respectively. The Figures shows that banding can be identified in all cases. However, considering the banding produced when the MBA algorithm was applied to the glass data sets (Fig. 5) the banding result

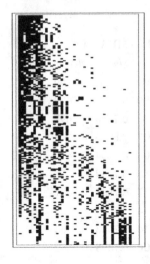 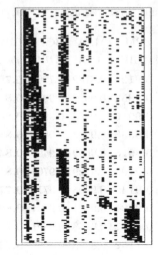 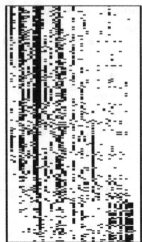

Fig. 4. BPM banding resulting using Glass dataset

Fig. 5. MBA banding resulting using Glass dataset

Fig. 6. BC banding resulting using Glass dataset

contains dots ("1s") at top-right and bottom-left corners while the BPM algorithm does not. Similarly, when the BC algorithm was applied to the glass data set (Fig. 6) the banding result is less dense than in the case of the proposed BPM algorithm (which features a smaller bandwidth).

6.3 Large Scale Study

To illustrate the utility of the proposed BPM algorithm and the previous BPM_A algorithm, the authors have applied both algorithms to a 5 dimensional data set constructed from the GB Cattle movement data base. The GB cattle movement database records all the movements of cattle registered within or imported into Great Britain. The database is maintained by the UK Department for Environment, Food and Rural Affairs (DEFRA). For the analysis reported in this work, data for the year 2003, for four counties (Aberdeenshire, Cornwall, Lancashire and Norfolk in Great Britain). In total we generated 16 data sets. The easting and northing dimensions were divided into ten sub-ranges and the temporal dimension divided into 3 intervals (each interval represented a month) was used. Each record comprises: (i) Animal Gender, (ii) Animal age, (iii) the cattle breed type, (iv) sender location in terms of easting and northing grid values, (v) the type of the sender location, (vi) receiver location type and (vii) the number of cattle moved. Discretization and Normalization were undertaken using the LUCS-KDD ARM DN Software[2] to convert the input data into the desired

---

[2] http://www.csc.liv.ac.uk/~frans/KDD/Software/LUCS_KDD_DN_ARM.

zero-one format. As a result the GB dataset comprised 110 items distributed over five dimensions: records, attributes, eastings, northings and time (months). Some statistics concerning the data sets are presented in Table 5.

The results obtained are presented in Table 6 (best results highligted in bold font). Note that the table includes results from using and not using a BPM M-table (as considered in Sect. 4) and results using the BPM_A algorithm presented previously in [1]. From the table the following can be noted: (i) using the BPM_A algorithm is more efficient than using BPM (Euclidean and Manhattan) (ii) not using BPM M-table requires less runtime than when using such a table (iii) using BPM with Manhattan weighting is more efficient than when using BPM with Euclidean weighting (because the use of Manhattan distances entail less calculation) and (iv) using BPM with Euclidean weighting produces the best bandings (*gbs* scores). Thus we can conclude that BPM with Euclidean weighting coupled with the use of a M-table is the most effective approach to the banded pattern mining problem. Closer inspection of the table also indicates that, as expected, there is a correlation between the number of records in the data sets and the run-time; as the number of records increases the processing time increases (this is to be expected). Note that the BPM_A algorithm does not necessarily find a best banding but only an approximation because it does not consider the entire data space when calculating bandings (it only consider dimension pairings). The BPM algorithm presented in this paper is designed to find an exact banding, instead of considering only dimension pairings as presented in [1], the bandings are derived with respect to the entire data space.

Table 5. Number of items in each dimension (after discretization) for the 16 5-D GB cattle movement data sets

Counties	Years	# Recs	# Atts	# Eings	# Nings	# Time
Aberdeenshire	Abd-Q1	42962	98	10	10	3
	Abd-Q2	46187	101	10	10	3
	Abd-Q3	41181	104	10	10	3
	Abd-Q4	47842	107	10	10	3
Cornwall	Corn-Q1	40501	101	10	10	3
	Corn-Q2	39626	104	10	10	3
	Corn-Q3	40226	107	10	10	3
	Corn-Q4	49890	110	10	10	3
Lancashire	Lanc-Q1	34325	97	10	10	3
	Lanc-Q2	40926	100	10	10	3
	Lanc-Q3	45765	103	10	10	3
	Lanc-Q4	47392	106	10	10	3
Norfolk	Nolf-Q1	11526	98	10	10	3
	Nolf-Q2	14311	101	10	10	3
	Nolf-Q3	9460	104	10	10	3
	Nolf-Q4	11680	107	10	10	3

Table 6. Runtime (RT) and *gbs* results obtained using: (i) Manhattan and Euclidean BPM and no M-Table (ii) Manhattan and Euclidean BMP and M-Table and (iii) BPM$_A$

Month	#	runtime (sec)					Global banding		
		BPM no M Tab.		BPM and M Tab.		BPM$_A$	*gbs* score		
ID	Recs	Manhat	Euclid	Manhat	Euclid		Manhat	Euclid	BPM$_A$
Abd-Q1	42962	**15.66**	48.68	**61.13**	212.03	**10.95**	0.6854	**0.6937**	0.4115
Abd-Q2	46187	**19.82**	50.95	**60.01**	219.04	**16.95**	0.6861	**0.6939**	0.4142
Abd-Q3	41181	**30.29**	43.86	**58.32**	211.89	**08.83**	0.6892	**0.6942**	0.4152
Abd-Q4	47842	**28.01**	80.86	**32.22**	231.59	**16.44**	0.6867	**0.6932**	0.4101
Corn-Q1	40501	**28.73**	93.82	**45.35**	163.42	**07.83**	0.6880	**0.6934**	0.4190
Corn-Q2	39626	**20.32**	69.33	**51.60**	185.91	**06.92**	0.6834	**0.6931**	0.4239
Corn-Q3	40226	**33.13**	94.41	**67.87**	201.86	**07.58**	0.6840	**0.6944**	0.4263
Corn-Q4	49890	**48.92**	121.11	**80.59**	210.54	**18.88**	0.6885	**0.6943**	0.4221
Lanc-Q1	34325	**27.66**	46.68	**51.02**	147.29	**05.13**	0.6856	**0.6936**	0.4346
Lanc-Q2	40926	**36.50**	50.95	**63.11**	182.74	**09.91**	0.6860	**0.6938**	0.4350
Lanc-Q3	45765	**25.74**	52.03	**59.85**	204.87	**13.86**	0.6859	**0.6936**	0.4352
Lanc-Q4	47392	**36.29**	55.52	**80.52**	228.89	**15.99**	0.6854	**0.6936**	0.4368
Nolf-Q1	11280	**05.32**	26.70	**19.65**	46.21	**01.58**	0.6830	**0.6934**	0.4124
Nolf-Q2	14557	**17.04**	25.85	**47.40**	86.82	**02.29**	0.6814	**0.6937**	0.4139
Nolf-Q3	9460	**10.48**	22.23	**45.17**	56.20	**01.27**	0.6852	**0.6942**	0.4202
Nolf-Q4	11680	**13.34**	25.84	**46.38**	63.15	**02.23**	0.6820	**0.6939**	0.4133

7   Conclusion

In this paper we have presented the BPM algorithm. Unlike the existing MBA and BC algorithms, the proposed algorithm does not consider large numbers of permutations but instead uses the concept of banding scores. In addition the proposed mechanism operates in N-D. The results presented indicate that the proposed BPM algorithm produces a better banding, using an independent Average Banding Width (ABW) measure, than in the case of MBA and BC. Results were also presented, using a large scale study (directed at the GB cattle movement database), indicating that the BPM algorithm and the BPM$_A$ algorithm (a previous variation of the proposed BPM algorithm) work well in 5-D (in terms of efficiency and effectiveness). The BPM mining algorithm is also more efficient than MBA and BC because it does not have to consider large numbers of permutations. For future work the authors intends to investigate High Performance Computing variation of the BPM algorithm to allow it to be applied in the context of Big Data Analytics.

References

1. Abdullahi, F.B., Coenen, F., Martin, R.: A scalable algorithm for banded pattern mining in multi-dimensional zero-one data. In: Bellatreche, L., Mohania, M.K. (eds.) DaWaK 2014. LNCS, vol. 8646, pp. 345–356. Springer, Heidelberg (2014)

2. Atkins, J., Boman, E., Hendrickson, B.: Spectral algorithm for seriation and the consecutive ones problem. SIAM J. Comput. **28**, 297–310 (1999)
3. Banerjee, A., Krumpelman, C., Ghosh, J., Basu, S., Mooney, R.: Model-based overlapping clustering. In: Proceedings of Knowledge Discovery and DataMining, pp. 532–537 (2005)
4. Coenen, F.: LUCS-KDD ARM DN software (2003). www.csc.liv.ac.uk/frans/KDD/Software/LUCS_KDD_DN_ARM/
5. Cuthill, A.E., McKee, J.: Reducing bandwidth of sparse symmetric matrices. In: Proceedings of the 1969 29th ACM national Conference, pp. 157–172 (1969)
6. Frans, C.: LUCS-KDD Data generator software (2003). www.csc.liv.ac.uk/frans/KDD/Software/LUCS_KDD_DataGen_Generator.html
7. Gemma, G.C., Junttila, E., Mannila, H.: Banded structures in binary matrices. Knowl. Discov. Inf. Syst. **28**, 197–226 (2011)
8. Junttila, E.: Pattern in Permuted Binary Matrices. Ph.D. thesis (2011)
9. Makinen, E., Siirtola, H.: The barycenter heuristic and the reorderable matrix. Informatica **29**, 357–363 (2005)
10. Mannila, H., Terzi, E.: Nestedness and segmented nestedness. In: Proceedings of the 13h ACM SIGKDD International Conference on Knowledge Discovery and Data Mining, New York, NY, USA, pp. 480–489 (2007)

Discrimination-Aware Association Rule Mining
for Unbiased Data Analytics

Ling Luo[1,2(✉)], Wei Liu[2,3], Irena Koprinska[1], and Fang Chen[2]

[1] School of Information Technologies, University of Sydney, Sydney, Australia
{ling.luo,irena.koprinska}@sydney.edu.au
[2] NICTA ATP Laboratory, Sydney, Australia
fang.chen@nicta.com.au
[3] Faculty of Engineering and IT, University of Technology, Sydney, Australia
wei.liu@uts.edu.au

Abstract. A discriminatory dataset refers to a dataset with undesirable correlation between sensitive attributes and the class label, which often leads to biased decision making in data analytics processes. This paper investigates how to build discrimination-aware models even when the available training set is intrinsically discriminating based on some sensitive attributes, such as race, gender or personal status. We propose a new classification method called Discrimination-Aware Association Rule classifier (DAAR), which integrates a new discrimination-aware measure and an association rule mining algorithm. We evaluate the performance of DAAR on three real datasets from different domains and compare it with two non-discrimination-aware classifiers (a standard association rule classification algorithm and the state-of-the-art association rule algorithm SPARCCC), and also with a recently proposed discrimination-aware decision tree method. The results show that DAAR is able to effectively filter out the discriminatory rules and decrease the discrimination on all datasets with insignificant impact on the predictive accuracy.

Keywords: Discrimination-aware data mining · Association rule classification · Unbiased decision making

1   Introduction

The rapid advances in data mining have facilitated the collection of a large amount of data and its uses for decision making in various applications. Although automatic data processing increases efficiency, it can bring potential ethical risks to users, such as discrimination and invasion of privacy. This paper focuses on building discrimination-aware classification models to eliminate potential bias against sensitive attributes such as gender and race.

Discrimination refers to the prejudicial treatment of individuals based on their actual or perceived affiliation to a group or class. People in the discriminated group are unfairly excluded from benefits or opportunities such as employment, salary or

© Springer International Publishing Switzerland 2015
S. Madria and T. Hara (Eds.): DaWaK 2015, LNCS 9263, pp. 108–120, 2015.
DOI: 10.1007/978-3-319-22729-0_9

education, which are open to other groups [1]. In order to reduce the unfair treatment, there are anti-discrimination legislations in different countries such as the Equal Pay Act of 1963 and the Fair Housing Act of 1968 in the US, and the Sex Discrimination Act 1975 in the UK [1]. Therefore, it is imperative to consider eliminating discrimination in applications such as decision support systems, otherwise the companies might be sued or penalized for acting against the law.

Our problem can be formally stated as follows. Suppose we are given a labeled dataset D with N instances, m nominal attributes $\{A_1, A_2,..., A_m\}$, from which the attribute $S = \{s_1, s_2,...,s_p\}$ has been identified as a sensitive attribute (e.g. race, gender, etc.), and a class attribute C. D is a *discriminatory dataset*, if there is an undesirable correlation between the sensitive attribute S and the class attribute C. For example, when performing credit history checks, if the probability P(*credit history = good | race = white*) is much higher than P(*credit history = good | race = black*), it is said that this dataset is biased against black people. The *discrimination severity of a classifier* is measured by computing the *discrimination score* DS (see Sect. 3.2) defined as:

$$DS = |P(C = C_t| S = S_1) - P(C = C_t| S = S_2)|, \text{ if } S \text{ is a binary nominal attribute,}$$

$$DS = \frac{1}{m} * (\sum_{i=1}^{m} \left| P(C = C_t|S = S_i) - P(C = C_t|S = S_{others}) \right|), \text{ if } S \text{ is a multi-value}$$

nominal attribute. This score is computed on the testing set using the predicted class labels. The goal is to learn a classifier with low discrimination score with respect to S, with minimal impact on the classification accuracy.

As an example assume that we are designing a recruitment system for a company to predict if a new candidate is suitable for a job or not. If the historical data contains more males than females, the prediction model may tend to favor the attribute gender. A prediction rule using gender or sensitive attribute like marital status, may achieve high accuracy, but it is not acceptable as it is discriminating, which is both unethical and against the law. Sensitive attributes such as gender, race and religion should be taken as an information carrier of a dataset, instead of distinguishing factors [2]. Females may be less suitable for a given job as on average they might have less work experience or lower educational level. It is acceptable to use work experience and educational level in the prediction model.

In this paper we investigate discrimination-aware classification that aims to decrease the discrimination severity for sensitive attributes, when the training data contains intrinsic discrimination. Our proposed method DAAR improves the traditional association rule classifier by removing the discriminatory rules while maintaining similar accuracy. DAAR also keeps the sensitive attributes during the classifier training of the classifier, which avoids information loss. Our contributions can be summarized as follows:

- We illustrate the discrimination problem and the importance of minimizing discrimination in real world applications.
- We propose a new measure, called Discrimination Correlation Indicator (DCI), which examines the discrimination severity of an association rule. DCI is applied as an effective criterion to rank and select useful rules in discrimination-aware association rule classification tasks.

- We extend the standard definition of the Discrimination Score measure (DS) from binary to multi-level nominal sensitive attributes.
- We propose DAAR, a new Discrimination-Aware Association Rule classification algorithm. We evaluate DAAR on three real datasets from different domains: traffic incident management, assessment of credit card holders and census income. We compare its performance with three methods: the standard association rule classifier [3], the state-of-the-art association rule classifier SPARCCC [4] and a discrimination-aware decision tree [5].

2 Related Work

2.1 Discrimination-Aware Methods

The discrimination-aware classification problem was introduced by Kamiran and Calders [6] and Pedreshi et al. [1], who formulated the direct and indirect discrimination definitions, and raised the attention of the Data Mining community to this problem. The existing discrimination-aware methods can be classified into two groups: methods that modify the dataset and methods that modify the algorithm.

The first group focuses on modifying the dataset during the pre-processing phase to eliminate the discrimination at the beginning. This includes removing the sensitive attribute, resampling [7] or relabeling some instances in the dataset to balance class labels for a certain sensitive attribute value [6, 8]. These methods typically lead to loss of important and useful information and undermine the quality of the predictive model that is learnt from the modified dataset. Additionally, just removing the sensitive attribute doesn't help due to the so called red-lining effect - the prediction model will still discriminate indirectly through other attributes that are highly correlated with the sensitive attribute [1, 9, 10].

The second group includes methods that integrate discrimination-aware mechanisms when building the classifier. Previous work [2, 5, 11] have adapted various widely used classification algorithms, including decision tree, naïve Bayes and support vector machine to deal with potential discrimination issues. The Discrimination-Aware Decision Tree (DADT) [5] uses a new splitting criterion, IGS, which is the information gain relative to the sensitive attribute, together with the standard information gain which is relative to the class (IGC). After generating a preliminary tree, the leaves are relabeled to decrease the discrimination severity to less than a non-discriminatory constraint $\varepsilon \in [0, 1]$ while losing as little accuracy as possible.

2.2 Association Rule Methods

An association rule takes the form $X \to Y$, where X and Y are disjoint item sets. X contains the set of antecedents of the rule, and Y is the consequent of the rule [12]. Given a dataset containing N instances and an association rule $X \to Y$, the support and confidence of this rule are defined as follows:

Support $(X \rightarrow Y) = \sigma(X \cup Y)/N$, Confidence $(X \rightarrow Y) = \sigma(X \cup Y)/\sigma(X)$

where $\sigma(\cdot)$ is the frequency of an item set $(\cdot)$. When learning association rules, we are interested in rules with high support and high confidence.

Firstly introduced in [3], Classification Based on Association rules (CBA) uses association analysis to solve classification problems. In CBA, only class attributes can appear in Y. When classifying a new instance, if there are multiple matching rules, the rule with the highest confidence will be used to determine the class label. This method will also be referred as "standard AR" in the rest of the paper. Our proposed method is based on CBA, as the rule-based classifier can produce easy-to-interpret models.

SPARCCC is a relative new variation of CBA, which adds a statistical test to discover rules positively associated with the class in imbalanced datasets [4]. SPARCCC introduced the use of p-value and Class Correlation Ratio (CCR) in the rule pruning and ranking. CCR is defined as:

$$CCR(X \rightarrow y) = corr(X \rightarrow y)/corr(X \rightarrow \neg y)$$

$$corr(X \rightarrow y) = (\sigma(X \cup y) * N)/(\sigma(X) * \sigma(y))$$

where $\sigma(\bullet)$ is the frequency of an item set $(\bullet)$. The method retains rules with *corr* $(X \rightarrow y) > 1$ and $CCR > 1$, which condition guarantees that they are statistically significant in the positive associative direction $X \rightarrow y$, rather than in the opposite direction $X \rightarrow \neg y$. SPARCCC has been shown to significantly outperform CCCS [4].

3   The Proposed Method DAAR

Our proposed method DAAR uses the new measure Discrimination Correlation Indicator (DCI), together with confidence and support, to efficiently select representative and non-discriminatory rules that can be used to classify new instances.

DAAR offers the following advantages: (1) unlike naïve methods which simply remove the sensitive attribute to deal with discrimination, DAAR keeps the sensitive attribute in the model construction to avoid losing useful information; (2) the new measure DCI is easy to compute and capable of filtering out discriminatory rules with minimal impact on the predictive accuracy; (3) DAAR generates a smaller set of rules than the standard AR and these discrimination-free rules are easy to use by the users.

3.1   DCI Measure

DCI is designed to measure the degree of discrimination for *each rule*. Given the rule $X \rightarrow y$, DCI is defined as

$$DCI = \begin{cases} \frac{|P(C=y| S=S_{rule})-P(C=y| S=S_{others})|}{(P(C=y| S=S_{rule})+P(C=y| S=S_{others}))} \\ 0 \qquad \textit{if either of the above } P(\cdot) \textit{ is } 0 \end{cases}$$

$P(C = y|S = S_{rule})$ is the probability of the class to be y given the value of the sensitive attribute S is S_{rule}. If S is a binary attribute, S_{rule} is the value of S in the target rule and

S_{others} is the other value of S. If S is a multi-value nominal attribute, S_{others} includes the set of all attribute values except the one which appears in the target rule.

For example, if the target rule is "gender = female, housing = rent → assessment = bad", where gender is the sensitive attribute, then S_{rule} is female and S_{others} is male. The DCI for this rule will be:

$$DCI = \frac{|P\,(C = low|gender = female) - P\,(C = low|gender = male)|}{P\,(C = low|gender = female) + P\,(C = low|gender = male\,)}$$

If the sensitive attribute does not appear in that rule at all, we define DCI to be 0. Therefore, the range of DCI is $[0, 1)$. When DCI equals to 0, which means the probability of the class value to be y is the same given different sensitive attribute values, the rule is considered to be free of discrimination. Otherwise, DCI is *monotonically increasing* with the discriminatory severity of a rule, which means that the larger DCI is, the more discriminatory the rule is with regard to the sensitive attribute S.

3.2 Discrimination Score

The Discrimination Score measure (DS) has been used in previous research [2, 5, 11] to *evaluate the discrimination severity of a classifier*. The conventional definition is only for the binary sensitive attribute case. If the sensitive attribute S is binary with values S_1 and S_2, DS is defined as:

$$DS = \left| P\left(C = C_{target}|S = S_1\right) - P\left(C = C_{target}|S = S_2\right) \right|$$

DS computes the difference between the probabilities of the target class C_{target} given the two values of the sensitive attribute $S = S_1$ or $S = S_2$, on the testing dataset. C_{target} can be any attribute value of the class label.

We extend this definition for the case with multi-value nominal attribute with m values, $m > 2$. We propose that DS is computed for each S_{value} and then averaged over the m scores. For each computation, it takes S_{value} as S_i and all the other values as S_{others}, and is defined as follows:

$$DS = 1/m* \left(\sum_{i=1}^{m} \left| P\left(C = C_{target}|S = S_i\right) - P\left(C = C_{target}|S = S_{others}\right) \right| \right)$$

The best case is when DS is zero, which means that the probabilities of the class value to be C_{target}, for all different values of the sensitive attribute, are the same, i.e. there is no discrimination. Otherwise, higher DS corresponds to higher discrimination severity. As the testing dataset has been labeled by the classifier, higher discrimination in the dataset indicates the classifier is biased, which should be prevented.

The purpose of DS and DCI is different. We note that DS cannot be used to filter discriminatory rules in DAAR instead of DCI, as DS is not for a single rule as required by DAAR. More specifically, DS is designed to measure the quality of a classifier (any classifier, not only AR) based on a testing dataset that has been labeled by the classifier. In contrast, DCI is computed for each rule (hence, it requires a rule-based classifier) and then compared against a threshold to check whether the rule is discriminatory or not. Another difference between DCI and DS is that DCI is a single ratio, as there is only one possible attribute value of S in one rule, so S_{rule} and S_{others} are fixed once we know the target rule. On the other hand, DS is the average score over m sub-scores, as it computes a sub score for each possible attribute value S_{value} in the dataset, which will be more than one for non-binary attribute S.

3.3 DAAR Algorithm

DAAR integrates DCI and the association rule classification to select discrimination-aware rules from all rules that have passed the minimum confidence and support thresholds. DAAR's algorithm is shown in Table 1.

Table 1. DAAR Algorithm

Algorithm Build Discrimination-Aware Classifier	
Input	dataset D with sensitive attribute S;
	max_length = k; thresholds *conf*, *spt* and *dci*;
Output	non-discriminatory rules
1	**for** i = 2 to k // generate *i*-item rules
2	**if** i = 2: generate 2-item rule, which is the base case;
3	**else** : merge (*i*-1)-item RuleSet and 2-item RuleSet;
4	prune rules to keep anti-monotonic;
5	**end if**
6	set up contingency tables to calculate confidence, support, DCI;
7	filter rules using thresholds *conf*, *spt* and *dci*;
8	store rules in *i*-item RuleSet;
9	**end for**
10	sort rules in *k*-item RuleSet by DCI in ascending order;
11	**return** *k*-item RuleSet as the classifier;

The algorithm defines the maximum length of the rule as k in the input, so as a result, all rules will contain at most $k - 1$ antecedents on the left and one class label on the right. In the loop, the algorithm merges the $(i-1)$-item rule set which was generated in the last round with the 2-item rule set (the base case), to get the i-item rule set. In line 10, the set of rules is sorted by DCI in ascending order for clear presentation to users, and this sorting does not affect the classification results. The majority voting is then used to classify new instances; if the vote is tied (e.g. the same numbers of rules support each class), the sum of DCI of all rules supporting each class is calculated and compared to determine the final class. As discussed in Sect. 3.1, the severity of discrimination is lower when DCI is smaller; therefore the voting will select the class value with lower sum value (i.e. which is less discriminatory).

4   Datasets and Experimental Setup

Three real datasets from different domains, such as public transport and finance management, are used to evaluate the performance of DAAR.

Traffic Incident Data was collected by the road authority of a major Australian city.[1] Each instance has 10 attributes, including the time, location and severity of the incident, the manager and other useful information. The *incident manager* is selected as the sensitive attribute *S*, and the class label is the *duration of the incident*, which takes two values: *long* and *short*. Our task is to predict whether an incident would be difficult to manage based on the available information. The *incident duration* is considered as a proxy to incident difficulty level – an incident with long duration corresponds to a difficult-to-manage incident and an incident with short duration corresponds to an easy-to-manage incident. Our experimental evaluation tests whether the proposed method can reduce the discrimination based on the sensitive attribute *incident manager* in predicting the *difficulty level of the incident*.

The data was preprocessed in two steps. Firstly, we noticed that there were more than 90 distinct manager values appearing in the full dataset, but most of them were only associated to less than 10 incidents. For simplicity, the managers were sorted by the number of associated incidents, and only instances handled by the top 5 managers were used in the experiment. Then, the majority class in the dataset was under-sampled to keep the dataset balanced with respect to the class attribute. This resulted in a dataset of 4,920 incidents, half of which were *difficult* and the other half were *easy* to manage.

German Credit Card Data is a public dataset from the UCI Machine Learning repository [13]. The dataset consists of 1,000 examples (700 *good* and 300 *bad* customers), described by 20 attributes (7 numerical and 13 categorical). The sensitive attribute is the *personal status and sex*, which shows the gender of a customer and whether he or she is single, married or divorced. Since it can be discriminatory to assess customers by their gender and marital status, we would like to decrease the discrimination based on this attribute when classifying new customers.

As the original dataset is strongly biased towards the class *good*, with ratio *good:bad* = 7:3, we randomly removed 400 good customers to keep the balance of the dataset. This resulted in 600 examples, 300 from each of the two classes.

Census Income Data is also a public dataset from the UCI Machine Learning repository. It contains 40 attributes (7 numerical and 33 categorical), which are used to predict the income level of a person. If the income is over $50 K, the person is classified as having *high income*, otherwise as having *low income*. The attribute *race* (with values: *white*, *black*, *asian or pacific islander*, *amer indian aleut or eskimo* and *other*) is the sensitive attribute. We randomly selected a smaller portion of the original dataset containing 1,200 examples; half with *high income* and half with *low income*.

---

[1] Data Collected by NSW Live Traffic: https://www.livetraffic.com/desktop.html#dev.

We evaluate the performance of DAAR and the other three methods in terms of predictive accuracy and discrimination score. The three baseline methods are CBA (the standard AR), SPARCCC and DADT. CBA was chosen as it is a standard association rule classifier. SPARCCC was selected as it is a state-of-the-art association rule classifier for imbalanced datasets. The discriminatory dataset can be considered as a special type of imbalanced dataset. In a discriminatory database, the bias is against a certain class label within a group, having the same value for a sensitive attribute, e.g. *race = black*, while in an imbalanced dataset the bias is against a class over the whole dataset. DADT was selected as it is a successful discrimination-aware classifier.

All three association rule mining methods (standard AR, DAAR and SPARCCC) use confidence and support thresholds to remove the uninteresting rules. These thresholds are controlled as a baseline, while the other measures, the CCR threshold in SPARCCC and the DCI threshold in DAAR, are varied to generate comparison conditions. For example, for the traffic data, we used the following pairs of confidence and support values (*conf* = 0.6, *spt* = 0.01; *conf* = 0.6, *spt* = 0.03; *conf* = 0.6, spt = 0.05; *conf* = 0.5, *spt* = 0.1).

The number of rules generated by the classifier is another important factor to consider. It affects both the accuracy and discrimination score, and is also very sensitive to the chosen thresholds for confidence, support, CCR and DCI. In order to compare the results fairly, it is important to make sure that the number of rules of comparable conditions are in the same range. Hence, once the thresholds for confidence and support are fixed for the standard AR, the thresholds for CCR and DCI (between [0,1)) are configured such that SPARCCC and DAAR can generate 4-5 conditions where the number of rules is of the same order of magnitude.

DADT, the discrimination-aware decision tree [5], uses the addition of the accuracy gain and the discrimination gain, IGC + IGS, as a splitting criterion, and relabeling of some of the tree nodes to reduce the discrimination. The non-discriminatory constraint $\varepsilon \in [0, 1]$ is tuned to generate comparison conditions.

5 Results and Discussion

All reported results are average values from 10-fold cross validation. The *p* value is the result of the independent two-sample *t*-test, which is used to statistically compare the differences in performance between DAAR and the other methods.

Traffic Incident Data. The sensitive attribute for this dataset is the *incident manager*, and the class label is the *incident difficulty level*. Table 2 shows the average predictive accuracy and discrimination score of the proposed DAAR method and the three methods used for comparison (standard AR, SPARCCC and DADT).

Table 2 shows that overall, considering both accuracy and discrimination, DAAR is the best performing algorithm – it has the second highest accuracy and the second lowest discrimination score, and also small standard deviations. DAAR is statistically significantly more accurate than SPARCCC (p = 0.041) and slightly more accurate than DADT (p > 0.05). Although AR is more accurate than DAAR, this difference is not statistically significant. In terms of discrimination score, DAAR has

Table 2. Accuracy and Discrimination Score for Traffic Incident Data

Methods	Accuracy		Discrimination Score	
	mean	std	mean	std
Standard AR	77.22%	0.021	0.236	0.010
SPARCCC	73.21%	0.059	0.266	0.064
DADT	74.19%	0.045	0.187	0.022
DAAR	76.32%	0.025	0.213	0.012

significantly lower discrimination score than both the standard AR (p = 0.009) and SPARCCC (p = 0.0016). DADT has the lowest discrimination score but its accuracy is impacted – it has the second lowest accuracy after SPARCCC. SPARCCC is the worst performing method – it has the lowest accuracy and highest discrimination score, and the largest variation for both measures.

Figure 1 presents a scatter plot of the accuracy and discrimination score. Ideally, we would like to see points in the top-left corner of the figure, which corresponds to high accuracy and low discrimination score simultaneously. However, there is a trade-off between the two measures, as the filtering out of discriminating rules will normally lower the predictive accuracy. Given this trade-off, our aim is to select the method which has lower discrimination score but no significant impact on accuracy.

The results in Fig. 1 are consistent with the results in Table 2. All individual results of DAAR (triangles) are clustered in the left part of the graph which corresponds to low discrimination score, and these scores are always lower than the results for the standard AR (diamonds). SPARCCC is more diverse – it has points with relatively low and very large discrimination score and some others with a large discrimination score, which explains the overall low discrimination score and the large standard deviation. DADT's results (circles) are also in the left part of the graph (low discrimination score) but the accuracy varies, which explains the overall lower accuracy and its higher standard deviation.

Table 3 presents some of the rules generated by the DAAR, together with their confidence and DCI values. These rules are easy to understand by users, which is one of the advantages of applying association rule classification.

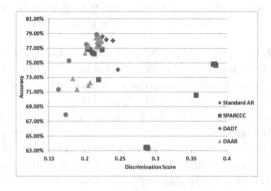

Fig. 1. Scatter plot for traffic incident data

Table 3. Examples of Rules Generate by DAAR for Traffic Incident Data

Rules	Confidence	DCI
type=road development, time = [20:00,23:59] → difficult incident	1	0
severity=2, time =[20:00,23:59], direction= north/south → difficult incident	0.95	0
day_of_week=7, direction = north /south → difficult incident	0.90	0
location= Cahill Expressway, Sydney → easy incident	0.76	0

Table 4 shows examples of rules that discriminate based on the manager with high confidence. These rules were filtered out by the proposed DAAR method.

Table 4. Examples of Discriminating Rules Filtered Out by DAAR

Rules	Confidence	DCI
manager=Frank, incident_severity=2 → difficult incident	0.87	0.273
manager=Charles →difficult incident	0.75	0.245
manager=Henry → easy incident	0.76	0.259

German Credit Card Data. For this data set, the aim is to eliminate the discrimination on *personal status and sex* when determining whether a customer is *good* or *bad*.

The results are shown in Table 5. We can see that the standard AR is the most accurate method, followed by SPARCCC, DAAR and DADT. The statistical testing results show that the accuracy of DAAR is not significantly lower than SPARCCC ($p > 0.05$) and that it is significantly higher than DADT ($p = 5.5e-5$). The accuracy range on this dataset (60-68 %) is lower compared to the traffic dataset (73-77 %). This might be due to the small size of the credit card dataset and the large number of its attributes. In terms of discrimination score, DADT is the best performing algorithm, followed by DAAR, SPARCCC and the standard AR. The t-test results show that our method DAAR has statistically significantly lower discrimination score than both the standard AR ($p = 8.0e-6$) and SPARCCC ($p = 0.0004$). Again, we can see that DAAR provides a good balance in terms of accuracy and discrimination score.

Table 5. Accuracy and Discrimination Score for German Credit Card Data

Methods	Accuracy		Discrimination Score	
	mean	std	mean	std
Standard AR	68.40%	0.012	0.329	0.006
SPARCCC	66.60%	0.029	0.305	0.071
DADT	60.13%	0.008	0.157	0.007
DAAR	64.90%	0.027	0.208	0.055

The scatter plot in Fig. 2 illustrates clearly the trade-off between accuracy and discrimination score. We can see that all DAAR's points (triangles) are on the left with respect to the standard AR points, but the accuracy of these points is lower than the accuracy of the standard AR points due to the removal of the discriminating rules. It is also interesting to observe that although SPARCCC has higher average accuracy than DAAR, the scatter plot demonstrates that for the same discrimination score (between 0.15 and 0.3), SPARCCC has lower accuracy than DAAR. The SPARCCC points are grouped into two main clusters: one in the middle that has similar discrimination score as DAAR and three points at the right corner that have high accuracy but large discrimination scores. DADT points are clustered at the bottom left of the graph, so that its accuracy lower than DAAR's for similar discrimination scores.

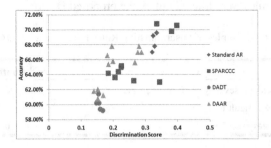

Fig. 2. Scatter Plot for German Credit Card Data

Census Income Data. The sensitive attribute in this dataset is *race* and the class label is *income level*. The aim is to avoid predicting the income level (*high* or *low*) of a person based on their race.

The results are presented in Table 6. We can see that in terms of average accuracy, all methods except DADT perform very similarly achieving accuracy of about 79-81 %, which is higher than the accuracy on the previous two datasets. SPARCCC is slightly more accurate and DAAR is slightly less accurate. The standard deviations of all three association rule methods are about 1 %. As to discrimination score, DADT again is the best performing method, followed by DAAR, which is consistent with the previous results. The standard AR comes next and the worst performing algorithm is SPARCCC. The t-test shows that the discrimination score of DAAR is significantly lower than both the standard AR (p = 0.016) and SPARCCC (p = 1.12e-6).

Table 6. Accuracy and discrimination score for Census Income Data

Methods	Accuracy		Discrimination Score	
	mean	std	mean	std
Standard AR	80.81%	0.011	0.285	0.009
SPARCCC	81.20%	0.010	0.289	0.010
DADT	68.57%	0.132	0.197	0.077
DAAR	79.65%	0.011	0.265	0.007

6   Conclusions

In this paper, we proposed DAAR, a discrimination-aware association rule classification algorithm that provides unbiased decision making support in data analytics processes. We have shown that DAAR is able to address the discrimination issues occurred on sensitive attributes, while having a minimal impact on the classification accuracy. DAAR uses DCI, a new discrimination measure, to prune rules that discriminate based on sensitive attributes, such as race and gender. The rules that pass the confidence-support-DCI filter will form the final DAAR rule set. To classify new instances, DAAR uses majority voting and a sum of DCI scores.

We empirically evaluated the performance of DAAR on three real datasets from traffic management and finance domains, and compared it with two non-discrimination-aware methods (a standard AR classifier and the state-of-the-art AR classifier SPARCCC), and also with the discrimination-aware decision tree DADT. The experimental results on all datasets consistently showed that DAAR is capable of providing a good trade-off between discrimination score and accuracy – it obtained low discrimination score while its accuracy was comparable with AR and SPARCCC, and higher than DADT. An additional advantage of DAAR is that it generates a smaller set of rules than the standard AR; these rules are easy to use by the users, in helping them make discrimination-free decisions.

Future work will include integrating DAAR in decision support applications such as assessment of social benefits. From a theoretical perspective, we plan to investigate the case with multiple sensitive attributes and the use of DCI in ensemble classifiers.

References

1. Pedreshi, D., Ruggieri, S., Turini, F.: Discrimination-aware data mining. In: Proceedings of the 14th ACM SIGKDD International Conference on Knowledge Discovery and Data Mining (KDD 2008), pp. 560–568. ACM (2008)
2. Calders, T., Verwer, S.: Three naive Bayes approaches for discrimination-free classification. Data Min. Knowl. Disc. **21**, 277–292 (2010)
3. Ma, Y., Liu, B., Yiming, W.H.: Integrating classification and association rule mining. In: Proceedings of the 4th ACM SIGKDD International Conference on Knowledge Discovery and Data Mining (KDD 1998), pp. 80–86 (1998)
4. Verhein, F., Chawla, S.: Using significant, positively associated and relatively class correlated rules for associative classification of imbalanced datasets. In: Proceedings of the 7th IEEE International Conference on Data Mining, pp. 679–684. IEEE (2007)
5. Kamiran, F., Calders, T., Pechenizkiy, M.: Discrimination aware decision tree learning. In: Proceedings of the 10th IEEE International Conference on Data Mining, pp. 869–874. IEEE (2010)
6. Kamiran, F., Calders, T.: Classifying without discriminating. In: International Conference on Computer, Control and Communication, pp. 1–6. IEEE (2009)
7. Kamiran, F., Calders, T.: Classification with no discrimination by preferential sampling. In: Proceedings of the Benelearn (2010)
8. Calders, T., Kamiran, F., Pechenizkiy, M.: Building classifiers with independency constraints. In: IEEE International Conference on Data Mining Workshops, pp. 13–18. IEEE (2009)

9. Hajian, S., Domingo-Ferrer, J.: A methodology for direct and indirect discrimination prevention in data mining. IEEE Trans. Knowl. Data Eng. **25**, 1445–1459 (2013)
10. Pedreschi, D., Ruggieri, S., Turini, F.: Integrating induction and deduction for finding evidence of discrimination. In: Proceedings of the 12th International Conference on Artificial Intelligence and Law, pp. 157–166. ACM, Barcelona, Spain (2009)
11. Ristanoski, G., Liu, W., Bailey, J.: Discrimination aware classification for imbalanced datasets. In: Proceedings of the 22nd ACM International Conference on Information and Knowledge Management, pp. 1529–1532. ACM (2013)
12. Simon, G.J., Kumar, V., Li, P.W.: A simple statistical model and association rule filtering for classification. In: Proceedings of the 17th ACM SIGKDD International Conference on Knowledge Discovery and Data Mining, pp. 823–831. ACM, 2020550 (2011)
13. University of California, Irvine, School of Information and Computer Sciences. http://archive.ics.uci.edu/ml

Social Computing

Big Data Analytics of Social Networks for the Discovery of "Following" Patterns

Carson Kai-Sang Leung$^{(\boxtimes)}$ and Fan Jiang

University of Manitoba, Winnipeg, MB, Canada
kleung@cs.umanitoba.ca

Abstract. In the current era of big data, high volumes of valuable data can be easily collected and generated. Social networks are examples of generating sources of these big data. Users (or social entities) in these social networks are often linked by some interdependency such as friendship or "following" relationships. As these big social networks keep growing, there are situations in which individual users or businesses want to find those frequently followed groups of social entities so that they can follow the same groups. In this paper, we present a big data analytics solution that uses the MapReduce model to mine social networks for discovering groups of frequently followed social entities. Evaluation results show the efficiency and practicality of our big data analytics solution in discovering "following" patterns from social networks.

1 Introduction and Related Works

Nowadays, high volumes of valuable data can be easily collected or generated from different sources such as social networks. Social networks are generally made of social entities (e.g., individuals, corporations, collective social units, or organizations) that are linked by some specific types of interdependencies (e.g., kinship, friendship, common interest, beliefs, or financial exchange). A social entity is connected to another entity as his next-of-kin, friend, collaborator, co-author, classmate, co-worker, team member, and/or business partner. Big data analytics of social networks computationally facilitates social studies and human-social dynamics in these big data networks, as well as designs and uses information and communication technologies for dealing with social context.

In the current era of big data (including big social network data), various social networking sites or services—such as Facebook, Google+, LinkedIn, Twitter, and Weibo [16,17]—are commonly in use. For instance, Facebook users can create a personal profile, add other Facebook users as friends, exchange messages, and join common-interest user groups. The number of (mutual) friends may vary from one Facebook user to another. It is not uncommon for a user A to have hundreds or thousands of friends. Note that, although many of the Facebook users are linked to some other Facebook users via their mutual friendship (i.e., if a user A is a friend of another user B, then B is also a friend of A), there are situations in which such a relationship is not mutual. To handle these situations, Facebook added the functionality of "follow", which allows a user

© Springer International Publishing Switzerland 2015
S. Madria and T. Hara (Eds.): DaWaK 2015, LNCS 9263, pp. 123–135, 2015.
DOI: 10.1007/978-3-319-22729-0_10

to subscribe or follow public postings of some other Facebook users without the need of adding them as friends. So, for any user C, if many of his friends followed some individual users or groups of users, then C might also be interested in following the same individual users or groups of users. Furthermore, the "like" button allows users to express their appreciation of content such as status updates, comments, photos, and advertisements. For example, when we liked the page "DEXA Society" (for information about DaWaK), many of our friends might also be interested in this page.

Similarly, Twitter users can read the tweets of other users by "following" them. Relationships between social entities are mostly defined by following (or subscribing) each other. Each user (social entity) can have multiple followers, and follows multiple users at the same time. The follow/subscribe relationship between follower and followee is not the same as the friendship relationship (in which each pair of users usually know each other before they setup the friendship relationship). In contrast, in the follow/subscribe relationship, a user D can follow another user E while E may not know D in person. For instance, a participant attending the DaWaK conference can follow @DEXASociety, but may not be followed by it. This creates a relationship with direction in a social network. We use D→E to represent the follow/subscribe (i.e., "following") relationship that D is following E.

In recent years, the number of users in the aforementioned social networking sites has grown rapidly (e.g., 1.44 billion monthly active Facebook users and 302 million monthly active Twitter users at the end of March 2015). This big number of users creates an even more massive number of "following" relationships. Over the past two decades, several data mining algorithms and techniques [1,5,8–10] have been proposed. Many of them [7,11,15] have been applied to mine social networks (e.g., discovery of special events [3], detection of communities [13,19], subgraph mining [20], as well as discovery of popular friends [6,11], influential friends [12] and strong friends [18]). In DaWaK 2014, we [4] proposed a *serial* algorithm to mine interesting patterns from social networks. While such an algorithm works well when mining a small focused portion of a social network due to its serial nature, there are situations in which one wants to mine a larger portion of a big social network. In response, we propose in the current DaWaK 2015 paper a new big data analytics and mining solution, which uses the MapReduce model to discover *interesting/popular "following" patterns* consisting of social entities (or their social networking pages) that are frequently followed by social entities. Such discovery of "following" patterns helps an individual user find popular groups of social entities so that he can follow the same groups. Moreover, many businesses have used social network media to either (i) reach the right audience and turn them into new customers or (ii) build a closer relationship with existing customers. Hence, discovering those who follow collections of popular social networking pages about a business (i.e., discovering those who care more about the products or services provided by a business) helps the business identify its targeted or preferred customers.

The remainder of this paper is organized as follows. The next section provides some background. Then, we present our new big data analytics and mining solution, which uses the MapReduce model to discover interesting "following" patterns from big social networks in Sect. 3. Evaluation results are shown in Sect. 4. Finally, conclusions are given in Sect. 5.

2 Background

High volumes of valuable data (e.g., web logs, texts, documents, business transactions, banking records, financial charts, medical images, surveillance videos, as well as streams of marketing, telecommunication, biological, life science, and social media data) can be easily collected or generated from different sources, in different formats, and at high velocity in many real-life applications in modern organizations and society. This leads us into the new era of *big data* [14], which refer to high-veracity, high-velocity, high-value, and/or high-variety data with volumes beyond the ability of commonly-used software to capture, manage, and process within a tolerable elapsed time. This drives and motivates research and practices in *data science*—which aims to develop systematic or quantitative processes to analyze and mine big data—for continuous or iterative exploration, investigation, and understanding of past business performance so as to gain new insight and drive science or business planning. By applying *big data analytics and mining* (which incorporates various techniques from a broad range of fields such as cloud computing, data analytics, data mining, machine learning, mathematics, and statistics), data scientists can extract implicit, previously unknown, and potentially useful information from big data (e.g., big social network data).

Over the past few years, researchers have used a high-level programming model—called *MapReduce* [2]—to process high volumes of big data by using parallel and distributed computing on large clusters or grids of nodes (i.e., commodity machines) or clouds, which consist of a master node and multiple worker nodes. As implied by its name, MapReduce involves two key functions: (i) the map function and (ii) the reduce function. Specifically, the input data are read, divided into several partitions (sub-problems), and assigned to different processors. Each processor executes the *map function* on each partition (sub-problem). The map function takes a pair of $\langle key_1, value_1 \rangle$ and returns a list of $\langle key_2, value_2 \rangle$ pairs as an intermediate result, where (i) key_1 and key_2 are keys in the same or different domains and (ii) $value_1$ and $value_2$ are the corresponding values in some domains. Afterwards, these pairs are shuffled and sorted. Each processor then executes the *reduce function* on (i) a single key key_2 from this intermediate result $\langle key_2, \text{list of } value_2 \rangle$ together with (ii) the list of all values that appear with this key in the intermediate result. The reduce function "reduces"—by combining, aggregating, summarizing, filtering, or transforming— the list of values associated with a given key key_2 (for all k keys) and returns a single (aggregated or summarized) value $value_3$, where (i) key_2 is a key in some domains and (ii) $value_2$ and $value_3$ are the corresponding values in some domains. An advantage of using the MapReduce model is that users only need to focus on (and specify) these "map" and "reduce" functions—without worrying about implementation details for (i) partitioning the input data, (ii) scheduling and executing the program across multiple machines, (iii) handling machine failures, or (iv) managing inter-machine communication. Examples of MapReduce applications include the construction of an inverted index as well as the word counting of a document for data processing [2].

3   Our Data Analytics Solution for Mining "Following" Patterns from Big Social Network Data

In this section, we present our new big data analytics and mining solution—called **BigFoP**—which mines **Big** social network data for interesting "**Fo**llowing"patterns using the MapReduce model.

3.1   "Following" Relationships in Big Social Networks

In social networking sites like Twitter, social entities (users) are linked by *"following" relationships* such as A→B indicating that a user A (i.e., *follower*) follows another user B (i.e., *followee*). Then, given a social network in which each social entity is *following* some other social entities, such a social network can be represented as a graph $G = (V, E)$ where (i) V is a set of vertices (i.e., social entities) and (ii) E is a set of directional edges connecting some of these vertices (i.e., "following" relationships). See Example 1.

Example 1. For illustrative purpose, let us consider a small portion of a big social network as shown in Fig. 1. It can be represented by $G = (V, E)$, where (i) set V of vertices = {Abel, Biel, Carlos, Desi, Elba, Fabio} and (ii) set E of edges = {⟨Abel, B⟩, ⟨Abel, E⟩, ⟨Biel, A⟩, ⟨Biel, C⟩, ⟨Biel, E⟩, ⟨Carlos, A⟩, ⟨Carlos, E⟩, ⟨Desi, B⟩, ⟨Desi, C⟩, ⟨Desi, E⟩, ⟨Elba, A⟩, ⟨Elba, B⟩, ⟨Elba, C⟩, ⟨Elba, D⟩, ⟨Fabio, A⟩, ⟨Fabio, B⟩, ⟨Fabio, C⟩, ⟨Fabio, E⟩}. Here, to avoid the confusion between followers and followees, we represent followers by their names and followees by their initials in these ⟨follower, followee⟩-pairs in the set E of edges.   □

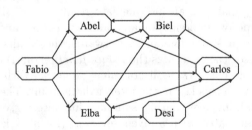

Fig. 1. A sample social network consisting of $|V| = 6$ users.

When compared with the mutual friendship relationships, the "following" relationships are different in that the latter are *directional*. For instance, a user B may be following another user C while C is not following B. As in Example 1, Biel is following Carlos, but Carlos is not following Biel. This property increases the complexity of the problem because of the following reasons. The group of users followed by B (e.g., Biel→{Abel, Carlos, Elba}) may not be same group of users as those who are following B (e.g., {Abel, Desi, Elba, Fabio}→Biel). Hence, we need to store directional edges (e.g., ⟨Abel, Biel⟩, ⟨Biel, Abel⟩) instead of undirectional edges (e.g., {Abel, Biel} indicating that Able and Biel are mutual friends). Given $|V|$ social entities, there are potentially $|V|(|V| - 1)$ directional edges for "following" relationships (cf. potentially $\frac{|V|(|V|-1)}{2}$ undirectional edges for mutual friendship relationships). Besides an increase in storage space, the computation time also increases because we need to check both directions to get relationships between pairs of users (e.g., cannot determine whether or not Carlos→Biel if we only know Biel→Carlos).

3.2   Discovery of "Following" Patterns

As the number of users in social networking sites (e.g., Twitter) is growing explosively nowadays, the number of "following" relationships between social network users is also growing. One of the important research problems with regard to this high volume of data is to discover interesting "following" patterns. A *"following" pattern* is a pattern representing the linkages when a significant number of users follow the same combination/group of users. For example, users who follow the twitter feed or tweets of NBA also follow the tweets of Adam Silver (current NBA commissioner). If there are large numbers of users who follow the tweets of both NBA and Adam Silver together, we can define this combination (NBA and Adam Silver) of followees as an interesting "following" pattern (i.e., a frequently followed group).

To discover interesting "following" patterns (i.e., collections of social network pages that are frequently followed by users), we propose a data science solution called **BigFoP** that mines **Big** social network data for interesting "**Following**"patterns by using sets of map and reduce functions.

3.3   The First Set of Map-Reduce Functions in BigFoP

Abstractly, BigFoP first applies a map function to each edge as follows:

$$\text{map}_1 : \langle \text{edge ID}, \text{"following"relationship captured by the edge} \rangle$$
$$\mapsto \langle \text{ follower, individual followee} \rangle, \tag{1}$$

in which the master node reads and divides big social network data in partitions. Specifically, the map_1 function can be specified as follows:

For each edge $e=\langle\text{follower, followee}\rangle \in E$ in social network $G=(V,E)$ **do**
   emit ⟨follower, followee, 1⟩.

This map function is applied to each edge $e=\langle$follower, followee$\rangle \in E$ in the social network represented by $G = (V, E)$, and results in a list of $\langle$follower, followee, 1$\rangle$ capturing all existing "following" relationships (between followers and followees) in the social network. See Example 2.

Example 2. After applying the map_1 function to the social network data in Example 1, our BigFoP returns a list containing $\langle$Abel, B, 1$\rangle$, $\langle$Abel, E, 1$\rangle$, $\langle$Biel, A, 1$\rangle$, $\langle$Biel, C, 1$\rangle$, $\langle$Biel, E, 1$\rangle$, $\langle$Carlos, A, 1$\rangle$, $\langle$Carlos, E, 1$\rangle$, $\langle$Desi, B, 1$\rangle$, $\langle$Desi, C, 1$\rangle$, $\langle$Desi, E, 1$\rangle$, $\langle$Elba, A, 1$\rangle$, $\langle$Elba, B, 1$\rangle$, $\langle$Elba, C, 1$\rangle$, $\langle$Elba, D, 1$\rangle$, $\langle$Fabio, A, 1$\rangle$, $\langle$Fabio, B, 1$\rangle$, $\langle$Fabio, C, 1$\rangle$, and $\langle$Fabio, E, 1$\rangle$. ☐

Afterwards, our big data analytics and mining solution BigFoP applies a reduce function to group and count the number of followers for each followee, as well as to list these followers for each followee. More specifically, $\langle$follower, followee, 1$\rangle$ pairs from the map_1 function are shuffled and sorted. Each processor then executes the reduce function on the shuffled and sorted pairs to count the number of followers and list them for each followee. To speed up this big social network data mining process, BigFoP also allows users to specify the *interestingness* of groups of social entities by a frequency threshold. Here, the users can indicate the minimum number of followers for a group of followees so that the group can be considered interesting. By incorporating this user preference, BigFoP returns (i) a list of followers only for those popular followees (i.e., followees who are frequently followed by at least the minimum number of followers) and (ii) the count for each followee. In other words, BigFoP applies the following reduce function:

$$reduce_1:\langle\text{followee, list of followers}\rangle$$
$$\mapsto \text{list of }\langle interesting \text{ followee, follower information}\rangle, \qquad (2)$$

with a detailed definition as follows:

> **For each** followee $\in \langle \_,$ followee, $\_\rangle$ emitted by map_1 **do**
> **set** counter[followee] = 0;
> **set** list[followee] = {};
> **for each** follower $\in \langle$follower, followee, 1$\rangle$ emitted by map_1 **do**
>   counter[followee] = counter[followee] + 1;
>   list[followee] = list[followee] $\cup$ {follower};
> **if** counter[followee] $\geq$ user-specified min frequency threshold
> **then emit** $\langle$followee, counter[followee], list[followee]$\rangle$.

This results in (i) a list of followers and (ii) its count for each *interesting/popular* followee. See Example 3.

Example 3. Continue with Example 2. Our BigFoP applies the $reduce_1$ function with user-specified minimum frequency threshold of 2 followers and returns $\langle$A, 4, {Biel, Carlos, Elba, Fabio}$\rangle$, $\langle$B, 4, {Abel, Desi, Elba, Fabio}$\rangle$, $\langle$C, 4, {Biel, Desi, Elba, Fabio}$\rangle$, and $\langle$E, 5, {Abel, Biel, Carlos, Desi, Fabio}$\rangle$. Note that our BigFoP does not return the lists for followees D or F because their corresponding counters were low (D and F were followed by only 1 and 0 followers, respectively).

To summarize, after applying the first set of map$_1$ and reduce$_1$ functions, our Big-FoP has so far discovered four interesting "following" patterns—in the form of *individual* frequently followed social entities—namely, {A}, {B}, {C} and {E}, who are followed by 4, 4, 4 and 5 followers respectively. In other words, each of these four individual followees is followed by at least 2 followers (the user-specified minimum frequency threshold).   □

3.4   The Second Set of Map-Reduce Functions in BigFoP

Thereafter, our BigFoP applies a next set of map and reduce functions to mine interesting "following" patterns in the form of *pairs* of frequently followed social entities based on the results from the first set of map$_1$ and reduce$_1$ functions. For instance, knowing that D and E are unpopular individual followees, it is guaranteed that any pairs containing followee D or E is also unpopular. By making use of this knowledge, the search space for mining interesting "following" patterns can then be pruned effectively. Specifically, the map$_2$ function, which returns ⟨follower, {p} ∪ {followee}, 1⟩ for every follower in the follower list of each popular/interesting individual followee p, can be specified as follows:

$$\text{map}_2 : \langle \text{interesting followee } p, \text{ its follower information} \rangle$$
$$\mapsto \langle \text{follower, followee pair} \rangle, \tag{3}$$

with a detailed definition as follows:

> **For each** $p \in \langle p, \_, \text{list}[p] \rangle$ emitted by reduce$_1$ **do**
>> **for each** follower $\in$ list$[p]$ **do**
>>> **for each** ⟨follower, followee⟩ $\in E$ of social network $G=(V,E)$ **do**
>>>> **if** isRelevant(followee, p)
>>>> **then emit** ⟨follower, {p} ∪ {followee}, 1⟩.

Here, isRelevant(followee, p) is a Boolean function checking the relevance (e.g., consistence to the mining order) of followee with respect to p. This results in lists of ⟨follower, followee, 1⟩, and a list for each popular individual followee p returned by the reduce$_1$ function. See Example 4.

Example 4. Continue with Example 3. Recall that the first set of map$_1$ and reduce$_1$ functions returns four popular followees A, B, C and E. So, for popular followee A (followed by four followers Biel, Carlos, Elba and Fabio), the map$_2$ function emits all *relevant* followees of these four followers: ⟨Biel, AC, 1⟩, ⟨Biel, AE, 1⟩, ⟨Carlos, AE, 1⟩, ⟨Elba, AB, 1⟩, ⟨Elba, AC, 1⟩, ⟨Fabio, AB, 1⟩, ⟨Fabio, AC, 1⟩, and ⟨Fabio, AE, 1⟩. Note that (i) followees of Abel are not emitted (because it is not meaningful for Abel to follow himself), (ii) followees of Desi are not emitted (because Desi does not follow A), (iii) four relationships in the form ⟨_, A, 1⟩ (e.g., ⟨Biel, A, 1⟩) are irrelevant with respect to p=A (because we already knew these four followers are following *single individual followee* A when we started this map$_2$ function and we aimed to find followers who follow *pairs of followees*), and (iv) ⟨Elba, AD, 1⟩ is also irrelevant (because followee D is unpopular).

Similarly, for popular followee B (followed by four followers Abel, Desi, Elba and Fabio), the map$_2$ function emits all *relevant* followee of these four followers: {⟨Abel, BE, 1⟩, ⟨Desi, BC, 1⟩, ⟨Desi, BE, 1⟩, ⟨Elba, BC, 1⟩, ⟨Fabio, BC, 1⟩, ⟨Fabio, BE, 1⟩}. Note that (i) followees of Biel are not emitted (because it is not meaningful for Biel to follow himself), (ii) followees of Carlos are not emitted (because Carlos does not follow B), (iii) four relationships in the form ⟨_, B, 1⟩ (e.g., ⟨Desi, B, 1⟩) are irrelevant with respect to p=B (because we already knew these four followers are following *single individual followee* B when we started this map$_2$ function and we aimed to find followers who follow *pairs of followees*), and (iv) ⟨Elba, BD, 1⟩ is also irrelevant (because followee D is unpopular). More important to note is that (v) relationships in the form ⟨_, AB, 1⟩ (e.g., ⟨Elba, AB, 1⟩, ⟨Fabio, AB, 1⟩) are irrelevant with respect to p=B (because these relationships are already processed by the map$_2$ function).

Then, for popular followee C (followed by four followers Biel, Desi, Elba and Fabio), the map$_2$ function emits all *relevant* followee of these four followers: {⟨Biel, CE, 1⟩, ⟨Desi, CE, 1⟩, ⟨Fabio, CE, 1⟩}.

Finally, for popular followee E (followed by five followers Abel, Biel, Carlos, Desi and Fabio), the map$_2$ function does not emit any followee because there is no *relevant* followee for these five followers.   □

Similar to reduce$_1$, the reduce$_2$ function shuffles and sorts ⟨follower, {p} ∪ {relevant followee}, 1⟩ to find and count followers for each followee pair P = ({p} ∪ {relevant followee}) as follows:

$$\text{reduce}_2 : \langle \text{followee pair, list of common followers} \rangle$$
$$\mapsto \text{list of } \langle \textit{interesting followee pair, follower information} \rangle, \quad (4)$$

with a detailed definition as follows:

> **For each** $P \in \langle$_, followee group P, _$\rangle$ emitted by map$_2$ **do**
>   **set** counter[P] = 0;
>   **set** list[P] = {};
>   **for each** follower ∈ ⟨follower, P, 1⟩ emitted by map$_2$ **do**
>   counter[P] = counter[P] + 1;
>   list[P] = list[P] ∪ {follower};
>   **if** counter[P] ≥ user-specified min frequency threshold
>   **then emit** ⟨P, counter[P], list[P]⟩.

This results in (i) a list of followers and (ii) its count for each *interesting/popular* followee pair P.

Example 5. Continue with Example 4. Our BigFoP applies the reduce$_2$ function with user-specified minimum frequency threshold = 2 followers and returns ⟨AB, 2, {Elba, Fabio}⟩, ⟨AC, 3, {Biel, Elba, Fabio}⟩, ⟨AE, 3, {Biel, Carlos, Fabio}⟩, ⟨BC, 3, {Desi, Elba, Fabio}⟩, ⟨BE, 3, {Abel, Desi, Fabio}⟩, and ⟨CE, 3, {Biel, Desi, Fabio}⟩. In other words, after applying this second set of map$_2$ and reduce$_2$ functions, our BigFoP algorithm discovered six interesting "following" patterns—in the form of pairs of frequently followed social entities—namely,

{A,B}, {A,C}, {A,E}, {B,C}, {B,E} and {C, E}, who are followed by 2, 3, 3, 3, 3 and 3 followers respectively. In other words, each of these six followee pairs is followed by at least 2 followers (the user-specified minimum frequency threshold). □

3.5   Subsequent Sets of Map-Reduce Functions in BigFoP

So far, our BigFoP has found interesting "following" patterns in the form of (i) *individual* frequently followed social entities as well as (ii) *pairs* of frequently followed social entities. BigFoP then applies *similar* sets of map and reduce functions to find triplets, quadruplets, quintuplets and higher (i.e., k-tuplets for $k \geq 3$) of frequently followed social entities:

$$\text{map}_{k \geq 3} : \langle \text{interesting followee}(k-1)\text{-tuplet } P, \text{ its follower information} \rangle$$
$$\mapsto \langle \text{follower, followee } k\text{-tuplet} \rangle, \qquad (5)$$

with a detailed definition as follows:

> **For each** $P \in \langle P, \_, \text{list}[P] \rangle$ emitted by reduce$_{k-1}$ **do**
>   **for each** follower $\in$ list$[P]$ **do**
>   **for each** $\langle$follower, followee$\rangle \in E$ of social network $G=(V,E)$ **do**
>   **if** isRelevant(followee, P)
>   **then emit** $\langle$follower, $P \cup \{$followee$\}$, 1$\rangle$.

Again, isRelevant(followee, P) is a Boolean function checking the relevance (e.g., consistence to the mining order) of followee with respect to P. Since reduce$_2$ can be considered as an instance of the reduce$_{k \geq 2}$ function, the latter can be defined in a way very similar to that for reduce$_2$ as shown below:

$$\text{reduce}_{k \geq 2} : \langle \text{followee group, list of common followers} \rangle$$
$$\mapsto \text{list of} \langle \textit{interesting} \text{ followee group, follower information} \rangle, \quad (6)$$

with a detailed definition as follows:

> **For each** $P \in \langle \_, \text{followee group } P, \_ \rangle$ emitted by map$_{k-1}$ **do**
>   **set** counter$[P]$ = 0;
>   **set** list$[P]$ = {};
>   **for each** follower $\in \langle$follower, P, 1$\rangle$ emitted by map$_{k-1}$ **do**
>   counter$[P]$ = counter$[P]$ + 1;
>   list$[P]$ = list$[P]$ $\cup$ {follower};
>   **if** counter$[P]$ $\geq$ user-specified min frequency threshold
>   **then emit** $\langle P$, counter$[P]$, list$[P] \rangle$.

This results in (i) a list of followers and (ii) its count for each *interesting/popular* followee group P. See Example 6.

Example 6. Continue with Example 5. For popular followee group AB (followed by two followers Elba and Fabio), the map$_3$ function emits three *relevant* followees: {$\langle$Elba, ABC, 1$\rangle$, $\langle$Fabio, ABC, 1$\rangle$, $\langle$Fabio, ABE, 1$\rangle$}. Then, for popular followee group AC (followed by three followers Biel, Elba and Fabio), the map_3 function emits two *relevant* followees: $\{\langle$Biel, ACE, $1\rangle, \langle$Fabio, ACE, $1\rangle\}$. Similarly, for popular followee group BC (followed by three followers Desi, Elba and Fabio), the map_3 function emits two *relevant* followees: $\{\langle$Desi, BCE, $1\rangle, \langle$Fabio, BCE, $1\rangle\}$.

Afterwards, by applying the $reduce_3$ function, our BigFoP discovers the following three interesting "following" patterns $\{$A, B, C$\}$, $\{$A, C, E$\}$ and $\{$B, C, E$\}$ with their associated lists and number of followees as $\langle$ABC, 2, $\{$Elba, Fabio$\}\rangle$, $\langle$ACE, 2, $\{$Biel, Fabio$\}\rangle$, and $\langle$BCE, 2, $\{$Desi, Fabio$\}\rangle$.

Based on the results returned by the $reduce_3$ function, BigFoP applies map_4 but returns nothing because there is no relevant quadruplet of frequently followed social entities. This completes the mining process for interesting "following" patterns from our illustrative example social network. Note that key concepts and steps illustrated in this example are applicable to any big social network. □

4   Observations, Evaluation and Discussion

To discover "following" patterns, our BigFoP takes advantages of the MapReduce model. The input social data are divided into several partitions (subproblems) and assigned to different processors. Each processor executes the map_k and $reduce_k$ functions (for $k \geq 1$). On the surface, one might worry that lots of communications or exchanges of information are required among processors. Fortunately, due to the divide-and-conquer nature of our big social network data analytics solution of discovering "following" patterns, once the original big social network is partitioned and assigned to each processor (e.g., one processor is assigned the followers of A, another is assigned the followers of B, a third one is assigned the followers of C), each processor handles the assigned data without any reliance on the results from other processors. As observed from the above examples, the processor assigned for the followers of a popular followee can apply the subsequent sets of map and reduce functions on data emitted by that processor. For example, a processor applies map_1 and $reduce_1$ to find popular followee A. That processor can then apply map_2 on the data emitted by $reduce_1$ from that processor to find popular followee group AB (i.e., group containing A). Similarly, the processor applies map_3 on the data emitted by $reduce_2$ from the same processor to find subsequent popular followee group ABC. Without the need of extra communications and exchanges of data among processors, our BigFoP discovers all interesting "following" patterns efficiently. Moreover, if a partition of the big social network is too big to be handled by a single processor, our BigFoP furthers sub-divide that partition so that the resulting sub-partitions can be handled by each of the multiple processors.

Furthermore, due to the divide-and-conquer nature of our big social network data analytics solution of discovering "following" patterns, the amount of data input for the map_k and $reduce_k$ functions monotonically decreases as the size of the popular group of k followees increases. Our BigFoP discovers all interesting "following" patterns in a space effective manner.

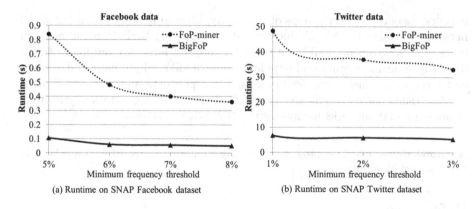

Fig. 2. Experimental results of BiigFoP on social network datasets.

As for runtime performance, we compared the performance of our BigFoP with related works (e.g., FoP-miner [4]). We used real-life social network datasets: The Stanford Network Analysis Project (SNAP) ego-Facebook dataset and ego-Twitter dataset (http://snap.stanford.edu/data/). The SNAP Facebook dataset contains 4,039 social entities and 88,234 connections ("following" relationships) between these social entities. The SNAP Twitter dataset contains 81,306 social entities and 1,768,149 connections between these social entities. All experiments were run using either (i) a single machine with an Intel Core i7 4-core processor (1.73 GHz) and 8 GB of main memory running a 64-bit Windows 7 operating system, or (ii) the Amazon Elastic Compute Cloud (EC2) cluster—specifically, 11 High-Memory Extra Large (m2.xlarge) computing nodes (http://aws.amazon.com/ec2/). We implemented both the existing FoP-miner algorithm and our proposed BigFoP in the Java programming language. The stock version of Apache Hadoop 0.20.0 was used. The results shown in Fig. 2, in which the x-axis shows the user-specified minimum frequency threshold (in percentage of the number of social entities) expressing the interestingness of the mined patterns, are based on the average of multiple runs. Runtime includes CPU and I/Os in the mining process of interesting "following" patterns. In particular, Fig. 2(a) shows that BigFoP provided a speedup of about 8 times when compared with FoP-miner when mining the SNAP Facebook dataset. Higher speedup is expected when using more processors. Figure 2(b) shows a similar result for the SNAP Twitter dataset. Moreover, our BigFoP is shown to be scalable with respect to the number of social entities in the big social network. As ongoing work, we are conducting more experiments, including an in-depth study on the quality of discovered "following" patterns.

5 Conclusions

In this paper, we proposed a big data analytics and mining algorithm—called *BigFoP*—for discovering interesting "following" patterns. BigFoP helps social network users to discover groups of frequently followed followees from big social networks by using the MapReduce model. By applying BigFoP, social network users (e.g., newcomers) could find popular groups of followees and follow them. Similarly, a business could find popular groups of followed products and services and incorporate customers' feedback on these products and services. Experimental results show the effectiveness of BigFoP in this big data analytics task of mining social networks for interesting "following" patterns.

Acknowledgement. This project is partially supported by NSERC (Canada) and University of Manitoba.

References

1. Cuzzocrea, A., Leung, C.K.-S., MacKinnon, R.K.: Mining constrained frequent itemsets from distributed uncertain data. Future Gener. Comput. Syst. **37**, 117–126 (2014)
2. Dean, J., Ghemawat, S.: MapReduce: simplified data processing on large clusters. Commun. ACM **51**(1), 107–113 (2008)
3. Dhahri, N., Trabelsi, C., Ben Yahia, S.: RssE-Miner: a new approach for efficient events mining from social media RSS feeds. In: Cuzzocrea, A., Dayal, U. (eds.) DaWaK 2012. LNCS, vol. 7448, pp. 253–264. Springer, Heidelberg (2012)
4. Jiang, F., Leung, C.K.-S.: Mining interesting "Following" patterns from social networks. In: Bellatreche, L., Mohania, M.K. (eds.) DaWaK 2014. LNCS, vol. 8646, pp. 308–319. Springer, Heidelberg (2014)
5. Jiang, F., Leung, C.K.-S.: Stream mining of frequent patterns from delayed batches of uncertain data. In: Bellatreche, L., Mohania, M.K. (eds.) DaWaK 2013. LNCS, vol. 8057, pp. 209–221. Springer, Heidelberg (2013)
6. Jiang, F., Leung, C.K.-S., Liu, D., Peddle, A.M.: Discovery of really popular friends from social networks. In: IEEE BDCloud 2014, pp. 342–349. IEEE, Los Alamitos (2014)
7. Kang, Y., Yu, B., Wang, W., Meng, D.: Spectral Clustering for Large-Scale Social Networks via a Pre-Coarsening Sampling based Nyström Method. In: Cao, T., Lim, E.-P., Zhou, Z.-H., Ho, T.-B., Cheung, D., Motoda, H. (eds.) PAKDD 2015, Part II. LNCS (LNAI), vol. 9078, pp. 106–118. Springer, Heidelberg (2015)
8. Leung, C.K.-S., Cuzzocrea, A., Jiang, F.: Discovering frequent patterns from uncertain data streams with time-fading and landmark models. LNCS TLDKS **8**, 174–196 (2013)
9. Leung, C.K.-S., MacKinnon, R.K.: BLIMP: a compact tree structure for uncertain frequent pattern mining. In: Bellatreche, L., Mohania, M.K. (eds.) DaWaK 2014. LNCS, vol. 8646, pp. 115–123. Springer, Heidelberg (2014)
10. Leung, C.K.-S., MacKinnon, R.K., Tanbeer, S.K.: Fast algorithms for frequent itemset mining from uncertain data. In: Kumar, R., Toivonen, H., Pei, J., Huang, J.Z., Wu, X. (eds.) IEEE ICDM 2014, pp. 893–898. IEEE, Los Alamitos (2014)
11. Leung, C.K.-S., Tanbeer, S.K.: Mining popular patterns from transactional databases. In: Cuzzocrea, A., Dayal, U. (eds.) DaWaK 2012. LNCS, vol. 7448, pp. 291–302. Springer, Heidelberg (2012)
12. Leung, C.K.-S., Tanbeer, S.K., Cameron, J.J.: Interactive discovery of influential friends from social networks. Soc. Netw. Anal. Min. **4**(1), Article 154 (2014)

13. Ma, L., Huang, H., He, Q., Chiew, K., Wu, J., Che, Y.: GMAC: a seed-insensitive approach to local community detection. In: Bellatreche, L., Mohania, M.K. (eds.) DaWaK 2013. LNCS, vol. 8057, pp. 297–308. Springer, Heidelberg (2013)

14. Madden, S.: From databases to big data. IEEE Internet Comput. **16**(3), 4–6 (2012)

15. Mumu, T.S., Ezeife, C.I.: Discovering community preference influence network by social network opinion posts Mining. In: Bellatreche, L., Mohania, M.K. (eds.) DaWaK 2014. LNCS, vol. 8646, pp. 136–145. Springer, Heidelberg (2014)

16. Rader, E., Gray, R.: Understanding user beliefs about algorithmic curation in the facebook news feed. In: Begole, B., Kim, J., Inkpen, K., Woo, W. (eds.) ACM CHI 2015, pp. 173–182. ACM, New York (2015)

17. Rajadesingan, A., Zafarani, R., Liu, H.: Sarcasm detection on Twitter: a behavioral modeling approach. In: Cheng, X., Li, H., Gabrilovich, E., Tang, J. (eds.) ACM WSDM 2015, pp. 97–106. ACM, New York (2015)

18. Tanbeer, S.K., Leung, C.K.-S., Cameron, J.J.: Interactive mining of strong friends from social networks and its applications in e-commerce. J. Organ. Comput. Electron. Commer. **24**(2–3), 157–173 (2014)

19. Wei, E.H.-C., Koh, Y.S., Dobbie, G.: Finding maximal overlapping communities. In: Bellatreche, L., Mohania, M.K. (eds.) DaWaK 2013. LNCS, vol. 8057, pp. 309–316. Springer, Heidelberg (2013)

20. Yu, W., Coenen, F., Zito, M., El Salhi, S.: Minimal vertex unique labelled subgraph mining. In: Bellatreche, L., Mohania, M.K. (eds.) DaWaK 2013. LNCS, vol. 8057, pp. 317–326. Springer, Heidelberg (2013)

Sentiment Extraction from Tweets: Multilingual Challenges

Nantia Makrynioti[(✉)] and Vasilis Vassalos

Athens University of Economics and Business, 76 Patission Street,
GR10434 Athens, Greece
{makriniotik,vassalos}@aueb.gr

Abstract. Every day users of social networks and microblogging services share their point of view about products, companies, movies and their emotions on a variety of topics. As social networks and microblogging services become more popular, the need to mine and analyze their content grows. We study the task of sentiment analysis in the well-known social network Twitter (https://twitter.com/). We present a case study on tweets written in Greek and propose an effective method that categorizes Greek tweets as positive, negative and neutral according to their sentiment. We validate our method's effectiveness on both Greek and English to check its robustness on multilingual challenges, and present the first multilingual comparative study with three pre-existing state of the art techniques for Twitter sentiment extraction on English tweets. Last but not least, we examine the importance of different preprocessing techniques in different languages. Our technique outperforms two out of the three methods we compared against and is on a par to the best of those methods, but it needs significantly less time for prediction and training.

1 Introduction

Users have integrated microblogging services and social networks in their daily routine, and tend to share through them increasingly more thoughts and experiences of their lives. As a result, platforms, such as Twitter, are a goldmine for the tasks of opinion mining and sentiment analysis, providing valuable information on topics of timeliness or not, by users of varying social, educational and demographic background.

In this paper, we examine sentiment analysis in Twitter with emphasis on tweets written in Greek and we suggest a method based on supervised learning. Sentiment analysis is defined as the task of classifying texts, in case of Twitter these correspond to tweets, into categories depending on whether they express

This research has been co-financed by the European Union (European Social Fund – ESF) and Greek national funds through the Operational Program "Education and Lifelong Learning" of the National Strategic Reference Framework (NSRF) – Research Funding Program: Thales. Investing in knowledge society through the European Social Fund.

© Springer International Publishing Switzerland 2015
S. Madria and T. Hara (Eds.): DaWaK 2015, LNCS 9263, pp. 136–148, 2015.
DOI: 10.1007/978-3-319-22729-0_11

positive or negative emotion or whether they enclose no emotion at all. As a consequence, sentiment analysis solves two classification tasks, the identification of objective and subjective tweets and the categorization of the latter according to their polarity. Given a number of tweets, our task is to categorize them in three classes, positive, negative and neutral depending on the presence of features that indicate emotion or not, as most of the times this is consistent with the sentiment of the message [12].

Although recently many papers study the task of sentiment analysis and many approaches have been proposed, almost all of them regard English text and work for other languages is limited. Moreover, many studies do not report results from comparisons with other pre-existent methods and each technique is usually evaluated on a single dataset. Evaluation on different datasets, including data of more than one languages, is an interesting process, which cross-checks the performance of the methods among languages.

The contributions of our paper are summarized below:

1. We propose a novel method for classification of tweets into three categories, positive, negative and neutral, and we evaluate our classifier on real Greek and English tweets. Our method outperforms two of the three compared approaches while giving statistically indistinguishable results to the third but with significant less time.
2. We present a case study of sentiment analysis in the context of the Greek language, unlike English that are much more studied and understood. For this purpose we collected and manually annotated a corpus of posts in Greek from Twitter, in order to be used as training and test data.[1]
3. We present extensive evaluation results and comparisons to three existing methods developed for English on a Greek as well as an English dataset. The purpose of these experiments is to provide the first comparative study of different state of the art techniques over Greek data, and examine their generalizability to address multilingual challenges. We also examine the contribution of specific preprocessing and postprocessing steps through ablation tests that demonstrate the degree to which certain steps of the proposed method improve the accuracy of the system with regard to Greek or English.

The rest of the paper is organized as follows. Section 2 presents some representative approaches on the problem of sentiment analysis and Sect. 3 analyzes the data used for training and testing. In Sect. 4 at first we give an overview of our method and then we describe in detail every step of it. Results from the evaluation of the classifier and the comparative analysis are reported in Sect. 5. Finally, Sect. 6 concludes and presents ideas for future work.

2   Related Work

The mining and analysis of unstructured data from social networks has attracted considerable attention in recent years. Go et al. [9] dealt with sentiment analysis

---

[1] Data are available by emailing the authors.

in Twitter, but their work was limited to positive and negative sentiments, and does not involve the recognition of objective (neutral) tweets. The machine learning algorithms that were applied are Multinomial Naive Bayes, Support Vector Machines (SVM) [24] and Maximum Entropy, whereas unigrams, bigrams as well as the combination of these two were used as features. Maximum accuracy reached 83 % and was achieved with Maximum Entropy and both unigrams and bigrams. Pak and Paroubek [20] emphasized the preprocessing of tweets before classification and adopted bigrams, trigrams, negation and part-of-speech tags as features. They used entropy and introduced a variant of it called "salience" to select the most representative features. Their results show that bigrams outperform trigrams and salience discriminated n-grams better than entropy. The method described in [5] divides the classification of tweets into two stages. The first stage classifies subjective and objective tweets, while the second categorizes subjective tweets into positive and negative. Part-of-speech tags are used as features in this paper too. Dictionaries of subjective terms and syntax features of Twitter, such as hashtags, links, punctuation and words in capital letters, were also employed. The classifier used SVM and maximum error rate for the first stage reached 18.1 %, whereas for the second stage it reached 18.7 %.

Even though the paper by Pang et al. [21] is not about Twitter, it is a benchmark and a comparison point with all the studies mentioned above. The paper addresses the task of sentiment analysis in movie reviews. Features include unigrams, bigrams and negation. Multinomial Naive Bayes and Maximum Entropy were tested, but SVM achieved 82.9 %, which was the maximum accuracy. Finally, a very recent approach by Mohammad et al. [17], which used a variety of features, including ngrams, syntax, lexicon and negation features, achieved the highest average F-score (69.02 %) with a SVM classifier in SemEval 2013 (International Workshop on Semantic Evaluation) and the task of sentiment analysis in Twitter [18]. Our work falls into the same category with the aforementioned studies, but apart from the certain difference of experimenting on Greek data, we apply a different combination of features and preprocessing steps, followed by a novel postprocessing negation identification step, which attempts to recognize the structure of negation in text and reverse the given prediction, rather than affect the features used for classification. Moreover, we reproduce published methods and present comparisons of them on a multilingual fashion, experimenting on datasets from two languages, Greek and English. All the above approaches belong to the category of supervised learning, but many studies have also performed unsupervised sentiment analysis. Due to limited space, we do not mention them here.

As stated earlier, there is lack of studies concerning other languages than English and the task is not sufficiently examined from this perspective. The paper by Atteveldt et al. [4] presents a system for automatically determining the polarity of relations between actors, e.g. politicians and parties, and issues, such as unemployment and healthcare, in Dutch text. To determine the polarity of relations, the authors use existing techniques for sentiment analysis in English and show that these methods can be translated to Dutch. Another study that addresses the multilingual perspective of the task is presented by Boiy and Moens [6]. The authors propose a supervised method for sentiment analysis and perform experiments on English, Dutch and French blog reviews and forum texts. There is also work about sentiment analysis on Modern Standard Arabic at the sentence level [2]. Arabic is a morphologically-rich language in contrast to English and the authors propose some Arabic-specific features along with the more commonly used and language-independent ones. Another work by Abbasi et al. [1] performs sentiment analysis on hate/extremist group forum postings in English and Arabic, and evaluates a variety of syntactic and stylistic features for this purpose. A method on Chinese data is also proposed by Zhao et al. in [26]. We are aware of a paper regarding reputation management on Greek data [22], but it presents a commercial product very briefly and in the abstract, and cannot be reproduced. Thus, our method not only is described extensively and in detail, but is also compared with other methods in the literature.

Finally, with regard to papers that compare methods and systems of sentiment analysis, such as [10] and [3], we take a step further and present comparisons in more than one languages.

3 Data

In this section we describe the datasets that are used for training and testing. Details about the size and contents of each dataset are given by Table 1. The Greek training data were collected between August 2012 and January 2015. Part of positive and negative tweets are based on subjective terms and around 20 % of neutral tweets were gathered from accounts of newspapers and news sites. The rest were streamed randomly. Respectively, Greek test set consists of random tweets posted between October 2013 and January 2015. We used Twitter Search and Streaming API[2] for the collection. Both training set and test set were labeled by three annotators. The calculated Fleiss' kappa [7] for the training set is 0.83, which is interpreted as almost perfect agreement, whereas for the test set is 0.691, which denotes substantial agreement. We will refer to the Greek training and test set as GR–train and GR–test.

For experiments on English we use training and test data provided by the organizers of SemEval 2013 [18] for the task of sentiment analysis in Twitter. The organizers collected tweets according to popular topics, which included

Table 1. Datasets

Dataset	Positive	Negative	Neutral	Total
GR–train	1870	2940	3190	8000
GR–test	261	249	378	888
Sem–train	3287	1601	4175	9063
Sem–test	1572	601	1640	3813

---

[2] https://dev.twitter.com/.

named entities previously extracted by a Twitter-tuned NER system [23], and used Mechanical Turk for annotation. We will refer to SemEval training and test set as Sem–train and Sem–test respectively.

4 Overview of Approach

The approach we adopt consists of three main steps: (1) Preprocessing of data. (2) Feature engineering. (3) Reversal of classifier's prediction for a tweet due to negation identification. The proposed method takes into account not only inflection but also word stress, both characteristics of morphologically-rich languages, and suggests a novel technique to reduce the negative effect of the combination of both in classification performance. Moreover, it treats identification of negation as a postprocessing step and attempts to capture its structure, which is a much different approach than adding a special suffix to bag-of-words features that most methods do until now. The aforementioned steps are described in detail in the following sections.

4.1 Preprocessing

Preprocessing is applied to both training and test set. The first step of preprocessing is the removal of noise from the data. Elements that do not indicate the polarity of a tweet are considered as noise. Such elements are listed below. (1) URL links. (2) Mentions of other users. (3) The abbreviation RT, which indicates that a tweet is a retweet of another one. (4) Stop words, including articles and pronouns. Stop words are extremely common words, which appear to be of little value in deciding the sentiment of a text.

Because users use plethora of emoticons/hashtags, we choose to replace positive emoticons[3] with the emoticon ":)" and negative emoticons[4] with the emoticon ":(". A number of hashtags, such as #fail and #win, are also replaced with the former two emoticons. The aim of this step is to group the emoticons/hashtags in two categories and to avoid the need of importing tweets in the training set for each one of them. In addition to the above replacements, possible repetitive vowels encountered in a word are reduced to one, whereas repetitive consonants are reduced to two.

Capitalization and removal of accent marks are the next steps. An accent mark over the vowel in the stressed syllable is used in Greek to denote where the stress goes, e.g. 'καλημέρα' (good morning). In order to avoid mistakes due to omission of stress marks and incorrect use of capital letters versus lowercase letters, we remove these marks from tweets and transform them to uppercase. Stemming is the third and last step, and is used mostly to compensate for data sparseness. Stems are generated by George Ntais' Greek stemmer [19] for Greek and by Lovins stemmer [15] via the Weka data mining software [11] for English.

---

[3] List of positive emoticons: :-), :), :o), :], :3, :c), :>, =], 8), =), :}, :^), <3, ^_^, ;>, (:, ;), (;, :d, :D.

[4] List of negative emoticons: >:[, :-(, :(, :-c, :c, :-<, :<, :-[, :[, :{, :'(, :/ .

The previous steps are applied to the test set too. However, the preprocessing of test set involves an additional step: part-of-speech tagging. It takes place before stemming and is an auxiliary step for the process of negation identification (Sect. 4.3), which follows classification. After the replacements, we annotate the words of each tweet with part-of-speech tags, which are not taken into account by the classifier to predict the class, but are used in patterns whose intention is to detect negation. A Greek part-of-speech tagger [13] is used for the tagging process in Greek, whereas in English the Carnegie Mellon University (CMU) Twitter Natural Language Processing (NLP) tool was selected [8].

4.2   Features

Feature engineering follows the bag-of-words representation with unigrams and term presence. Due to the limit in the number of characters that compose a tweet, a unigram is enough to denote the sentiment in most of the cases. For some unigrams there is a dependency with a particular class, while others do not give any information under any circumstances about the polarity. We decided to keep only a subset of them in order to eliminate noisy features and build a simpler model. We experimented with two methods, Information Gain and Chi Squared [14]. They both gave equally good results, so Information Gain is chosen arbitrarily for the experiments displayed in sections below.

Apart from word ngrams, lexicons of subjective terms, which contain terms with association to positive and negative sentiments, may provide various features for sentiment analysis. There are plenty of subjective lexicons in English, but we are not aware of any such lexicons in Greek. Nonetheless, we attempted to create manually two simple Greek subjective lexicons, one with positive words and one with negative words according to their prior polarity. Words were derived from random tweets, not contained in GR–train or GR–test, or translated from subjective English lexicons. The positive lexicon contains 199 words and the negative one consists of 292 words. We use two simple features, the presence of positive/negative terms of such lexicons in Greek tweets and more sophisticated features, such as those proposed in [17], for English data. In the aforementioned paper, lexicon-based features proved to be useful for the task of sentiment analysis. We present our conclusions about this kind of features in Sect. 5.

4.3   Negation Identification and Polarity Reversal

Negation identification is based on patterns of part-of-speech tags combined with negation words. We attempt to identify these patterns in each tweet and store the token that is negated. For example, the Greek word 'δεν' (not) followed by a verb and an adjective constitutes a negation pattern. If a tweet contains the phrase 'Η ταινία δεν ήταν καλή' (The movie wasn't good), the former negation pattern will be identified due to the word 'δεν', the verb 'ήταν' (both correspond to wasn't) and the adjective 'καλή' (good). Then the token 'καλή', which is the one that is negated, will be stored. Nine frequent patterns are recognized for Greek and eight for English. The detection of negation aims to reverse the prediction given by the classifier for a tweet from positive to negative or from negative to positive. If the prediction is neutral, no change is made. So following the decision of the classifier, we first check if a negated token is stored for the tweet. If yes, then we examine if this token belongs to the features that are present in the tweet. Suppose we have the aforementioned tweet for which we have kept the token 'καλή' as the negated token. If the unigram 'καλή' is one of the features and its value is 1, which indicates that this feature is present in the tweet, then the appropriate reversal of polarity will be performed. Otherwise, it will not.

4.4 Challenges of the Greek Language

The Greek language has a highly inflective nature that reduces the effectiveness of usual bag-of-words features. Greek verbs and adjectives are inflected for person, number and gender, which affects mostly the suffixes of the words. The various suffixes due to inflection increase ngram features, many of which are not contained in the training set. Hence, classification performance decreases. As a countermeasure to the inflective nature of Greek, the words of each tweet are replaced by their stems, assuming that stems are enough to denote the sentiment of a tweet in most cases.

Except from inflected verbs and adjectives, stress marks used in Greek make things even more complicated. Twitter users often forget to add these marks or they add them at the wrong syllable, creating this way a number of different versions of the same word (e.g. 'καλημέρα' is a different unigram from 'καλημερα'). As stated earlier, we chose to remove accent marks in order to reduce ngram features, but in case of stemming this choice may lead to mistaken predictions. Specifically, although stemming operates positively and helps the method to generalize better on unseen data (a conclusion that is drawn from the ablation tests included in Sect. 5.4), there is a case where it operates negatively: the stem of two words is the same whereas their polarities are different. For example the Greek words 'συμφωνώ' (agree) and 'σύμφωνα' (according to) have completely different meaning. The first word has positive polarity, whereas the second is neutral. Since the stem of both words is 'συμφων', there is no information to reveal the original word before stemming. As unigrams are used for predictions after stemming, the above case may be handled incorrectly. The described phenomenon can be frequently seen in Greek, even with words that are spelled exactly the same, but because they are stressed differently, their meaning changes. Note that these words are not homonyms as the word "like" for example, which serves both as verb and as proposition. In fact, purely homonyms with different polarity are extremely rare if non existent in the Greek language.

In order to handle the particular cases properly, a database is created with each record storing the following information: (i) stem, (ii) part-of-speech tag, (iii) polarity. The presence of unigrams (if any), on which the classifier based each prediction, are stored. If one of them, along with its part-of-speech tag, exists in the database, it is replaced with another one that has the same polarity. Specifically, if the unigram that exists in the database is positive, it will be replaced with the emoticon ":)", whereas if it is negative, it will be replaced with the emoticon ":(". Finally, if the unigram is neutral, an article will replace it. At the moment, seven such cases are identified and stored in the database. However, this database can be continually improved by a user feedback mechanism.

The described particularities show that depending on language, different pre-processing steps may improve the performance of the classifier and thus it is not trivial to suggest a method that proves to be best for every language.

5 Experiments

There are two versions of the proposed method that are developed for the experiments. The first one uses SVM as the classification algorithm and we will refer to it as #Sentiment_v1. The second version is called #Sentiment_v2 and uses Logistic Regression. SVM uses linear kernel and the value of parameter C is 1.0. The implementations of both algorithms are provided by the Weka data mining software. The section of experiments is divided in two parts. The first part presents the results of the evaluation on the Greek dataset collected by us, whereas the second includes results of experiments on English data provided by SemEval 2013. In these subsections we also compare the proposed method to three pre-existing methods developed for English [5,9,17], which we followed and implemented according to the descriptions in the corresponding papers. We will refer to these methods as Go_method, Mohammad_method and Barbosa_method according to the first author. Due to space restrictions we do not describe these methods, but of course we provide the corresponding references for details.

The evaluation metrics we report in the experiments are average precision, recall and F-score, i.e. the sum of the corresponding metrics for each class divided by the number of classes. We also use McNemar's test [16] to check the statistical significance of the difference in performance between systems in each experiment.

5.1 Greek Data

The experiments of this section concern the evaluation on Greek data gathered by us, i.e. the GR–test. We present a comparison between the two versions of our system and the three pre-existing methods described above. For Barbosa_method we implement only two lexicon features, number of positive and number of negative words, as all other lexicon features depend on the structure of the MPQA lexicon [25], which is separated into strong subjective and weak subjective terms (this distinction does not exist in current Greek lexicons). Figure 1 displays the evaluation results of the five systems, #Sentiment_v1, #Sentiment_v2, Go_method, Mohammad_method and Barbosa_method on Greek data. As far as our method is concerned, the difference in F-score between #Sentiment_v1 and #Sentiment_v2 on GR–test is statistically significant, which means that SVM outperforms Logistic Regression. However, our performance is statistically indistinguishable from Mohammad_method. The other two methods by Go and Barbosa achieve much lower average F-scores.

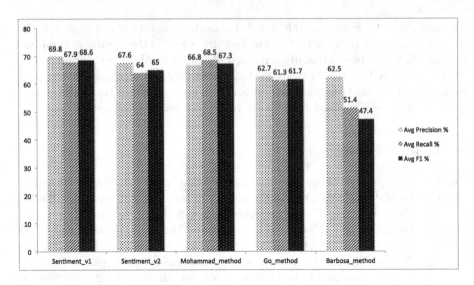

Fig. 1. Results on Greek data (GR–train for training and GR–test for testing)

The main conclusion of this experiment is that just the use of unigrams as features is not enough to achieve high accuracy in a classification problem with three classes. The Go_method was originally tested on a two-class classification of English tweets and generated good results, but the extension of the method to three classes and on another language seems not so simple and would need further preprocessing steps/features to work. This is demonstrated by #Sentiment_v1, #Sentiment_v2 and Mohammad_method, which also support the use of unigrams, but extend it with lexicon features, more preprocessing, such as stemming or feature selection, and achieve to reach higher average F-scores. Mohammad_method addresses inflection by keeping all ngram features, which however means a much larger model and more training time.

5.2   English Data

This section is dedicated to experiments on English data provided by the organizers of SemEval 2013. Again we present a comparison between the system proposed (only #Sentiment_v1, which is the best version according to the previous experiments) and the other three methods.

The evaluation results on the SemEval dataset (Sem–train and Sem–test) are displayed in Fig. 2. #Sentiment_v1 and Mohammad_method are again statistically indistinguishable and give the highest F-scores. Again methods that include both ngram and lexicon features, along with preprocessing and feature selection techniques, perform better on English data.

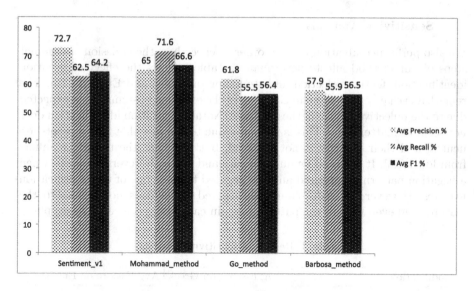

Fig. 2. Results on SemEval data (Sem–train for training and Sem–test for testing)

5.3 Time Consumption

In this section we present time performance results of the methods using the number of predicted tweets per second and training time. All methods ran on a single machine with an Intel Core i5 processor at 2.6 GHz and 16 GB of RAM. Details about time consumption of each method are given by Table 2.

Table 2. Time consumption

Method	Predicted tweets/sec	Training time (minutes)
#Sentiment_v1	16 tweets/sec	8.45 min
Mohammad_method	9 tweets/sec	14.91 min
Go_method	807 tweets/sec	5.9 min
Barbosa_method	8 tweets/sec	15 min

Although #Sentiment_v1 and Mohammad_method are indistinguishable in terms of F-score, #Sentiment_v1 needs 43 % less prediction and training time. This difference in time performance is reasonable, since Mohammad_method generates more features, such as part-of-speech and Twitter syntax features (RTs, hashtags, e.g.), which based on the experimental results they do not contribute that much to accuracy, but they increase processing time. Go_method is by far the fastest method. This is because it only involves unigram features, which are quickly generated. Nevertheless, they fail to predict test data effectively as experimental results demonstrated.

5.4 Sensitivity Analysis

We also performed ablation tests in order to check how the omission of different steps of our method affects performance. Table 3 shows the effect of negation identification, feature selection and stemming on Greek and English data. The remarkable change in F-score in Greek after the omission of stemming is expected due to the inflective nature of the language. Notably, negation identification does not seem to matter a lot. This is probably due to the fact that many tweets are neutral and their polarity is not reversed, but also that the technique suffers from low recall. It tends to be quite precise and correctly reverse polarity when a negation pattern is captured and the negated token is one of the classification features. However, in many cases the negated token does not belong to the features and even though the pattern is again captured, no reversal takes place.

Table 3. Results of sensitivity analysis

Modification	Avg F-score on Greek	Avg F-score on English
No modification	68.6 %	64.2 %
Without negation identification	68.7 %	64.1 %
Without feature selection	66.7 %	62.2 %
Without stemming	63.1 %	62.2 %

6 Conclusion and Future Work

We present a method for sentiment analysis in Twitter focused on the Greek language. We perform the first multilingual comparative analysis and report comparison results to three leading existing methods, from experiments on two different datasets (Greek and English). Our method clearly outperforms two of the three methods we compared against in sentiment extraction, while being statistically indistinguishable from the third. However, the proposed method needs 43 % less time for predictions and training. These experiments reveal that the generalization of a method to different languages or from a two to a three class classification problem is not trivial. Moreover, they give evidence about the effect of different preprocessing steps and features, such as stemming, in performance for Greek and English. An interesting idea to pursue in the future is the assignment of sentiment to the correct entity in the tweet.

References

1. Abbasi, A., Chen, H., Salem, A.: Sentiment analysis in multiple languages: feature selection for opinion classification in web forums. ACM Trans. Inf. Syst. **26**(3), 12:1–12:34 (2008)

2. Abdul-Mageed, M., Diab, M.T., Korayem, M.: Subjectivity and sentiment analysis of modern standard arabic. In: Proceedings of the 49th Annual Meeting of the Association for Computational Linguistics: Human Language Technologies: Short Papers . HLT 2011, vol. 2, pp. 587–591. Association for Computational Linguistics, Stroudsburg, PA, USA (2011)
3. Annett, M., Kondrak, G.: A comparison of sentiment analysis techniques: polarizing movie blogs. In: Bergler, S. (ed.) Canadian AI. LNCS (LNAI), vol. 5032, pp. 25–35. Springer, Heidelberg (2008)
4. Atteveldt, W.V., Ruigrok, N., Schlobach, S.: Good news or bad news? conducting sentiment analysis on dutch text to distinguish between positive and negative relations. J. Inf. Technol. Polit. 5(1), 73–94 (2008)
5. Barbosa, L., Feng, J.: Robust sentiment detection on twitter from biased and noisy data. In: Proceedings of the 23rd International Conference on Computational Linguistics: Posters, pp. 36–44. Association for Computational Linguistics (2010)
6. Boiy, E., Moens, M.F.: A machine learning approach to sentiment analysis in multilingual web texts. Inf. Retrieval 12(5), 526–558 (2009)
7. Fleiss, J., et al.: Measuring nominal scale agreement among many raters. Psychol. Bull. 76, 378–382 (1971)
8. Gimpel, K., Schneider, N., O'Connor, B., Das, D., Mills, D., Eisenstein, J., Heilman, M., Yogatama, D., Flanigan, J., Smith, N.A.: Part-of-speech tagging for twitter: annotation, features, and experiments. In: Proceedings of the 49th Annual Meeting of the Association for Computational Linguistics: Human Language Technologies: short papers. HLT 2011, vol. 2, pp. 42–47 (2011)
9. Go, A., Bhayani, R., Huang, L.: Twitter sentiment classification using distant supervision. Processing 150(12), 1–6 (2009)
10. Gonçalves, P., Araújo, M., Benevenuto, F., Cha, M.: Comparing and combining sentiment analysis methods. In: Proceedings of the First ACM Conference on Online Social Networks. pp. 27–38. COSN '13, ACM, New York, NY, USA (2013)
11. Hall, M., Frank, E., Holmes, G., Pfahringer, B., Reutemann, P., Witten, I.H.: The weka data mining software: an update. SIGKDD Explor. Newsl. 11(1), 10–18 (2009)
12. Hu, X., Tang, J., Gao, H., Liu, H.: Unsupervised sentiment analysis with emotional signals. In: Proceedings of the 22nd International Conference on World Wide Web. WWW 2013 (2013)
13. Koleli, E.: A new Greek part-of-speech tagger, based on a maximum entropy classifier. Master's thesis, Athens University of Economics and Business (2011)
14. Liu, H., Setiono, R.: Chi2: Feature selection and discretization of numeric attributes. In: 1995 Proceedings of Seventh International Conference on Tools with Artificial Intelligence, pp. 388–391. IEEE (1995)
15. Lovins, J.B.: Development of a stemming algorithm. Mech. Translation Comput. Linguist. 11, 22–31 (1968)
16. McNemar, Q.: Note on the sampling error of the difference between correlated proportions or percentages. Psychometrika 12(2), 153–157 (1947)
17. Mohammad, S., Kiritchenko, S., Zhu, X.: Nrc-canada: building the state-of-the-art in sentiment analysis of tweets. In: Second Joint Conference on Lexical and Computational Semantics (*SEM), Proceedings of the Seventh International Workshop on Semantic Evaluation (SemEval 2013), vol. 2, pp. 321–327 (2013)

18. Nakov, P., Rosenthal, S., Kozareva, Z., Stoyanov, V., Ritter, A., Wilson, T.: Semeval-2013 task 2: sentiment analysis in twitter. In: Second Joint Conference on Lexical and Computational Semantics (*SEM), Proceedings of the Seventh International Workshop on Semantic Evaluation (SemEval 2013), vol. 2, pp. 312–320 (2013)

19. Ntais, G.: Development of a Stemmer for the greek Language. Master's thesis, Stockholm's University (2006)

20. Pak, A., Paroubek, P.: Twitter as a corpus for sentiment analysis and opinion mining. In: Proceedings of the Seventh Conference on International Language Resources and Evaluation (LREC 2010). European Language Resources Association (ELRA) (2010)

21. Pang, B., Lee, L., Vaithyanathan, S.: Thumbs up?: sentiment classification using machine learning techniques. In: Proceedings of the ACL-02 Conference on Empirical Methods in Natural Language Processing, vol. 10, pp. 79–86. Association for Computational Linguistics (2002)

22. Petasis, G., Spiliotopoulos, D., Tsirakis, N., Tsantilas, P.: Sentiment analysis for reputation management: mining the Greek web. In: Likas, A., Blekas, K., Kalles, D. (eds.) SETN 2014. LNCS, vol. 8445, pp. 327–340. Springer, Heidelberg (2014)

23. Ritter, A., Clark, S., Mausam, Etzioni, O.: Named entity recognition in tweets: an experimental study. In: Proceedings of the Conference on Empirical Methods in Natural Language Processing, pp. 1524–1534. EMNLP 2011 (2011)

24. Vapnik, V.: Statistical Learning Theory. Wiley, New York (1998)

25. Wilson, T., Wiebe, J., Hoffmann, P.: Recognizing contextual polarity in phrase-level sentiment analysis. In: Proceedings of the Conference on Human Language Technology and Empirical Methods in Natural Language Processing. HLT 2005, pp. 347–354 (2005)

26. Zhao, J., Dong, L., Wu, J., Xu, K.: Moodlens: an emoticon-based sentiment analysis system for chinese tweets. In: Proceedings of the 18th ACM SIGKDD International Conference on Knowledge Discovery and Data mining, pp. 1528–1531. KDD 2012 (2012)

TiDE: Template-Independent Discourse Data Extraction

Jayendra Barua[1(✉)], Dhaval Patel[1], and Vikram Goyal[2]

[1] Indian Institute of Technology, Roorkee, India
{jbarudec,patelfec}@iitr.ac.in
[2] Indraprastha Institute of Information Technology, Delhi, India
vikram@iiitd.ac.in

Abstract. The problem of Discourse Data Extraction focuses on identifying comments and reviews from social networking websites. Existing approaches for Discourse Data extraction are either template-dependent or they are limited to comment-posting-structure discovery. We are not aware of any technique that extracts the detailed comment information like comment text, commenter and discussion structure from the comment page. In this paper, we present a template-independent two step approach, namely TiDE, which extracts the discourse data such as comments, reviews, posts and structural relationship among them. In the first step, we parse the input comment page to prepare a Document Object Model tree and then find the location of discourse data in the tree using the concept of Path-Strings. The outputs of the first step are Comment Blocks and these Comment Blocks are leveraged in second step to extract the comments, commenter and discussion structure. Experimental studies on 19 well known Discourse websites having different templates show that our Comment Block discovery is more adaptable than the existing posting-structure discovery technique. We are able to extract 97 % of comment-text and 79 % commenter information which is significant compared to state of the art techniques. We also show the usefulness of TiDE by building a news comment crawler.

1 Introduction

The evolution and popularity of online social networks (OSN) have changed the perspective of Internet among people. Initially, Internet was just a source of information for users. Presently, web users share their experiences and thoughts about various entities such as news articles, travel, restaurants, hotels, movies, videos etc. via OSN such as Twitter, Facebook, YouTube, Yelp and Tripadvisor. E.g. Yelp contains 67 million reviews of about 89000 thousand business entities, the popular Indian news media "The Hindu" on an average gets 15-20 user comments per news article, etc. We notice that the some OSN uses third party commenting systems such as Disqus, and Vuukle to provide the discussion/commenting facilities about their entity of interest. E g. two popular Indian news media "The Hindu" and "The Telegraph" are using "Vuukle" and "Disqus" commenting systems respectively in order to allow the users to post the comments on news articles. Hereafter, we refer all the OSN which allow user to write comments as "Discourse websites" and user comment and reviews are referred as "discourse data".

© Springer International Publishing Switzerland 2015
S. Madria and T. Hara (Eds.): DaWaK 2015, LNCS 9263, pp. 149–162, 2015.
DOI: 10.1007/978-3-319-22729-0_12

The discourse data is useful in many data analytics applications, Wang. et al. in [1] present a technique to suggest the news articles to a user by analyzing his comments on previously published news articles. In another application, Krishna et al. in [2] have done sentiment analysis of the YouTube comments and learned the polarity trends of public sentiments on videos. Yu X. et al. in [10], proposed a LeaderRank algorithm to identify the opinion leaders in Bulletin-Board systems using blog-posts of a university website. In summary, noticeable work has been done on different aspects of discourse data such as, "analyzing all the comments on a news article, video, etc." or "examine all the comments of a specific commenter". However, less attention has been made on extraction of discourse data from Discourse websites.

Existing techniques for discourse data extraction is largely focused on writing template dependent scraper. Yu X. et al. in [10] constructed a blog-post crawler to retrieve the discourse data from a university Bulletin Board System. Since, the Different Discourse websites are following different templates to publish comments; we need to inspect the source of the Discourse website for data extraction. Thus, developing template dependent method is a tedious task and only a computer professional can do it. Moreover, the discourse data about an online entity may be available at multiple websites. For e.g. More than 100 reviews about "Taj Campton Place, San Fancisco" are available at Yelp (http://bit.ly/1xQObAy) and Tripadvisior (http://bit.ly/1NBiU8b). Similarly, more than 100 users has commented the news "India lost semifinal in CWC-2015" at different Indian news media Hindustan Times (http://bit.ly/1bLjUZP), The Hindu(http://bit.ly/1G9QQ9q)., etc. In summary, people may discuss about same entity at multiple Discourse website and these Discourse websites may follow different templates to publish the comments. In such scenario, it is not feasible to apply the template dependent approach as we have to build a separate scraper for each of the website. Thus, development of a template independent approach is necessary.

Some existing approaches like Posting Structure Detection [7], Blog Post extraction [9] and MDR [12] extract the regularly structured web data records from web pages. Applying these techniques on the Discourse website may discover each comment as a record. However, these techniques do not extract detailed information like comment text, commenter and discussion structure among comments. Alternatively, article extraction techniques such as Boilerplate [3], URL Tree [4] and ECON [5] can be used to extract the comments from the web page. However, these approaches output all comments as a single large textual content and such output is not useful for the applications where individual comments along with their structural relationship (comment-reply) are required. In summary, there is a need of developing a template independent Discourse data extraction technique that can extract the individual comments along with the discussion structure. In this paper, the discussion structure refers to a comment-reply relationship in-between the comments.

Technical Contribution: First, we introduce the concept of Comment Page Structure that is used to model the comments/discussion on the given input comment page (Sect. 3). Next, we propose a generic two steps technique, namely TiDE (Template-Independent Discourse data Extraction), to extract the discourse data from a given comment webpage (Sect. 4). TiDE employs a set of heuristics, such as Path-Strings, text-count of HTML tags and least common ancestor to discover the Comment Page Structure and then extracts the comment texts, commenters and discussion structure from the identified Comment Page Structure. Our experimental results on 19 well known Discourse websites having different templates, for publishing comments suggest 79 % extraction accuracy of comment text along with commenter and 97 % extraction accuracy of comment text alone (Sect. 5). Finally, a news comment crawler is also developed using TiDE to accumulate the discourse data about the news articles published by Indian news media.

2   Related Work

Bank et al. in [7] provides an algorithm for detection of posting-structures in Internet forums. Posting-structure refers to the HTML block of a comment which contains all the information of the underlying comment such as comment text, commenter etc. They presented a two-step approach where they first detect the main content of the page using the concept of site-style tree and in the next step, they find out the XPath expression of the common posting-structure. However, they only detect posting-structure, but do not extract the more precise information such as commenter, comment text and discussion structure.

Subercaze et al. in [6] has proposed a complex seven step approach for lifting comments from a given webpage. Their approach extracts the identical frequent sub-trees from the DOM tree of the given web page. This approach includes clustering, merging, pattern expansion and winner selection in order to extract comments. However, these steps incur high computational costs as well as require parameters tuning. Further, experimental results do not reveal any information about dataset and algorithmic parameters.

Cao et al. in [9] provides a blog post and comment extraction approach based on information quantity. They assume that a web page is divided into two parts: (1) Post and (2) Comment, and provide an algorithm that separates the blog posts from the comments. However, they do not provide any details about how the comments, commenter and discussion structure can be extracted from the input web page.

3   Preliminaries

In a Discourse website, we assume that there is a separate comment page for each entity. The comment page is an HTML document that contains the information about "who commented what and who replied to whom" in the form of HTML tags. An HTML tag has tag name and attribute-value pairs. For example, consider a tag,

    <div class = "Grid" data-component-term = "tweet" role = "presentation"> .

The name of HTML tag is *div*. It has three attributes, namely *class*, *data-component-term*, and *role*, having values *"Grid"*, *"tweet"* and *"presentation"* respectively.

The HTML document is parsed using HTML parser such as Jsoup, NekoHTML etc. to create a Document Object Model (DOM) tree. Each node in the DOM tree represents an HTML tag and is identified by a tag name. We compare the HTML structure of various Discourse websites using DOM tree, and observed that these Discourse websites uses different HTML tags but follow some common layout while publishing comments. As a result, we introduce the concept of Comment Page Structure to model the layout of comment web page.

A **Comment Page Structure** of a comment page consists of three components, namely, (1) Comment Block, (2) Parent Comment Block, and (3) Reply Comment Block. In Fig. 1, we highlight the structural relationship between these three components as well as their locations in the HTML structure.

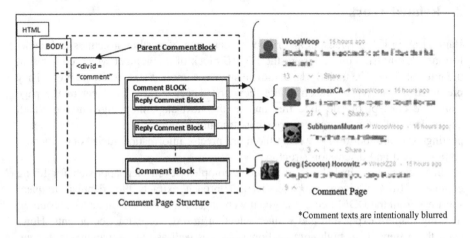

Fig. 1. Components of "Comment Page Structure"

Definition 1 (Comment Block). A Comment Block is the smallest HTML unit on the Comment Page Structure that mainly contains two important information (1) comment text and (2) commenter. The comment text refers to the textual part of the comment, and commenter is the author of the comment text.

Definition 2 (Comment Tag). An HTML tag that contains the comment text is known as a Comment Tag.

Definition 3 (Author Tag). An HTML tag that contains the commenter name is known as an Author Tag.

Definition 4 (Parent Comment Block). A Parent Comment Block is an HTML tag of a comment page such that all the Comment Blocks are its immediate children.

Definition 5 (Reply Comment Block). A Reply Comment Block is also a Comment Block, but it is not an immediate child of Parent Comment Block.

In Fig. 1, Node < div id = "comment" > is the Parent Comment Block. Reply Comment Blocks in Fig. 1 are contained inside the Comment Blocks. Now, we introduce a concept of Path-String for nodes in a DOM tree. The Path-Strings are used to extract the components of Comment Page Structure.

Definition 6 (Path-String). A Path-String of a node n in given DOM tree is a path from the root node to the node n along with the positional information of each node on the path. In our case, the position information of a node is obtained with respect to its parent. The **path-length** of a Path-String is the number of tags inside that Path-String. For example, consider a Path-String for a node with tag <p>:

$$\langle \text{html} : 0 \rangle - \langle \text{body} : 1 \rangle - \langle \text{div} : 12 \rangle - \langle \text{div} : 0 \rangle - \langle \text{p} : 3 \rangle$$

This Path-String is for tag < p > in the DOM tree since this tag appears last in the string. Value "3" just after "p" in the string indicates that node "p" is the 4th child of "div" tag. "div" tag is the second last in the string and parent of tag "p" and so on. The path-length of the above Path-String is 5.

4 Template-Independent Discourse Data Extraction (TiDE)

TiDE follows a two-step approach. In the first step, Comment Blocks are discovered from the input comment page (Sect. 4.1). Next, comment text, commenter and discussion structure are extracted from each Comment Blocks (Sect. 4.2).

4.1 Locate Comment Blocks

Since, Comment Block contains user comments in the form of Comment Tag, our idea is to first identify the Comment Tag. Note that, the user comments are published in a form of textual content, i.e., text count value of Comment Tag should be high. Based on this intuition, we use *text-count* heuristic to identify the Comment Tags. Next, we use the least common ancestor of Comment Tags to identify the Parent Comment Block. Finally, the immediate children of Parent Comment Block are outputted as Comment Blocks. This process is explained in detail as follows.

First, we parse the input comment page and create a DOM tree. Recall, each node in the DOM tree is identified by an HTML tag. Next, we traverse each node t in the DOM tree and obtain the value of *text-count$_t$* and *key$_t$* for node t. Here, *text-count$_t$* is the number of characters in the textual content of node t (excluding the *text-count* of children node of t) and *key$_t$* is the combination of tag name and tag attributes (with values). The following equation explains the derivation of *key$_t$*:

$$key_t = \begin{cases} TagName[\text{class} = \text{value}] & \textit{if attribute ``class''is present in node t} \\ TagName[\text{atr}_1 = \text{val}_1][\text{atr}_2 = \text{val}_2]..[\text{atr}_k = \text{val}_k] & \textit{Otherwise} \end{cases}$$

Example 1 (Key Generation). Suppose, we encounter a node in DOM tree having HTML Tag < div class = "*Grid*" data-component-term = "*tweet*" role = "*presentation*" > . For this node, "div[class = *Grid*]" is generated as the key. Similarly, we generate "div[id = *contentArea*][role = *main*]" if we encounter a node with HTML tag < div id = "*contentArea*" role = "*main*" > . Note that, "*class*" attribute in HTML is known as a tag-identifier attribute. This attribute is used for identifying the tags in an

HTML document that belongs to same class. As a result, we ignore other attributes, if the class attribute is present in the HTML tag while generating the key.

Next, we insert (key_t, $text\text{-}count_t$) of each node t in a hash-table HT. If key_t is already present in hash-table HT, then we increase the value (i.e. $text\text{-}count$) of key_t stored in hash-table by $text\text{-}count_t$, i.e. $HT[key_t] = HT[key_t] + text\text{-}count_t$. Once, we finish inserting all (key_t, $text\text{-}count_t$) pairs, we identify a key from hash-table having maximum value. The key corresponding to the maximum value represents the Comment Tag. This is our maximum text-count heuristic for Comment Tag discovery.

Once Comment Tag is identified, we again traverse the DOM tree and identify the nodes having tag name and tag attributes (with values) same as of "Comment Tag". For each such node, we find out the Path-String and put it into a *CommentPathList*. Next, we discover the maximum common prefix of Path-Strings that are stored in *CommentPathList*. The maximum common prefix is also a Path-String and points to Parent Comment Block in a DOM tree. Note that, the maximum common prefix discovers the least common ancestor of Comment Tags and thus, we term this heuristic as least common ancestor heuristic. The immediate children of Parent Comment Block in DOM tree are outputted as a Comment Blocks.

Example 2 (Comment Block Discovery). Assume, for a given Comment Tag, we have identified its five occurrences in a comment page. Figure 2 shows the five Path-Strings, one for each occurrence of Comment Tag. In this example, ⟨html : 0⟩ − ⟨body : 1⟩ − ⟨div : 6⟩ − ⟨section : 4⟩ − ⟨div : 1⟩ − ⟨ul : 3⟩ is the maximum common prefix. This Path-String is pointing to a Parent Comment Block, a node having an HTML tag < ul > . We notice in Fig. 2 that, the immediate children of this Parent Comment Block are four < li > tags, < li:21 > ,<li:22 > , < li:23 > , < li:24 > . These tags represent Comment Blocks.

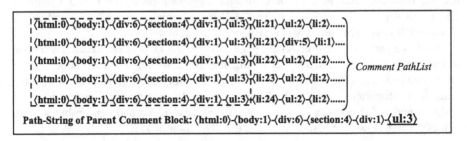

Fig. 2. Example of Comment Block discovery

We test our algorithm on comment pages of three different Discourse websites for identifying the tags that are used to represent the Comment Tag, Parent Comment Block, and Comment Block. The results are shown in Table 1. Observe that, different Discourse websites uses different tags and attributes and we are able to detect them correctly.

Time Complexity Analysis. The time complexity of identifying Comment Tags, Parent Comment Block and Comment Block is $O(T)$, $O(N)$, and $O(P)$ respectively. Where T is the number of tags in DOM tree, N is the number of occurrences of the

Table 1. Identified Comment Tag, Parent Comment Block and Comment Block

Web Site	Comment Tag	Parent Comment Block	Comment Block
YouTube	<div class = "comment-text-content">	<div class = "comments-list">	<div class = "comment-entry">
Yelp	<p itemprop = "description">	<ul class = "ylist ylist-bordered reviews">	<li>
TripAdvisor	<p class = "partial_entry">	<div id = "REVIEWS">	<div class = "reviewSelector ">

Comment Tag in DOM tree, and P is the number of children of a Parent Comment Block and $P \leq N < T$. The overall time complexity of step 1 is $O(T)$.

4.2   Extraction of Comments, Discussion Structure and Commenter

Having identified the Comment Blocks in the previous step, we process them further to discover the comment text, discussion structure and commenter. Given a Comment Block X, our first task is to identify what are the comments (i.e., comment text) and what is the relationship between these comments in form of the discussion structure (Sect. 4.2.1). Once, comment texts and discussion structures are identified; our second task is to identify their authors, i.e., commenter (Sect. 4.2.2).

4.2.1   Discover Comment Text with Discussion Structure

Recall, comment text is embedded inside the Comment Tag, and we have already discovered the Comment Tag is the previous step. As a result, comment texts can be identified by searching all the occurrences of the Comment Tag in the given Comment Block. However, this straightforward approach does not capture the discussion structures among comment text. Now, we explain how the discussion structure is discovered along with the comment text.

Example 3 (Discussion Structure). Consider three different comment pages with three comments A, B and C as shown in Fig. 3. Here, comment pages (b) and (c) involve discussion. In the comment page (a), there are three Comment Blocks, one for each comment. In the comment page (b), comment B is a reply to comment A, and Comment C is the reply of Comment B. Similarly in (c), comments B and C are reply of comment A.

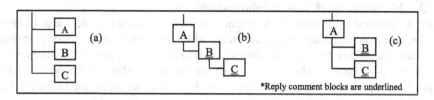

Fig. 3. Comment pages with varying discussion structures

In order to capture the discussion structure from the Comment Block X, we assign a unique-id and parent-id to each comment text of X. For example, for the Comment Block A in the comment page (a) of Fig. 3, the output should be (Comment Tag = A, unique-id = 1, parent-id = NULL). Similarly, for the Comment Block A in the comment page (b), the output should be three comments (Comment Tag = A, unique-id = 1, parent-id = Null) and (Comment Tag = B, unique-id = 2, parent-id = 1) and (Comment Tag = C, unique-id = 3, parent-id = 2).

Now, we discuss how comment text and discussion structure are extracted from Comment Block X. Recall, Comment Block X points to a single node in a DOM tree. First, we traverse the sub-tree rooted at Comment Block X in DFS manner, locate all the occurrences of Comment Tag and merge[1] two Comment Tags if they are siblings of each other. We assign an auto-incrementing unique id to each Comment Tag during traversal. Moreover, we also obtain Path-String for each Comment Tag and append it into a temporary list, namely *path-list*.

Once *path-list* is prepared, we traverse it from left to right and process each Path-String p_i to identify the comment text ct_i and parent-id of comment text *parent_i*. The comment text ct_i is the textual-content of the node that is pointed by Path-String p_i in the DOM tree. In our case, each ct_i is given a unique id, denoted as *unique id(ct_i)*. The parent id of the first comment text ct_1 is set to *NULL*. The parent id of the remaining comment text ct_i is obtained using Path-Strings p_i and p_{i-1} using following equation:

$$parent_i = \begin{cases} uniqueid(ct_{i-1}) & \text{if path} - \text{length}(p_i) > \text{path} - \text{length}(p_{i-1}) \\ parent_{i-1} & \text{if path} - \text{length}(p_i) = \text{path} - \text{length}(p_{i-1}) \\ parent_j & \text{if path} - \text{length}(p_i) < \text{path} - \text{length}(p_{i-1}) \end{cases}$$

where $j \in [2, i\text{-}1]$ such that Path-String p_j is the nearest Path-String to p_j in path-list and *path-length(p_i) = path-length(p_j)*.

Example 4 (Discussion Structure Discovery). Consider a Comment Block shown in Fig. 4. The Comment Block contains four Reply Comment Blocks. The Path-String corresponding to Comment Tag is also shown in Fig. 4 and is assigned a unique id starting from 1 to 5. After creating *path-list*, we apply the aforementioned process for identifying discussion structure. Following results were obtained: *parent_1 = NULL, parent_2 = 1, parent_3 = 2, parent_4 = 3, parent_5 = 3*.

4.2.2   Identification of Author Information

Now, we proceed to discover the commenter for each identified comment text. In most of the Discourse website, the commenter information is published using anchor tag < a>. The anchor tag < a > navigates reader to the profile page of the commenter. Generally, the "**First**" anchor tag < a > of the Comment Block or Reply-Comment Block is used to encode the author information. By "*First*" we mean "the *first < a > tag obtained while traversing the DOM-tree of Comment Block in DFS manner*".

---

[1] *merge*: append text of all the sibling tags into first sibling and except first, remove all sibling tags. Thus, output the Path-String of first sibling only.

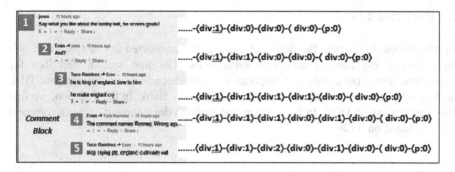

Fig. 4. Identification of discussion structure.

Given a Comment Block X, we search the first anchor tag < a > in the DOM tree rooted at X. This process discovers the author of comment text that is associated with X. As discussed, Comment Block may contain Reply Comment Block and there is an author for each Reply Comment Block. However, Reply Comment Blocks are yet to be identified. We address this issue, using *path-list* and discussion structure discovered in previous Sect. 4.2.1.

Given a Path-String p_i from a *path-list*, we obtain another Path-String p_j such that p_j is parent of p_i. Now, using p_i and p_j, we identify the maximum common prefix $mcp_{i,j}$. The immediate tag in Path-String p_i after the occurrence of $mcp_{i,j}$ in p_i is pointing to the Reply Comment Block. Once Reply Comment Block is obtained, we again start searching for the first anchor tag < a > in the DOM tree rooted at Reply Comment Block. The discovered anchor tag will be author tag for Path-String p_i. We process all the Path-Strings from the *path-list* in the similar way except the first Path-String.

Example 5 (Reply Comment Block Discovery). Let the Path-Strings associated with two Comment Tags of Comment Block A be P_A and P_B. Assume, P_A is a parent of P_B and given as follow

$$P_A : \langle html : 0 \rangle - \langle body : 1 \rangle - \langle div : 6 \rangle - \langle div : 1 \rangle - \langle ul : 3 \rangle - \langle div : 1 \rangle - \langle div : 0 \rangle$$
$$- \langle div : 0 \rangle - \langle div : 0 \rangle - \langle p : 0 \rangle$$
$$P_B : \langle html : 0 \rangle - \langle body : 1 \rangle - \langle div : 6 \rangle - \langle div : 1 \rangle - \langle ul : 3 \rangle - \langle div : 1 \rangle - \langle div : 1 \rangle$$
$$- \langle div : 0 \rangle - \langle div : 0 \rangle - \langle div : 0 \rangle - \langle p : 0 \rangle$$

The maximum common prefix between P_A and P_B, $mcp_{A,B}$, is $\langle html : 0 \rangle - \langle body : 1 \rangle - \langle div : 6 \rangle - \langle div : 1 \rangle - \langle ul : 3 \rangle - \langle div : 1 \rangle$. The immediate tag in P_B after $mcp_{A,B}$ is $\langle div : 1 \rangle$. This tag is the root node in the sub-tree of Reply Comment Block of Comment Block A.

Time Complexity Analysis. In a comment page, time complexity for identification of comment text and discussion structure is $O(N^2)$, where N is the number of Comment Tag. The time complexity for identification of commenter is $O(N)$.

5 Experiments

In this section, we present the details of experiments conducted to evaluate TiDE. We have divided experiments into four subsections. The first section describes the implementation detail and dataset preparation. In the second section, we compare TiDE with posting-structure discovery algorithm [7], namely Bank. In third section, we find the effectiveness of TiDE. In last section, we describe development of news comment crawler based on TiDE.

5.1 Experimental Setup

We implement TiDE algorithm in Java programming language using Jsoup API (implementation available at http://bit.ly/1Fcmemp). We prepare a Dataset using web pages from 19 popular Discourse websites. The first column in Table 2 lists the selected Discourse Website. Among these 19 websites, there are 15 Review websites which do not contain discussions. For each of these Review websites, we have selected 20 comment pages. The remaining 4 Discourse websites contain discussion structure among which one is YouTube and remaining 3 are third-party commenting systems Disqus, Facebook, and Vuukle. For YouTube we have selected comment page of 20 videos. We have selected 80 comment pages from each third-party commenting systems. In total, our dataset consist of $(16 \times 20 + 3 \times 80 = 560)$ comment web pages with an average 22 comments per page.

Table 2. Comparision between TiDE and bank approach

S. No.	Review Website	Accuracy		S. No.	Review Website	Accuracy	
		Bank	TiDE			Bank	TiDE
1	Citysearch	0.00%	95.00%	11	Trustpilot	0.00%	100.00%
2	Dealerrater	0.00%	100.00%	12	Urbanspoon	0.00%	100.00%
3	Foursquare	0.00%	95.00%	13	Virtualtourist	50.00%	100.00%
4	Insiderpages	10.00%	80.00%	14	Yelp	100.00%	100.00%
5	Merchant circle	68.18%	100.00%	15	Twitter	90.00%	95.00%
6	Metacritic	0.00%	100.00%	16	Disqus	0.00%	100.00%
7	Rottentomatoes	0.00%	100.00%	17	Vuukle	0.00%	100.00%
8	Travbuddy	94.44%	100.00%	18	Facebook	0.00%	92.00%
9	Traveller	0.00%	100.00%	19	Youtube	70.00%	85.00%
10	Tripadvisor	0.00%	100.00%				
Overall Accuracy: Bank = 25.40% , TiDE = 96.95%							

We implemented a template-dependent web scrapper for each of the Discourse website. Using these scrappers, we prepare the ground-truth for each comment page. In our ground-truth, we have Comment Block information along with comment text, commenter, and discussion structure.

5.2 Comparative Study of Comment Block Discovery Techniques

We compare the Comment Blocks discovered by our technique with the baseline algorithm Bank. We have implemented Bank algorithm for the comparison. Both the algorithms, TiDE and Bank, are applied on each comment page of the experimental dataset. TiDE outputs the Comment Blocks while Bank algorithm outputs the XPath expression of posting-structure. Using the XPath expression, we extract the posting-structures from input comment page. Finally, we compare extracted Comment Blocks/Posting Structure with the Comment Blocks of ground-truth. Table 2 shows the accuracy of extracted Comment Blocks and Posting structures for each Discourse Websites. The accuracy in Table 2 is the percentage of comment pages from which at least 80 % Comment Blocks/Posting Structure can be extracted successfully.

We can see in Table 2 that our approach is able to extract Comment Blocks accurately from all the Discourse websites whereas Bank approach [7] works efficiently only on 5 Discourse websites. The Overall accuracy achieved by TiDE is 97 % which outperforms the algorithm given by Bank et al. in [7]. The Bank algorithm is unable to work on most of the Discourse websites because of the varying posting structures of modern websites. For example modern websites publishes comments nested inside each other due to which there may be multiple XPath for posting structures, whereas Banks algorithm outputs only a single XPath for all posting-structures.

5.3 Effectivness of TiDE

In this experiment, we have tested the effectiveness of TiDE against the template-dependent technique. Template dependent techniques are accurate. However, it is not possible to gather discourse data from multiple Discourse websites automatically using template-dependent techniques.

We have applied TiDE on the prepared dataset which outputs comment text, a discussion structure and commenter information. The output is evaluated against the ground-truth. Recall that, we have used template-dependent scrappers for preparing the ground-truth. We have checked the presence of every comment on each comment page in the corresponding ground-truth. Table 3 shows the average number of correct comments and commenters extracted from each Discourse website.

In the results, we can see that our TiDE is performing perfectly on Travbuddy, Trustpilot, Virtualtourist and Yelp. Our approach for extraction of comment text is very effective and extracted an average of 97 % of comments on all 560 comment pages of 19 Online Discourse websites. However, our approach for extraction of commenters did not work on Dealerrater and Tripadvisor because their webpages structure does not have an author tag as first anchor tag < a > of Comment Block. Despite it, we have achieved an accuracy of about 79 %, which is quite fair.

Recall, that comment pages of four websites Disqus, Vuukle, Facebook and You-Tube of our dataset have discussion structure. In comparison with ground-truth, TiDE is able to discover 100 % comment-reply relationship in comments of Facebook, Vuukle and YouTube, however, 75.62 % comment-reply relations are discovered in Disqus comments. So, our approach discovered an average of 91 % comment-reply relations correctly from four Discourse websites.

Table 3. Extraction results of comment and commentor information

| S. No. | Review Website | Average | | S. No. | Review Website | Average | |
		%Comments Extracted	%Correct Author			%Comments Extracted	%Correct Author
1	Metacritic	83.60%	94.86%	11	Traveller	100.00%	77.33%
2	Insiderpages	88.14%	42.02%	12	Tripadvisor	100.00%	6.18%
3	Citysearch	91.36%	88.46%	13	Urbanspoon	100.00%	57.64%
4	Foursquare	95.00%	94.68%	14	Virtualtourist	100.00%	100.00%
5	Twitter	95.24%	95.24%	15	Yelp	100.00%	100.00%
6	Trustpilot	97.11%	100.00%	16	Disqus	98.69%	99.58%
7	Travbuddy	99.38%	100.00%	17	Vuukle	100.00%	96.58%
8	Dealerrater	100.00%	0.00%	18	Facebook	100.00%	99.88%
9	Merchant circle	100.00%	52.15%	19	Youtube	99.69%	98.80%
10	Rottentomatoes	100.00%	98.91%				
Overall Average: %Comment Extracted=97.27%, %Correct author = 79.07%							

5.4   News Comment Crawler Based on TiDE

We build a news comment crawler (NCC) using TiDE. First, NCC discovers the newly published news articles from the predefined set of 10 news websites. Next, NCC uses TiDE to extract the discourse data from the comment pages of newly discovered articles. In particular, the NCC monitors the news article at an interval of one hour in order to fetch the newly arrived user comments. We ran the NCC for three months and accumulated 309319 comments by crawling 52260 news articles. Due to space constraint, the statistics about obtained discourse data is available at (http://bit.ly/1ND5Mzy). We dig the data accumulated by the NCC to obtain the top three commented news articles, top three commenters (see Table 4) and a Tag Cloud (see Fig. 5). From Table 4, we deduce that people majorly reacts on incidents like polling, and statements from rival nations. Figure 5 evince that in India, people mostly comments on topics related to politics, Bollywood, and economy.

Table 4. Most commented articles and top commenters

S. No	News article	News source	Number of comments	Top commenters
1	N Korea threatens 'merciless' response over Seth Rogen film	The Guardian	360	TejPal - 845
2	David Cameron faces defeat in Juncker row as EU summit begins	The Guardian	245	Indian - 749
3	Bypoll results LIVE: Lalu-Nitish `grand alliance` set to beat BJP in Bihar	Zee News	179	Amit - 689

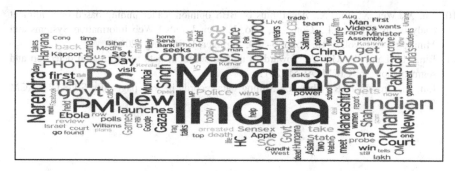

Fig. 5. Tag cloud of news headlines that received user comments

6 Conclusion

We proposed a novel approach TiDE for extraction of discourse data in Discourse websites. First, we present a Comment Page Structure as common layout of comment pages to model the user comments. Then, we describe a two-step method based on heuristics such as Path-Strings, least common ancestor and maximum common prefix, to discover the aforementioned Comment Page Structure. Our experiments suggest 79 % extraction accuracy of comment text along with commenter and 97 % accuracy of comment text alone. Development of NCC shows that TiDE is very much useful in the applications where automatic aggregation of user comments is required.

References

1. Wang, J., Li, Q., Yuanzhu, P., Liu, J., Zhang, C., Lin, Z.: News recommendation in forum-based social media. In: AAAI Conference. AAAI Press (2010)
2. Krishna, A., Zambreno, J., Krishnan, S.: Polarity trend analysis of public sentiment on YouTube. In: COMAD. ACM (2013)
3. Kohlschütter, C., Fankhauser, P., Nejdl, W.: Boilerplate detection using shallow text features. In: WSDM. ACM (2010)
4. Sluban, B., Grčar, M.: URL tree: efficient unsupervised content extraction from streams of web documents. In: CIKM. ACM (2013)
5. Guo, Y., Tang, H., Song, L., Wang, Y., Ding, G.: ECON: an approach to extract content from web news page. In: APWEB. Springer (2010)
6. Subercaze, J., Gravier, C.: Lifting user generated comments to SIOC. In: KECSM. CEUR-WS.org (2012)
7. Bank, M., Mattes, M.: Automatic user comment detection in flat internet fora. In: DEXA Workshops. IEEE (2009)
8. Yi, L., Bing, L., Li, X.: Eliminating noisy information in web pages for data mining. In: SIGKDD. ACM (2003)
9. Cao, D., Liao, X., Xu, H., Bai, S.: Blog post and comment extraction using information quantity of web format. In: Li, H., Liu, T., Ma, W.-Y., Sakai, T., Wong, K.-F., Zhou, G. (eds.) AIRS 2008. LNCS, vol. 4993, pp. 298–309. Springer, Heidelberg (2008)

10. Yu, X., Wei, X., Lin, X.: Algorithms of BBS opinion leader mining based on sentiment analysis. In: Wang, F.L., Gong, Z., Luo, X., Lei, J. (eds.) Web Information Systems and Mining. LNCS, vol. 6318, pp. 360–369. Springer, Heidelberg (2010)
11. Bing, L., Lam, W., Gu, Y.: Towards a unified solution: data record region detection and segmentation. In: CIKM. ACM (2011)
12. Liu, B., Grossman, R., Zhai, Y.: Mining data records in web pages. In: SIGKDD. ACM (2003)

Heterogeneous Networks and Data

A New Relevance Measure for Heterogeneous Networks

Mukul Gupta[✉], Pradeep Kumar, and Bharat Bhasker

Indian Institute of Management, Lucknow, India
{fpm13008,pradeepkumar,bhasker}@iiml.ac.in

Abstract. Measuring relatedness between objects (nodes) in a heterogeneous network is a challenging and an interesting problem. Many people transform a heterogeneous network into a homogeneous network before applying a similarity measure. However, such transformation results in information loss as path semantics are lost. In this paper, we study the problem of measuring relatedness between objects in a heterogeneous network using only link information and propose a meta-path based novel measure for relevance measurement in a general heterogeneous network with a specified network schema. The proposed measure is semi-metric and incorporates the path semantics by following the specified meta-path. For relevance measurement, using the specified meta-path, the given heterogeneous network is converted into a bipartite network consisting only of source and target type objects between which relatedness is to be measured. In order to validate the effectiveness of the proposed measure, we compared its performance with existing relevance measures which are semi-metric and applicable to heterogeneous networks. To show the viability and the effectiveness of the proposed measure, experiments were performed on real world bibliographic dataset DBLP. Experimental results show that the proposed measure effectively measures the relatedness between objects in a heterogeneous network and it outperforms earlier measures in clustering and query task.

Keywords: Heterogeneous network · Meta-path · Clustering · Query task · Relevance measure

1 Introduction

Measuring relatedness between objects (represented as nodes) in a heterogeneous network for different mining tasks has gained the attention of researchers and practitioners due to the information rich results. To mine a heterogeneous network, many researchers transform the heterogeneous network into the corresponding homogenous network before applying different mining techniques [1–3]. However, mining of heterogeneous-transformed homogeneous network involves information loss as path semantics are lost [1]. Due to this, there is a surge of studies for measuring relatedness between objects. Relevance measurement for objects in a heterogeneous network has become important for various mining tasks like clustering, classification and query [1, 2].

© Springer International Publishing Switzerland 2015
S. Madria and T. Hara (Eds.): DaWaK 2015, LNCS 9263, pp. 165–177, 2015.
DOI: 10.1007/978-3-319-22729-0_13

Mining of heterogeneous information networks has become important as they are ubiquitous and play a critical role in modern information infrastructure [1, 3, 4]. For example, a bibliographic network can be modelled as a heterogeneous network and typically has nodes representing papers, authors, conferences and keywords. The relationship of nodes are represented by links in the network. Since in a heterogeneous network different typed objects and different typed relationships co-exist, so measuring relatedness between different typed objects, following different relationship paths, can give inter-

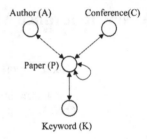

Fig. 1. Network schema of DBLP heterogeneous network data

esting results and may reflect the real nature of data. Figure 1 shows the network schema of DBLP database [2]. In Fig. 1, a bidirectional link exists between Paper and Author indicating that every paper is associated with author(s) and vice versa. Similarly bidirectional links exist between Paper and Keyword as well as Paper and Conference. A paper may be cited by another paper hence a self-loop exists at Paper node.

Different mining and query tasks can be performed on a heterogeneous network to answer different questions. For example, in case of DBLP network, we can answer questions like *"Who are the peer researchers of a specified author?"*, *"Who are the leading researchers in Information Retrieval area?"*, *"How are Computer Science research areas structured?"* and many others. To answer these questions, we need to perform different mining tasks like clustering, classification, ranking etc. over different meta-paths (meta-level description of paths) [2–4]. For these mining tasks, measuring relatedness between objects is an important step. For example, in case of DBLP dataset, in order to answer the question like *"Who are the leading researchers in Information Retrieval area?"* we need to measure the relevance of Authors (A) with the Conferences (C) related to information retrieval area. Also, in case of clustering, we need to measure the relatedness between same typed objects following a relationship path involving different typed objects.

In this paper, we propose a meta-path based novel relevance measure which can measure the relatedness between same as well as different typed objects in a heterogeneous network. The proposed measure is semi-metric meaning it has important properties of reflexivity, symmetry and limited range [5]. In addition, the proposed measure does not require decomposition of an atomic relation when source and target objects are of different type and, thus, reduces the computational requirement [4]. Further, the proposed measure can be used to measure the relatedness between objects following any arbitrary meta-path. To validate the effectiveness of the proposed measure, we use real world bibliographic dataset from DBLP and perform clustering and query task [4, 6]. For comparison, we use other relevance measures which are semi-metric and applicable to heterogeneous networks, namely, HeteSim and PathSim [3, 4]. For clustering task, we use both PathSim and HeteSim and compare the performance with the proposed measure. But for query task, we compare the performance of the proposed measure only with HeteSim as PathSim cannot measure similarity between different typed objects [3].

The rest of the paper is organized as follows. Section 2 introduces the related work. In Sect. 3, we present our proposed relevance measure. An illustration is given in Sect. 4. Section 5 presents experimental setup and results. Finally, in Sect. 6 conclusion and future research directions are given.

2 Related Work

Information networks can be broadly classified as homogeneous and heterogeneous networks [1]. Mining of homogeneous networks has been attempted by many researchers. Research done by Jeh and Widom [7] and Page et al. [8] are two such examples. However, mining of heterogeneous networks is relatively an emerging research area and has been attempted by few researchers [1, 2]. Heterogeneous networks are richer in information as compared to their heterogeneous-transformed homogeneous counterpart [1]. Earlier, researchers have used various similarity/distance measures to compute the relatedness between objects [5, 9]. Conventional similarity/distance measures like Jaccard coefficient, Cosine similarity and Euclidean distance are features based measures and not suitable to use directly on heterogeneous networks [4, 5, 10]. Link based similarity measures such as PageRank [8] and SimRank [7] are widely accepted but these are limited to use in homogeneous networks.

Measuring relatedness between objects in a heterogeneous information network has gained momentum recently and there are only few measures which are directly applicable to heterogeneous networks and can incorporate the path semantics like PCRW (Path Constrained Random Walk), PathSim, HeteSim and AvgSim. PCRW, HeteSim and AvgSim use random walk based methods while PathSim uses path count for relevance measurement [3, 4, 11, 12]. For all the four measures, meta-path based approach is applied for measuring relatedness between objects [3, 4, 11, 12].

PCRW, proposed by Lao and Cohen [11], is not a symmetric measure, which means relatedness between two objects will not be equal in forward and reverse directions and it is a major limitation with PCRW [4]. Sun et al. [3] proposed PathSim to measure the relatedness between objects in a heterogeneous network. But the limitation with PathSim is that it can measure relatedness only between same typed objects following an even length symmetric meta-path. Since, in many practical applications it may be required to measure the relatedness between different typed objects like mining of movies database, Flickr, Social network data, in that situation, PathSim would not be applicable.

In order to measure the relatedness between different as well as same typed objects, Shi et al. [4] proposed HeteSim. HeteSim has shown improved performance as compared to both PCRW and PathSim [4]. Although better than PCRW and PathSim, HeteSim has its own limitation. The limitation with HeteSim is that in order to measure the relatedness between objects, if the length of meta-path is odd, we need to do the decomposition of atomic relations to make the length of meta-path even which is computationally intensive. Both PathSim and HeteSim are semi-metric. Another work done by Meng, X. et al. [12] proposed a random walk based measure AvgSim which, although, has symmetric property but is not semi-metric which makes it suitable only for limited mining techniques/algorithms.

The prior overview shows that relevance measurement for objects in a heterogeneous network is an area which has lot of potential for research. This motivated us to design a new relevance measure which is more effective as compared to earlier measures and also addresses their limitations.

3   A Novel Relevance Measure

Mining of a heterogeneous network can give information rich results and by following different paths, we can capture different semantics and subtleties which is not possible in case of heterogeneous-transformed homogeneous network. In this section, we propose a novel relevance measure to compute the relatedness between same as well as different typed objects for any arbitrary meta-path. Building upon the framework described by Sun et al. [3], first, we define information network and network schema, middle object type, middle relation type and weighted path matrix which will be used in formulating the new measure.

Definition 1 (Information Network and Network Schema). *An information network is defined as a directed graph* $G = (V, E)$ *with an object type mapping function* $\emptyset:V \rightarrow A$ *and a link type mapping function* $\psi:E \rightarrow R$, *where each object* $v \in V$ *belongs to one particular object type* $\emptyset(v) \in A$, *and each link* $e \in E$ *belongs to a particular relation type* $\psi(e) \in R$. *However, the network schema is the meta-level representation of* $G = (V, E)$ *which is a directed graph over object types* A *and relation types* R, *denoted as* $T_G = (A, R)$.

When the types of objects $|A| > 1$ or the types of relations $|R| > 1$, the network is heterogeneous information network; otherwise, it is homogeneous information network.

Definition   2   (Middle   Object   Type). *For   an   even   length   meta-path* $A_1 \overset{R_1}{\rightarrow} A_2 \overset{R_2}{\rightarrow} ... \overset{R_{m-1}}{\rightarrow} A_m \overset{R_m}{\rightarrow} ... \overset{R_{l-1}}{\rightarrow} A_l \overset{R_l}{\rightarrow} A_{l+1}$, *where l is even, the middle object type* A_m *is the object type which is equidistant from source object type* A_1 *as well as target object type* A_{l+1}.

For DBLP network schema shown in Fig. 1, consider the meta-paths *APA* ("Author-Paper-Author") and *APCPA* ("Author-Paper-Conference-Paper-Author") which have middle object types *P* ("Paper") and *C* ("Conference") respectively. All symmetric meta-paths of length more than one are essentially even length path and, therefore, have middle object type. For meta-path *APKPC*, which is not a symmetric path, the middle object type is *K* and for meta-path *APCP* there is no middle object type as the length of meta-path is odd. For a meta-path, using middle object type, we can divide the path into two equal length paths i.e.one from source to middle object type and other from middle to target object type. For example, path *APCPA* can be divided into *APC* and *CPA*.

Definition   3   (Middle   Relation   Type). *For   an   odd   length   meta-path* $A_1 \overset{R_1}{\rightarrow} A_2 ... \overset{R_m}{\rightarrow} ... A_l \overset{R_l}{\rightarrow} A_{l+1}$, *where l is odd, the middle relation type* R_m *is the relation type which has equal number of preceding and succeeding relation types from source to target object type.*

For meta-path *APCP* in DBLP network schema which has odd length, the middle relation type is *PC*.

For measuring relatedness between source and target object, we can transform a heterogeneous network into a bipartite network consisting of only source and target type objects. For that, we need to compute the weighted path matrix as given in Definition 4.

Definition 4 (Weighted Path Matrix). *For a heterogeneous network and its schema level representation, a weighted path matrix M for meta-path $P = (A_1 A_2 \ldots A_{l+1})$ is defined as $M = W_{A_1 A_2} \times W_{A_2 A_3} \times \ldots \times W_{A_l A_{l+1}}$, where $W_{A_i A_j}$ is the adjacency matrix between object types A_i and A_j. $M(x_i, y_j)$ represents the number of path instances between objects $x_i \in A_1$ and $y_j \in A_{l+1}$ following meta- path P.*

Now, we present the proposed relevance measure for measuring relevance for objects in a heterogeneous network.

Definition 5. *Given a meta-path $P = (A_1 A_2 \ldots A_{l+1})$, for bipartite representation of a heterogeneous network that has only source and target type objects, relatedness between source object $a_i \in A_1$ and target object $b_j \in A_{l+1}$ is:*

$$Rel\left(a_i, b_j | P\right) = \frac{w_{a_i b_j} \left(\frac{1}{\deg(a_i)} + \frac{1}{\deg(b_j)}\right)}{\frac{1}{\deg(a_i)} \sum_j w_{a_i b_j} + \frac{1}{\deg(b_j)} \sum_i w_{a_i b_j}} \tag{1}$$

where $w_{a_i b_j}$ is the value $M(a_i, b_j)$ from the weighted path matrix i.e. the number of paths connecting objects a_i and b_j. $\deg(a_i)$ and $\deg(b_j)$ are the degree of objects a_i and b_j respectively in bipartite representation.

There can be three different cases for computing relatedness using proposed measure based on type of objects and length of meta-path. Below, we present all the three cases and present the formula for calculating the relevance using Definition 5.

Case I: Relevance Measurement for Different Typed Objects. When source and target objects are of different type, then regardless of length of the meta-path, the relatedness between source object $a_i \in A_1$ and target object $b_j \in A_{l+1}$ following meta-path $P = (A_1 A_2 \ldots A_{l+1})$ is calculated by first computing the bipartite representation using Definition 4. Then we can calculate the relatedness $DPRel\left(a_i, b_j | P\right)$ between source and target objects using Definition 5.

$$DPRel\left(a_i, b_j | P\right) = Rel\left(a_i, b_j | P\right) \tag{2}$$

Case II: Relevance Measurement for Same Typed Objects When Path Length is Even. When source and target objects are of same type and length of meta-path is even then the relatedness between source object $a_i \in A_1$ and target object $a_j \in A_{l+1}$ is calculated as follows. First, we find the middle object type of meta-path P using Definition 2. In this case, the length of meta-path is even so $P = P_L P_R = (A_1 A_2 \ldots A_m)(A_m A_{m+1} \ldots A_{l+1})$. For $P_L = A_1 A_2 \ldots A_m$, we calculate

$Rel(a_i, b_k|P_L)$ for source object a_i and all middle objects $b_k \in A_m$, $\forall k$. Similarly, for target object a_j, we calculate $Rel(a_j, b_k|P_R^{-1})$ following $P_R^{-1} = A_{l+1}A_l \ldots A_m$. Then, we calculate relatedness $DPRel\left(a_i, a_j|P\right)$ between a_i and a_j as follows using Tanimoto coefficient [5]. Here middle objects are taken as attribute objects.

$$X = Rel(a_i, b_k|P_L) \tag{3}$$

$$Y = Rel(a_j, b_k|P_R^{-1}) \tag{4}$$

$$DPRel(a_i, a_j|P) = \frac{X.Y}{|X|^2 + |Y|^2 - X.Y} \tag{5}$$

where X and Y are the vectors of relevance of a_i and a_j respectively with all middle objects $b_k \in A_m$, $\forall k$ in the meta-path.

Case III: Relevance Measurement for Same Typed Objects When Path Length is Odd. This situation occurs rarely when length of the meta-path is odd for same type objects [3, 4]. In this case, we first find the middle relation type using Definition 3 and decompose instances of that middle relation type to create middle objects as described by Shi et al. [4]. This will result in a meta-path of even length. After that we will follow the same process as in case 2.

Thus, only in third case i.e. relevance measurement for same typed objects when path length is odd, we are required to do decomposition of atomic relations.

3.1 Properties of the Proposed Measure

Our proposed measure is semi-metric which makes it useful for many applications involving heterogeneous networks. Before applying the proposed measure, we need to convert the heterogeneous network into a bipartite network consisting only of source and target object types using Definition 4. Since in case of same source and target object types, we convert the odd length meta-path case into even length case by decomposing atomic middle relations, so there is no need to give a separate proof for this case. The proof of semi-metric properties [5] are given below for the two cases i.e. when the source and target objects are of same type and of different types.

Case I: Relevance Measurement for Different Typed Objects
Property 1 (Limited Range). $0 \leq DPRel(a_i, b_j|P) \leq 1$

Proof: Relatedness between two objects $a_i \in A_1$, and $b_j \in A_{l+1}$ following meta-path $P = A_1A_2 \ldots A_{l+1}$

$$DPRel\left(a_i, b_j|P\right) = \frac{w_{a_ib_j}\left(\frac{1}{\deg(a_i)} + \frac{1}{\deg(b_j)}\right)}{\frac{1}{\deg(a_i)}\sum_j w_{a_ib_j} + \frac{1}{\deg(b_j)}\sum_i w_{a_ib_j}}$$

If there is no path between a_i and b_j, then, $w_{a_ib_j} = 0$ and $Rel\left(a_i, b_j|P\right) = 0$

If a_i and b_j are connected to only each other but with no other node then, $\deg(a_i) = \deg(b_j) = 1$ and $\sum_j w_{a_i b_j} = \sum_i w_{a_i b_j} = w_{a_i b_j}$, therefore,

$$DPRel\left(a_i, b_j|P\right) = 2 \times \frac{w_{a_i b_j}}{w_{a_i b_j} + w_{a_i b_j}} = 1$$

The relevance value can never be negative since degree of a node and weight $w_{a_i b_j}$ can never be negative in case of meta-path based framework. Also, the degree of a node can never be zero because no isolated node will be present in the network.

Thus,

$$0 \le DPRel(a_i, b_j|P) \le 1$$

Property 2 (Reflexivity). $DPRel\left(a_i, b_j|P\right) = 1 \, iff \, a_i = b_j$

Proof: Since a_i and b_j are of different type therefore $a_i = b_j$ means that their connection patterns are same. This can happen only when they are connected to only each other. In this situation, $\sum_j w_{a_i b_j} = \sum_i w_{a_i b_j} = w_{a_i b_j}$ and $\deg(a_i) = \deg(b_j) = 1$. Therefore,

$$DPRel\left(a_i, b_j|P\right) = 2 \times \frac{w_{a_i b_j}}{w_{a_i b_j} + w_{a_i b_j}} = 1$$

Property 3 (Symmetry). $DPRel\left(a_i, b_j|P\right) = DPRel\left(b_j, a_i|P^{-1}\right)$

Proof:

$$DPRel\left(a_i, b_j|P\right) = \frac{w_{a_i b_j}\left(\frac{1}{\deg(a_i)} + \frac{1}{\deg(b_j)}\right)}{\left(\frac{1}{\deg(a_i)} \sum_j w_{a_i b_j} + \frac{1}{\deg(b_j)} \sum_i w_{a_i b_j}\right)}$$

$$= \frac{w_{b_j a_i}\left(\frac{1}{\deg(b_j)} + \frac{1}{\deg(a_i)}\right)}{\left(\frac{1}{\deg(b_j)} \sum_i w_{b_j a_i} + \frac{1}{\deg(a_i)} \sum_j w_{b_j a_i}\right)} = DPRel\left(b_j, a_i|P^{-1}\right)$$

Case II: Relevance Measurement for Same Typed Objects. Since, in case of same typed objects, we are converting the odd path length case into the even path length case by decomposing atomic middle relations, therefore, there is no need to give separate proof for the odd length case. For same typed objects, we calculate relatedness by calculating first the relevance of the source object a_i with middle objects $b_k, \forall k$ i.e. $Rel(a_i, b_k|P_L)$ and relevance of the target object a_j with middle objects $b_k, \forall k$ i.e. $Rel(a_j, b_k|P_R^{-1})$ where meta-path $P = P_L P_R = (A_1 A_2 \ldots A_m)(A_m A_{m+1} \ldots A_{l+1})$. Then we use Tanimoto coefficient [5] to measure the relatedness between a_i and a_j following meta-path P.

$$X = Rel(a_i, b_k | P_L)$$

$$Y = Rel\left(a_j, b_k | P_R^{-1}\right)$$

$$DPRel(a_i, a_j | P) = \frac{X.Y}{|X|^2 + |Y|^2 - X.Y}$$

Since the Tanimoto coefficient has all the three properties [5] i.e. reflexivity, symmetry and limited range, therefore, in case of same type objects all three properties are automatically proven.

4   Illustration

In order to explain the working of the proposed measure, we take the following heterogeneous network as an example shown in Fig. 2(a). In this network, there are three types of nodes i.e. Author (A), Paper (P), and Subject (S). The semantic relationship between author and paper is different from relationship between paper and subject. The semantic relationships are bidirectional. For example, an author "*writes*" paper or a paper is "*written by*" authors. Therefore, we have taken undirected links in our example. The network schema and the bipartite representation following meta-path *APS* for this example are shown in Fig. 2(b) and (c) respectively. The calculation of weighted path matrix is shown in Fig. 3 (a).

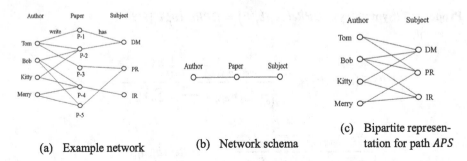

(a)   Example network   (b)   Network schema

(c)   Bipartite representation for path *APS*

Fig. 2. Example network, its schema, and bipartite representation

Now, we calculate the relatedness between different typed objects *Tom* and *DM* following meta-path *APS* i.e. "Author-Paper-Subject". In this case, regardless of the length of meta-path, decomposition of atomic relations is not required as we are calculating relatedness between different type objects.

$$DPRel\left(Tom, DM | APS\right) = \frac{w_{Tom,DM}\left(\frac{1}{\deg(Tom)} + \frac{1}{\deg(DM)}\right)}{\left(\frac{w_{Tom,DM} + w_{Tom,PR} + w_{Tom,IR}}{\deg(Tom)} + \frac{w_{Tom,DM} + w_{Bob,DM} + w_{Kitty,DM} + w_{Merry,DM}}{\deg(DM)}\right)}$$

	P-1	P-2	P-3	P-4	P-5
Tom	1	1	1	0	0
Bob	0	1	0	1	1
Kitty	0	1	0	1	0
Merry	0	0	0	1	1

$\times$

	DM	PR	IR
P-1	1	0	0
P-2	1	0	0
P-3	0	1	0
P-4	0	0	1
P-5	0	1	0

$=$

	DM	PR	IR
Tom	2	1	0
Bob	1	1	1
Kitty	1	0	1
Merry	0	1	1

(a)   Calculation of weighted path matrix for meta-path APS

	DM	PR	IR
Tom	0.588	0.333	0
Bob	0.286	0.333	0.333
Kitty	0.357	0	0.417
Merry	0	0.417	0.417

	Tom	Bob	Kitty	Merry
Tom	1	0.579	0.383	0.209
Bob	0.579	1	0.662	0.744
Kitty	0.383	0.662	1	0.366
Merry	0.209	0.744	0.366	1

(b)   Relatedness between Authors and Subjects following meta-path APS

(c)   Relatedness between Authors following meta-path $APSPA$

Fig. 3. Calculation of relevance

$$= \frac{2(\frac{1}{2} + \frac{1}{3})}{(\frac{2+1+0}{2} + \frac{2+1+1+0}{3})} = 0.588$$

The results of rest of these computations in matrix form are shown in Fig. 3 (b).

Next, we show how to calculate relatedness between same typed objects. Now, we calculate relatedness between authors by following meta-path $P = APSPA$. Since, we are measuring relatedness between same typed objects and the length of meta-path is even, therefore, there is no need to decompose atomic relations in this case. We first calculate the relatedness of authors with middle object type of this meta-path i.e. Subject (S). Then we use Tanimoto coefficient to calculate the relatedness between authors. Here $P = APSPA = P_L P_R = (APS)(SPA)$. Therefore, $P_L = APS$ and $P_R^{-1} = APS$.

$$X = R(Tom, \{DM, PR, IR\}|APS) = \{0.588, 0.333, 0\}$$

$$Y = R(Bob, \{DM, PR, IR\}|APS) = \{0.286, 0.333, 0.333\}$$

$$DPRel(Tom, Bob|APSPA) = \frac{X.Y}{|X|^2 + |Y|^2 - X.Y} = 0.579$$

The results of rest of these computations in matrix form are shown in Fig. 3(c).

5   Experimental Setup and Results

To show the viability and the effectiveness of the proposed measure DPRel, we take four area DBLP dataset collected from the website http://web.engr.illinois.edu/~mingji1/ [6]. For comparison, we take only PathSim and HeteSim as these are the only two meta-path based measures which are semi-metric and directly applicable to heterogeneous networks. All experiments were performed on a system with Intel Core i5 processor and 8 GB RAM using R version 3.0.3.

174   M. Gupta et al.

5.1 Dataset

The DBLP dataset used in our experiment is a subset
of data available on DBLP website [6]. The dataset
used in our experiment involves major conferences
in four research areas: Database, Data Mining, Information Retrieval and Artificial Intelligence which
naturally forms four classes. The dataset used in our
experiment contains 14376 papers, 20 conferences,
14475 authors and 8920 keywords (terms) with
170794 links in total and the dataset is stored in plain
text file. In this dataset, 4057 authors, all 20 confer-

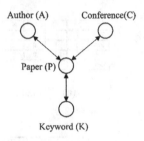

Fig. 4. Network schema of DBLP
heterogeneous network data

ences and 100 papers are labelled with one of the four research area classes. The network
schema for DBLP is shown in Fig. 4. Since citation information is not present in this
dataset, therefore, paper node in the schema has no self-loop.

5.2 Performance Comparison for Clustering

For clustering task, we use three meta-paths *APCPA,CPAPC* and *PAPCPAP* for clustering of authors, conferences and papers respectively [4]. We use Partition Around
Medoid (PAM) [9, 10] and Affinity Propagation (AP) [13] for clustering and take value
of $k = 4$ for PAM as there are four natural classes. The results of comparison are given
in Tables 1 and 2. For performance evaluation, we use F-Measure, Normalized Mutual
Information (NMI), Cluster Purity and Adjusted Rand Index (ARI) [9]. In clustering,
we do not perform clustering of keywords since the four areas are very much overlapping
in terminology used. Therefore, the clustering accuracy would be too low in case of
clustering of keywords and would not be able to capture the essence of comparison of
performance of three measures.

From the results, it is clear that DPRel performs better as compared to HeteSim and
PathSim for clustering of authors, conferences and papers in case of PAM. For AP, in
case of paper and conference, DPRel performs better. We also see that for clustering of
authors and conferences, performance of DPRel is high but for clustering of papers
performance is low. The reason might be the selection of meta-path. The accuracy of
clustering and other mining tasks depend upon the meta-path selected. The relatedness
between conferences are measured using meta-path *CPAPC* which means conferences
sharing same authors. Likewise, meta-path *APCPA* which means authors publishing
papers in same conferences effectively capture the similarity of authors. Since, in both
cases of authors and conferences, the similarity is captured appropriately by meta-paths,
the performance of PAM and AP are high in both cases. However, in case of papers
since similarity is captured by referenced authors (i.e., the *APCPA* path) which is not
effectively measuring the similarity of papers, therefore, the performance of PAM and
AP are low in this case. This shows that the performance highly depends upon the
selection of meta-path apart from the accuracy of clustering algorithm.

Table 1. Comparison for clustering task using PAM

	Path	Precision	Recall	F-Measure	NMI	Cluster purity	ARI
DPRel	*APCPA*	0.8383	0.8392	**0.8387**	**0.737**	**0.912**	**0.7965**
HeteSim		0.6829	0.7615	0.72	0.6062	0.7693	0.5285
PathSim		0.8319	0.8384	0.8351	0.7322	0.9095	0.7907
DPRel	*CPAPC*	0.9	0.9	**0.9**	**0.9058**	**0.95**	**0.8588**
HeteSim		0.8167	0.8	0.8082	0.8073	0.9	0.7072
PathSim		0.7571	0.75	0.7536	0.7585	0.85	0.5983
DPRel	*PAPCPAP*	0.628	0.6199	**0.6239**	**0.5064**	**0.77**	**0.4632**
HeteSim		0.5358	0.5104	0.5228	0.3899	0.7	0.3326
PathSim		0.4872	0.5269	0.5063	0.3479	0.65	0.2746

Table 2. Comparison for clustering task using Affinity Propagation clustering

	Path	Precision	Recall	F-Measure	NMI	Cluster Purity	ARI
DPRel	*APCPA*	0.785	0.8014	0.7931	0.6748	0.8812	0.7349
HeteSim		0.8654	0.8693	**0.8674**	**0.7759**	**0.9283**	**0.8289**
PathSim		0.7834	0.7999	0.7916	0.674	0.8802	0.7342
DPRel	*CPAPC*	0.9	0.9	**0.9**	**0.9058**	**0.95**	**0.8588**
HeteSim		0.8167	0.8	0.8082	0.8073	0.9	0.7072
PathSim		0.8167	0.8	0.8082	0.8073	0.9	0.7072
DPRel	*PAPCPAP*	0.6334	0.6233	**0.6283**	**0.5161**	**0.77**	**0.4723**
HeteSim		0.5151	0.4899	0.5022	0.335	0.68	0.2559
PathSim		0.5791	0.556	0.5673	0.4054	0.73	0.4135

5.3 Performance Comparison for Query Task

Using query task, we can evaluate the effectiveness of DPRel for different typed objects in heterogeneous network. Since PathSim cannot measure the relatedness between different typed objects [3], therefore, we compare the performance of DPRel only with HeteSim. Using labelled subset of DBLP dataset, we measure the relatedness of conferences with authors following two meta-paths: *CPA* and *CPAPA*. For each conference, we rank authors according to its relevance value. We compute the AUC (Area Under ROC Curve) score based on the labels of authors and conferences to evaluate the performance of DPRel and HeteSim. For comparison, we take 7 representative conferences out of 20 and their AUC score values are listed in Table 3 for DPRel and HeteSim. We can see that for both meta-paths, performance of DPRel is better for all 7 conferences as compared to HeteSim (Table 3).

Table 3. Comparison for query task using AUC

Conference	CPA		CPAPA	
	DPRel	HeteSim	DPRel	HeteSim
KDD	**0.8117**	0.8111	**0.8296**	0.827
SIGIR	**0.9522**	0.9507	**0.9456**	0.9402
SIGMOD	**0.7674**	0.7662	**0.8016**	0.7934
VLDB	**0.8282**	0.8262	**0.8589**	0.8477
ICDE	**0.7296**	0.7282	**0.7709**	0.7648
AAAI	**0.812**	0.8109	**0.8215**	0.8113
IJCAI	**0.8771**	0.8754	**0.8911**	0.8785

5.4 Time Complexity Analysis

Let n be the average number of objects of one type in the network. Then, for DPRel, the space complexity would be $O(n^2)$ to store the relevance matrix. Let d be the average degree of a node in the network. Then, for a specified meta-path of length l, the time complexity for HeteSim would be $O(ld^2n^2)$, since for all node pairs (i.e. n^2), the relevance is calculated along the relevance path before the matrix pair multiplication [4]. However, in case of DPRel, relevance is calculated after getting the bipartite representation (i.e. after doing the multiplication of matrices). Therefore, the time complexity of calculating DPRel would be $O(ln^2 + dn^2)$ which is far less than the time complexity of HeteSim. Also, in case of DPRel we need to do the decomposition operation only in the case of same typed objects when path length is odd. This property further improves the efficiency of the DPRel as compared to HeteSim.

6 Conclusion and Future Research Directions

In this paper, we proposed a novel meta-path based measure, DPRel, to compute the relatedness between objects in a heterogeneous information network. The proposed measure addresses the limitations of earlier measures and is able to measure the relatedness between same as well as different typed objects. In this paper, we comparatively and systematically examined the performance of DPRel and compared the effectiveness with two well-known relevance measures, PathSim and HeteSim. From the experiments performed on real world bibliographic dataset DBLP, it is clear that DPRel outperforms PathSim and HeteSim in clustering and query tasks. Our work has following main contributions: First, in this work, we study different relevance measures in heterogeneous networks and address the problems of earlier meta-path based measures. Second, we propose a novel measure for relevance measurement in general heterogeneous networks. Our proposed measure is semi-metric, therefore, has applicability in real world applications. Third, the proposed measure can measure the relatedness between objects of different as well as same typed objects following any arbitrary meta-path. Fourth, the proposed measure has no need to decompose atomic relation while computing relatedness between different typed objects following any arbitrary path. Finally, our proposed measure performs better than other measures.

Future research directions include a dynamic programming based approach for fast computation of DPRel to compute the relevance in heterogeneous networks. Also, apart from DBLP dataset, the proposed measure can also be tested on other heterogeneous datasets emerging from social networking sites like, Facebook, Twitter etc.

References

1. Huang, Y., Gao, X.: Clustering on heterogeneous networks. Wiley Interdisc. Rev. Data Min. Knowl. Discov. **4**(3), 213–233 (2014)
2. Sun, Y., Han, J.: Mining heterogeneous information networks: a structural analysis approach. ACM SIGKDD Explor. Newsl. **14**(2), 20–28 (2013)
3. Sun, Y., Han, J., Yan, X., Yu, P.S., Wu, T.: Pathsim: meta path-based top-k similarity search in heterogeneous information networks. In: VLDB (2011)
4. Shi, C., Kong, X., Huang, Y., Philip, S.Y., Wu, B.: HeteSim: a general framework for relevance measure in heterogeneous networks. IEEE Trans. Knowl. Data Eng. **26**(10), 2479–2492 (2014)
5. Theodoridis, S., Koutroumbas, K.: Pattern Recognition. Academic Press, London (2009)
6. Ji, M., Sun, Y., Danilevsky, M., Han, J., Gao, J.: Graph regularized transductive classification on heterogeneous information networks. In: Balcázar, J.L., Bonchi, F., Gionis, A., Sebag, M. (eds.) ECML PKDD 2010, Part I. LNCS, vol. 6321, pp. 570–586. Springer, Heidelberg (2010)
7. Jeh, G., Widom, J.: SimRank: a measure of structural-context similarity. In: Proceedings of the Eighth ACM SIGKDD International Conference on Knowledge Discovery and Data Mining, pp. 538–543. ACM (2002)
8. Page, L., Brin, S., Motwani, R., Winograd, T.: The PageRank citation ranking: bringing order to the web. Technical report, Stanford University Database Group (1998)
9. Manning, C.D., Raghavan, P., Schütze, H.: Introduction to Information Retrieval. Cambridge University Press, Cambridge (2008)
10. Kumar, P., Raju, B.S., Radha Krishna, P.: A new similarity metric for sequential data. Int. J. Data Warehouse. Min. **6**(4), 16–32 (2010)
11. Lao, N., Cohen, W.W.: Relational retrieval using a combination of path-constrained random walks. Mach. Learn. **81**(1), 53–67 (2010)
12. Meng, X., Shi, C., Li, Y., Zhang, L., Wu, B.: Relevance measure in large-scale heterogeneous networks. In: Chen, L., Jia, Y., Sellis, T., Liu, G. (eds.) APWeb 2014. LNCS, vol. 8709, pp. 636–643. Springer, Heidelberg (2014)
13. Frey, B.J., Dueck, D.: Clustering by passing messages between data points. Science **315**(5814), 972–976 (2007)

UFOMQ: An Algorithm for Querying for Similar Individuals in Heterogeneous Ontologies

Yinuo Zhang[1]([⊠]), Anand Panangadan[2], and Viktor K. Prasanna[2]

[1] Department of Computer Science,
University of Southern California, Los Angeles, CA, USA
`yinuozha@usc.edu`
[2] Ming Hsieh Department of Electrical Engineering,
University of Southern California, Los Angeles, CA, USA
{`anandvp,prasanna`}`@usc.edu`

Abstract. The chief challenge in identifying similar individuals across multiple ontologies is the high computational cost of evaluating similarity between every pair of entities. We present an approach to querying for similar individuals across multiple ontologies that makes use of the correspondences discovered during ontology alignment in order to reduce this cost. The query algorithm is designed using the framework of fuzzy logic and extends fuzzy ontology alignment. The algorithm identifies entities that are related to the given entity directly from a single alignment link or by following multiple alignment links. We evaluate the approach using both publicly available ontologies and from an enterprise-scale dataset. Experiments show that it is possible to trade-off a small decrease in precision of the query results with a large savings in execution time.

1 Introduction

With the increasing use of ontologies for storing large amounts of heterogeneous data in enterprise-scale applications, there has been a corresponding interest in automatically discovering links between such ontologies [11]. Links between ontologies are represented as *alignments* between the entities contained in them. While a variety of approaches have been proposed in recent years to discover such alignments [3,11], much less work has been carried out in identifying similar individuals in different ontologies. The chief challenge in identifying similar individuals is the scale of the search space. Potentially, every individual represented in the ontologies has to be evaluated for its similarity to the query individual along with all of its properties. Such exhaustive evaluation of the search space is not feasible in enterprise-scale datasets.

We propose an approach to querying for similar individuals across multiple ontologies that makes use of the alignments discovered during the ontology linking process. Specifically, if the alignments represent the degree to which two entities in different ontologies share a particular type of relationship, then a query algorithm that returns individuals in decreasing order of their similarity to the target individual only needs to follow alignments starting from all of that

© Springer International Publishing Switzerland 2015
S. Madria and T. Hara (Eds.): DaWaK 2015, LNCS 9263, pp. 178–189, 2015.
DOI: 10.1007/978-3-319-22729-0_14

target individual's properties. Such a query mechanism can be designed using the framework of fuzzy logic. In prior work [15], we developed UFOM, a unified framework to generate ontology alignments for computing different types of correspondences. UFOM is based on a fuzzy representation of ontology matching. In this work, we present algorithms that utilize the fuzzy correspondences discovered by UFOM to efficiently query for all entities in an ontology that are similar to a given entity.

The algorithms can identify entities that are related to the given entity directly from a single alignment link (*direct matching*) or by following multiple alignment links (*indirect matching*). The algorithms are specified using a fuzzy extension of SPARQL (f-SPARQL [2]). The fuzzy SPARQL queries are then converted to crisp queries for execution. We evaluate this approach using both publicly available ontologies provided by the Ontology Alignment Evaluation Initiative (OAEI) campaigns and ontologies of an enterprise-scale dataset. Our experiments show that it is possible to trade-off precision of the similarity of the identified entities and the running time of the proposed query algorithms. Compared with a baseline approach (traversing all properties in an ontology), our proposed approach reduces execution time by 99 %.

The main contributions of this paper are (1) algorithms to efficiently identify similar entities in ontologies using links discovered during ontology alignment, (2) use of fuzzy SPARQL extensions to compactly represent similarity queries, and (3) quantitative evaluation of the approach on real-world datasets. The rest of the article is organized as follows. Section 2 gives an overview of the related work. In Sect. 3, we present background concepts and the problem definition. Section 4 summarizes our previous work on the UFOM ontology alignment approach. Section 5 describes how queries can be executed efficiently using the computed ontology alignment. In Sect. 6, we present the experimental evaluation of this query approach. We conclude in Sect. 7.

2   Related Work

Ontology matching is one of the key research topics in the Semantic Web community and has been widely investigated in recent years [9,11]. An ontology matching system discovers correspondences between entities in two ontologies. Most existing systems adopt multiple strategies such as terminological, structural, and instance-based approaches to fully utilize information contained in the ontologies.

However, exact matches do not always exist due to the heterogeneity of ontologies. Fuzzy set theory has recently been used for ontology matching in order to address this issue. Todorov et al. [14] propose an ontology matching framework for multiple domain ontologies. They represent each domain concept as a fuzzy set of reference concepts. The matches between domain concepts are derived based on their corresponding fuzzy sets. In [6], a rule base provided by domain experts is used in the matching process. In that system, both Jaccard coefficient and linguistic similarity are first calculated for each pair of entities. Then, the system uses the rule base to generate the final similarity measure of each correspondence. The fuzzy set is used as a link between the preliminary similarities and the final one. Both of the above-described works adopt fuzzy set theory for the ontology matching task; however they do not provide a formal definition of a fuzzy representation of correspondence. Moreover, both systems are specific to equivalence type relations of instances. There is no generic framework to identify correspondences for different types of relations. [12] propose an approach for automatically aligning instances, relations and classes. They measure degree of matching based on probability estimate. They focus on discovering different relations between classes and relations. For instances, only equivalence relation is considered. [7] introduce a robust ontology alignment framework which can largely improve the efficiency of retrieving equivalent instances compared with [12]. In [15], we proposed a unified fuzzy ontology matching (UFOM) framework to address these two issues. Specifically, UFOM uses fuzzy set theory as a general framework for discovering different types of alignment across ontologies. It enables representation of multiple types of correspondence relations and characterization of the uncertainty in the correspondence discovery process. Section 4 summarizes the UFOM framework.

Federated ontologies facilitate the process of querying and searching for relevant information. Nowadays, hundreds of public SPARQL endpoints have been deployed on the web. However, most of these are works in progress in terms of discoverability, interoperability, efficiency, and availability [1]. Mora and Corcho [8] investigate the evaluation of ontology-based query rewriting systems. They provide a unified set of metrics as a starting point towards a systematic benchmarking process for such systems. [4] formalizes the completeness of semantic data sources. They describe different data sources using completeness statements expressed in RDF and then use them for efficient query answering. [10] presents a duplicate-aware method to querying over the Semantic Web. Specifically, they generate data summaries to reduce the number of queries sent to each endpoint. [13] proposes a query-specific matching method which generates path correspondences to facilitate query reformulation. All of these works aim to design a unified framework for query execution and information integration. However, none of them consider the uncertainty and ambiguity inherent in ontology correspondences. In our work, we present a query execution algorithm in which query execution is facilitated by using ontology correspondences with fuzzy relation types.

3   Problem Definition

The problem of ontology matching problem is to find an *alignment A* for a pair of ontologies O_1 and O_2. An ontology alignment is defined as follows [11],

Definition 1. *An **ontology alignment** A is a set of* correspondences *between entities of the matched ontologies O_1 and O_2.*

Definition 2. *An **ontology correspondence** is a 4-tuple: $< id, e_1, e_2, r >$, where id is the identifier for the given correspondence, e_1 and e_2 are the entities of the correspondence (e.g., properties in the ontology), and r is the relation type between e_1 and e_2 (e.g., equivalence and disjointedness).*

Note that in the above definition, the correspondence is exact, i.e., the relation r strictly holds between the ontology entities e_1 and e_2. However, in systems that have to automatically determine the set membership from real-world data, it is natural that a degree should be associated with the relation between the entities. The higher the degree is, the higher the likelihood that the relation holds between them. In order to represent the uncertainty in the correspondences, we presented a fuzzy variant of ontology alignment in [15].

Definition 3. *A* ***fuzzy ontology alignment*** *is a set of* ***fuzzy correspondences*** *in the form of 6-tuple:* $< id, e_1, e_2, r, s, c >$

where $s = \mu_r(e_1, e_2)$ is the *relation score* denoting the membership of (e_1, e_2) in relation r, and c is the *confidence score* computed while generating the correspondence. With this definition of fuzzy correspondence, we can extend the relation type set with other useful types such as *Relevance* [15].

A query execution is a process of retrieving a set of *individuals* $I = \{i_1, i_2, ..., i_n\}$ which are relevant to a given individual t. I and t can belong to either one ontology or multiple ontologies. In this paper, we study the case in which I and t come from two heterogeneous ontologies O_1 and O_2. The one-ontology case is a specialization of the multi-ontology one.

A brute force method is to compare t with all property values in O_2. The time complexity of this approach is $O(|O_2||E_2|)$ where $|O_2|$ is the number of individuals in O_2 and $|E_2|$ is the number of properties in O_2. The brute force approach is inefficient in terms of search time. In this paper, we describe how we can improve the query performance using fuzzy alignment information.

4 Unified Fuzzy Ontology Matching (UFOM)

The foundation of UFOMQ algorithm is the fuzzy ontology alignment derived using the unified fuzzy ontology matching (UFOM) framework [15]. UFOM computes fuzzy ontology correspondences based on both a *relation score* and a *confidence score* between every possible entity pair in the ontologies. In order to provide an extensible framework, every relation score (i.e., for every type of relation of interest) is computed from a set of pre-defined similarity functions. The framework is illustrated in Fig. 1.

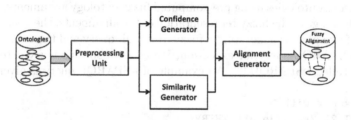

Fig. 1. Components of the UFOM system for computing fuzzy ontology alignment

UFOM takes two ontologies as input and outputs a fuzzy alignment between them. UFOM consists of four components: Preprocessing Unit (PU), Confidence Generator (CG), Similarity Generator (SG), and Alignment Generator (AG). PU classifies the entities in the ontologies based on their types. An entity can be either a class or a property. If it is a property, it is further identified as one of the following types: ObjectProperty, String DatatypeProperty, Datetime Datatype-Property, and Numerical DatatypeProperty. Different score computation strategies are developed for different types of entities. CG computes a confidence score for each correspondence which reflects the sufficiency of the underlying data resources used to generate this correspondence. Volume and variety are the two major metrics considered in this step. SG generates a vector of similarities for each pair of entities. The similarities form the basis for computing correspondences with different types of relation. In UFOM, we consider four types of similarity: Name-based Similarity, Mutual Information Similarity, Containment Similarity, and Structural Similarity.

Name-based Similarity is calculated based on both the semantic similarity and syntactic similarity of between the names of the two properties. Mutual Information Similarity models the mutual information that exists between the *individuals* of one entity and the *domain* represented by the second entity. Containment Similarity models the average level of alignment between an instance of property e_1 and the most similar instance in property e_2. Structural Similarity represents the degree of structural similarity between two properties as they are represented within their ontologies. The detailed formulation is presented in [15]

The component AG is responsible for generating a set of fuzzy correspondences in the form of 6-tuples: $< id, e_1, e_2, r, s, c >$. It calculates the relation score r using a fuzzy membership function for each relation type. After both s and c are derived, AG prunes the correspondences with s and c less than predefined cutoff thresholds s_δ and c_δ. For instance, the following are examples of fuzzy correspondences.

$< 1, isbn : author, bkst : desc, relevance, 0.73, 0.92 >$
$< 2, isbn : author, bkst : desc, equivalence, 0.58, 0.92 >$
$< 3, isbn : author, bkst : pub, disjoint, 0.83, 0.75 >$

5   Query Execution

We now describe the UFOMQ algorithm to efficiently execute queries over two heterogeneous ontologies using pre-computed fuzzy ontology alignments. In order to take advantage of the fuzzy representation of the alignments, the query process consists of two phases: generating a fuzzy SPARQL query and then converting it to a crisp SPARQL query for execution. We adopt a specific fuzzy extension of SPARQL called f-SPARQL [2]. An example of f-SPARQL query is given below.

```
#top-k FQ# with 20
SELECT ?X ?Age ?Height WHERE{
    ?X rdf:type Student
```

```
?X ex:hasAge ?Age with 0.3.
FILTER (?Age=not very young && ?Age=not very old) with 0.9.
?X ex:hasHeight ?Height with 0.7.
FILTER (?Height close to 175 cm) with 0.8.
}
```

In this example, "not very young" and "not very old" are fuzzy terms, and "close to" is a fuzzy operator. Each condition is associated with a user-defined weight (e.g., 0.3 for age and 0.7 for height) and a threshold (e.g., 0.9 for age and 0.8 for height). The top 20 results are returned based on the score function [2].

---

1   Identify a set of properties $E_{direct} = \{e_1, e_2, ..., e_m\}$ in O_2 where e_j is in a fuzzy correspondence $< id, e^t, e_j, r, s, c >$ with $s \geq S$, $s \geq C$ and $r \in \{equivalence, relevance\}$, S and C are user-defined thresholds and e^t is t's identifier property;

2   Identify a set of property triples $E_{indirect} = \{\{e_1^1, e_1^2, e_1^3\}, ..., \{e_n^1, e_n^2, e_n^3\}\}$ from O_2 where e_j^1 is in a fuzzy correspondence $< id, e^t, e_j^1, r, s, c >$ with $s \geq S$, $s \geq C$ and $r \in \{equivalence, relevance\}$, and e_j^1 and e_j^2 are the properties of the same class (intermediate class) where e_j^2 is its identifier, and e_j^3 is the target property equivalent to e_j^2;

3   **for** *each e_j in E_{direct}* **do**

4   Generate a fuzzy SPARQL using the direct matching generation rule;

5   Calculate a seed vector $\vec{s} = \{s_{syn}, s_{sem}, s_{con}\}$ for each pair (t, v_x) where t is the given individual and v_x is a value in e_j;

6   Generate individuals with grades calculated by a relevance function of $\vec{s}$ in the instance ontology $onto_{ins}$;

7   **for** *each $\{e_j^1, e_j^2, e_j^3\}$ in $E_{indirect}$* **do**

8   Generate a fuzzy SPARQL using the indirect matching generation rule;

9   Calculate a seed vector $\vec{s} = \{s_{syn}, s_{sem}, s_{con}\}$ for each pair (t, v_x) where t is the given individual and v_x is a value in e_j^1;

10   Generate individuals with grades calculated by a relevance function of $\vec{s}$ in the instance ontology $onto_{ins}$;

11   Generate a crisp SPARQL by computing the α-cut of the fuzzy terms based on the membership function and each graph pattern corresponds to a value in E_{direct} or $E_{indirect}$;

12   Return the individual set $I = \{i_1, i_2, ..., i_n\}$ by executing the crisp SPARQL over $onto_{ins}$;

Algorithm 1. UFOMQ - A query algorithm for UFOM

The UFOMQ algorithm is shown in Algorithm 1. The inputs to the algorithm are two ontologies O_1 and O_2, a set of fuzzy correspondences pre-computed using UFOM, and the target individual $t \in O_1$. The algorithm returns a set of individuals from O_2 that are similar to t. In our description, we state that the correspondences are either *equivalence* or *relevance*. However, the approach can be extended to other types of relations provided the corresponding alignments are discovered by UFOM.

Steps 1 and 2 in Algorithm 1 identify related properties using fuzzy correspondences generated by UFOM. These properties are computed using two methods: *direct matching* and *indirect matching* (Fig. 2).

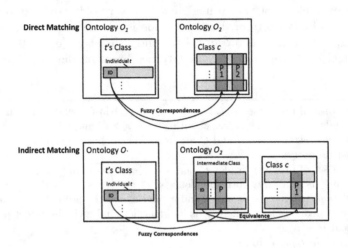

Fig. 2. Illustration of direct matching and indirect matching

For direct matching, we retrieve properties in O_2 which have fuzzy relations *equivalence* and *relevance*) with t's identifier property (e.g., *ID*) using the fuzzy alignment derived by UFOM. For example, the following SPARQL code retrieves only the *relevant* properties of *id* based on direct matching. Thresholds for relation score and confidence score (e.g., 0.5 and 0.7) are also specified in the query.

```
SELECT ?prop WHERE {
    ?prop ufom:type ''onto2_prop''.
    ?corr ufom:hasProp1 onto1:id.
    ?corr ufom:hasProp2 ?prop.
    ?corr ufom:relation ''Relevance''.
    ?corr ufom:score ?s.
    FILTER (?s > 0.5).
    ?corr ufom:conf ?c.
    FILTER (?c > 0.7).
}
```

Indirect matching is used to identify entities that do not share a single correspondence with t but are related via intermediate properties, i.e., more than one correspondence. We first identify all intermediate classes in O_2. The properties of such classes have a fuzzy relation with t's identifier property (e.g., *id*).

From these intermediate classes, we discover the properties which are equivalent to the identifier of the intermediate class. This equivalence relation is found by checking Object Properties in O_2. In contrast to direct matching which outputs a set of properties, indirect matching produces a collection of triples in the form of (e^1, e^2, e^3), where e^1 is the intermediate class' property with fuzzy relations with t's identifier property, e^2 is the intermediate class' identifier property, and e^3 is the target property equivalent to e^2. An example of the indirect matching approach for the relevance relation as expressed in SPARQL is shown below. *prop1*, *prop2*, and *prop3* correspond to e^1, e^2 and e^3 respectively.

```
SELECT ?prop1 ?prop2 ?prop3 WHERE {
    ?prop1 ufom:type ''onto2_prop''.
    ?prop1 rdfs:domain ?class.
    ?prop2 ufom:idof ?class.
    ?prop3 ufom:type ''onto2_prop''.
    ?prop3 rdfs:range ?class.
    ?corr ufom:hasProp1 onto1:id.
    ?corr ufom:hasProp2 ?prop1.
    ?corr ufom:relation ''Relevance''.
    ?corr ufom:score ?s.
    FILTER (?s > 0.5).
    ?corr ufom:conf ?c.
    FILTER (?c > 0.7).
}
```

Given the properties discovered by direct matching (*prop*) and indirect matching (*prop1*, *prop2* and *prop3*), we can build fuzzy SPARQL queries based on the rules expressed in Steps 4 and 8:

```
#top-k FQ# with 20
SELECT ?d WHERE {
    ?x onto2:id ?d.
    ?x onto2:prop ?p.
    FILTER (?p relevant to t) with 0.75.
}
```

```
#top-k FQ# with 20
SELECT ?d WHERE {
    ?x onto2:prop1 ?p.
    ?x onto2:prop2 ?c.
    ?y onto2:prop3 ?c.
    ?y onto2:id ?d.
    FILTER (?p relevant to t) with 0.75.
}
```

In the above rules, t is the given individual (the identifier used to represent the individual) and "relevant-to" is the fuzzy operator.

Since, eventually the fuzzy queries will have to be converted to crisp ones, we calculate a seed vector $\vec{s} = \{s_{syn}, s_{sem}, s_{con}\}$ for each value pair (t, v_x) where t is the given value (e.g., identifier of the given individual) and v_x is the value in the matched properties (e.g., "onto2:prop") (Steps 5 and 9). $\vec{s}$ represents multiple similarity metrics including syntactic, semantic, and containment similarities as described in [15]. The results are used to calculate the relevance scores which are stored as individuals in the instance ontology $onto_{ins}$ (Steps 6 and 10). In Step 11, we compute the α-cut of the fuzzy terms based on the membership function in order to convert to a crisp query. The resulting crisp SPARQL consists of multiple graph patterns and each of these corresponds to a matched property. The individual set I is derived by executing this crisp SPARQL query. An example of such a crisp SPARQL query returning individuals ranked based on their membership grades is shown below.

```
SELECT ?d WHERE {
    ?x onto2:id ?d.
    ?x onto2:prop ?p.
    ?ins onto_ins:value1 ?p.
    ?ins onto_ins:value2 t.
    ?ins onto_ins:type ''relevance''.
    ?ins onto_ins:grade ?g.
    FILTER (?g ≥ 0.75).
}
ORDER BY DESC(?g)
```

Using UFOMQ, the computation cost for retrieving relevant instances can be reduced. The time complexity of the UFOMQ algorithm is $O(|O_2||E_2(t)|)$ where $|E_2(t)|$ is the number of properties in O_2 which have fuzzy correspondences with t's identifier property. We evaluate the computation cost of UFOMQ on datasets in Sect. 6.

6   Experimental Evaluation

In this section, we present the results of applying the UFOMQ approach to two datasets. The first is publicly available ontologies from the Ontology Alignment Evaluation Initiative (OAEI) campaigns [5]. The second dataset comprises of ontologies of an enterprise-scale dataset.

OAEI Datasets. We first performed a set of experiments to evaluate the query execution process. The dataset is the Instance Matching (IM) ontology[1] from OAEI 2013. The dataset has 5 ontologies and 1744 instances. The fuzzy alignment is generated first using UFOM [15]. Then, we initialize 10 individuals from

---

[1] http://islab.di.unimi.it/im_oaei_2014/index.html.

one of the ontologies and retrieve related individuals from the other ontologies. WordNet[2] and DBPedia[3] are used to retrieve the synset and similar entities of a given individual. The membership grade threshold is set to 0.75. Figure 3 shows the performance of our query execution component on the IM ontology. Each data point is generated by averaging the results of 10 individuals.

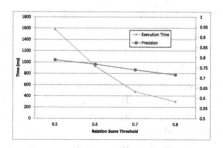

Fig. 3. Precision and execution time on applying UFOM query execution to the instance matching ontology

As the relation score threshold increases, both precision and running time for query execution decrease. This is because the number of correspondences decreases when we raise the relation score threshold. As a result, we have fewer correspondences to consider when we generate crisp queries and therefore the computational time is reduced. The reason precision also decreases is that we lose some true correspondences when we increase the relation score threshold. Those correspondences can help in finding more related individuals. However, as shown in Fig. 3, an increase in the threshold from 0.5 to 0.8 causes precision to decrease by only 9.3 % while execution time is reduced by 81.1 %. This indicates that the elimination of correspondences caused by increasing the threshold do not affect retrieving correct related individuals significantly. This is because the remaining correspondences are still sufficient to find connections between classes. In terms of querying for similar individuals, the correspondences between the same pair of classes have functional overlap.

Enterprise-Scale Dataset. For the evaluation of querying for related individuals, we considered two ontologies from an enterprise-scale information repository. Each ontology focuses on a different application area, but they are related in terms of the entities that they reference. Ontology O_1 has 125,865 triples. Ontology O_2 has 651,860 triples. Due to privacy concerns, we do not expose the real names of the properties and ontologies. We considered two classes, C_1 and C_2, in O_1 and two classes, C_3 an C_4, in O_2.

We identified 29 fuzzy correspondences between these two ontologies using UFOM. To evaluate query performance, we selected 10 representative individuals

---

[2] http://wordnet.princeton.edu/.
[3] http://dbpedia.org/.

Table 1. Query Execution Time (UFOM vs Baseline)

Scenario	UFOM(ms)	Baseline(ms)
C_1 to C_3	259	35974
C_2 to C_3	173	25706
C_1 to C_4	487	53937
C_2 to C_4	401	45752

from C_1 or C_2 and retrieve their relevant instances from C_3 or C_4 using the fuzzy alignment. Both precision and recall achieve 1.0 after we verified the results with the ground truth obtained by manually examining the ontologies for each of the automatically retrieved entities. We also generated the average execution time and the results are shown in Table 1. Compared with the baseline approach which traverses the values of all properties in O_2, our proposed approach reduces the execution by 99 % on average.

7   Conclusion

We presented the UFOMQ algorithm which enables scalable and efficient querying for related entities over heterogeneous ontologies. UFOMQ uses the fuzzy correspondences discovered during ontology alignment as computed by the previously developed UFOM framework. The query algorithm exploits the redundancy in the correspondences between classes — similar entities can be identified by following the strongest correspondences, not necessarily all the correspondences. In experiments performed on publicly available datasets, the query algorithm achieves a trade-off between precision of the returned query result and the computational cost of the query execution process. We also demonstrated the efficiency of UFOMQ in large enterprise-scale datasets.

For future work, we will develop query optimization strategies to facilitate efficient query execution. We will also adopt different entity identification techniques to improve the usability of the ontology alignment and query framework.

Acknowledgment. This work is supported by Chevron U.S.A. Inc. under the joint project, Center for Interactive Smart Oilfield Technologies (CiSoft), at the University of Southern California.

References

1. Buil-Aranda, C., Hogan, A., Umbrich, J., Vandenbussche, P.-Y.: SPARQL web-querying infrastructure: ready for action? In: Alani, H., Kagal, L., Fokoue, A., Groth, P., Biemann, C., Parreira, J.X., Aroyo, L., Noy, N., Welty, C., Janowicz, K. (eds.) ISWC 2013, Part II. LNCS, vol. 8219, pp. 277–293. Springer, Heidelberg (2013)

2. Cheng, J., Ma, Z.M., Yan, L.: f-SPARQL: a flexible extension of SPARQL. In: Bringas, P.G., Hameurlain, A., Quirchmayr, G. (eds.) DEXA 2010, Part I. LNCS, vol. 6261, pp. 487–494. Springer, Heidelberg (2010)
3. Choi, N., Song, I.-Y., Han, H.: A survey on ontology mapping. SIGMOD Rec. **35**(3), 34–41 (2006)
4. Darari, F., Nutt, W., Pirrò, G., Razniewski, S.: Completeness statements about RDF data sources and their use for query answering. In: Alani, H., Kagal, L., Fokoue, A., Groth, P., Biemann, C., Parreira, J.X., Aroyo, L., Noy, N., Welty, C., Janowicz, K. (eds.) ISWC 2013, Part I. LNCS, vol. 8218, pp. 66–83. Springer, Heidelberg (2013)
5. Euzenat, J., Meilicke, C., Stuckenschmidt, H., Shvaiko, P., Trojahn, C.: Ontology alignment evaluation initiative: six years of experience. In: Spaccapietra, S. (ed.) Journal on Data Semantics XV. LNCS, vol. 6720, pp. 158–192. Springer, Heidelberg (2011)
6. Fernández, S., Velasco, J.R., Marsá-Maestre, I., López-Carmona, M.A.: Fuzzyalign - a fuzzy method for ontology alignment. In: KEOD, pp. 98–107 (2012)
7. Lee, S., Lee, J., Hwang, S.-W.: Fria: fast and robust instance alignment. In: 22nd International World Wide Web Conference, WWW 2013, Rio de Janeiro, Brazil, 13–17 May 2013, Companion Volume, pp. 175–176 (2013)
8. Mora, J., Corcho, O.: Towards a systematic benchmarking of ontology-based query rewriting systems. In: Alani, H., Kagal, L., Fokoue, A., Groth, P., Biemann, C., Parreira, J.X., Aroyo, L., Noy, N., Welty, C., Janowicz, K. (eds.) ISWC 2013, Part II. LNCS, vol. 8219, pp. 376–391. Springer, Heidelberg (2013)
9. Rahm, E.: Towards large-scale schema and ontology matching. In: Bellahsene, Z., Bonifati, A., Rahm, E. (eds.) Schema Matching and Mapping. Data-Centric Systems and Applications, pp. 3–27. Springer, Heidelberg (2011)
10. Saleem, M., Ngonga Ngomo, A.-C., Xavier Parreira, J., Deus, H.F., Hauswirth, M.: DAW: Duplicate-AWare federated query processing over the web of data. In: Alani, H., Kagal, L., Fokoue, A., Groth, P., Biemann, C., Parreira, J.X., Aroyo, L., Noy, N., Welty, C., Janowicz, K. (eds.) ISWC 2013, Part I. LNCS, vol. 8218, pp. 574–590. Springer, Heidelberg (2013)
11. Shvaiko, P., Euzenat, J.: Ontology matching: state of the art and future challenges. IEEE Trans. Knowl. Data Eng. **25**(1), 158–176 (2013)
12. Suchanek, F.M., Abiteboul, S., Senellart, P.: PARIS: probabilistic alignment of relations, instances, and schema. PVLDB **5**(3), 157–168 (2011)
13. Tian, A., Sequeda, J.F., Miranker, D.P.: QODI: query as context in automatic data integration. In: Alani, H., Kagal, L., Fokoue, A., Groth, P., Biemann, C., Parreira, J.X., Aroyo, L., Noy, N., Welty, C., Janowicz, K. (eds.) ISWC 2013, Part I. LNCS, vol. 8218, pp. 624–639. Springer, Heidelberg (2013)
14. Todorov, K., Geibel, P., Hudelot, C.: A framework for a fuzzy matching between multiple domain ontologies. In: König, A., Dengel, A., Hinkelmann, K., Kise, K., Howlett, R.J., Jain, L.C. (eds.) KES 2011, Part I. LNCS, vol. 6881, pp. 538–547. Springer, Heidelberg (2011)
15. Zhang, Y., Panangadan, A., Prasanna, V.K.: Ufom: unified fuzzy ontology matching. In: IRI 2014 - Proceedings of the 15th International Conference on Information Reuse and Integration, San Francisco, CA, USA, 13–15 August 2014

Semantics-Based Multidimensional Query Over Sparse Data Marts

Claudia Diamantini, Domenico Potena, and Emanuele Storti[✉]

Dipartimento di Ingegneria dell'Informazione,
Università Politecnica delle Marche - Via Brecce Bianche, 60131 Ancona, Italy
{c.diamantini,d.potena,e.storti}@univpm.it

Abstract. Measurement of Performances Indicators (PIs) in highly distributed environments, especially in networked organisations, is particularly critical because of heterogeneity issues and sparsity of data. In this paper we present a semantics-based approach for dynamic calculation of PIs in the context of sparse distributed data marts. In particular, we propose to enrich the multidimensional model with the formal description of the structure of an indicator given in terms of its algebraic formula and aggregation function. Upon such a model, a set of reasoning-based functionalities are capable to mathematically manipulate formulas for dynamic aggregation of data and computation of indicators on-the-fly, by means of recursive application of rewriting rules based on logic programming.

Keywords: Logic-based formalisation of performance indicators · Multidimensional model · Sparse data marts · Query rewriting · Virtual enterprises

1 Introduction

More and more organisations recognise today the need to monitor their business performances to detect problems and proactively make strategic and tactical decisions. In this context, one of the biggest challenges is represented by consolidating performance data from disparate sources into a coherent system. Due to the lack of enforced integrity and relationship, Performance Indicators (PIs) based on this data are often incomplete, conflicting, or limited to a particular department/function within the organisation. Such a problem is more critical in a Virtual Enterprise (VE) environment, where the data providers can be very different. Heterogeneous organisations are in fact characterised by high levels of autonomy, that determines several kinds of data integration problems.

In the context of indicator management, such heterogeneities derive from internal business rules of the enterprise as well as the structure of the information system and hence cannot be simply resolved. Indeed, structural heterogeneity refers to the granularity level at which data are aggregated. For instance, some enterprises can have their revenues calculated globally for all its products,

© Springer International Publishing Switzerland 2015
S. Madria and T. Hara (Eds.): DaWaK 2015, LNCS 9263, pp. 190–202, 2015.
DOI: 10.1007/978-3-319-22729-0_15

or disaggregated product by product. Structural heterogeneity also applies to the formula adopted to calculate an indicator: again, the total revenue can be given as unique number, or the enterprise can store the total sales and costs separately. As a consequence, the interaction between granularity and formula structural heterogeneities needs to be carefully managed. Moreover, after data reconciliation, in traditional data warehouses query answering is simply realised by aggregating over the available values, and this can generate unreliable results (e.g. obtaining a cost of 100 at the 1st semester 2013 and of 500 at the 2nd semester 2013 can be interpreted erroneously as a sudden increase of costs, while it turns out that the cost for the 1st semester has been calculated by summing over the first two months only). For such reasons we believe that awareness of the completeness of results is a fundamental feature of a supporting environment. This is especially true at VE level, where besides heterogeneities in dimension hierarchies and members, usually there is no agreement about which level of aggregation for data will be used, and also corporate policies about monitoring may change over time (e.g., a partner can change frequency for data storage from monthly to weekly basis at any time). As a consequence, sparsity comes from the fact that, even for the same indicator, data will likely refer to different members, levels or dimensions.

In this paper we address these problems by presenting a semantic-based approach for multidimensional query over various data marts, capable to cope with sparse data marts by dynamic calculation of indicators values. The work has been developed during the EU project BIVEE[1], aimed at supporting innovation in a VE environment, that is therefore the target scenario of the paper. In our approach, local data are described in global terms by means of a *semantic multidimensional model*, in which we extend the data cube model with description of the structure of an indicator in terms of its algebraic formula. The model allows to reconcile dimensional heterogeneities by referring to the same representation of dimensions. Moreover, having the semantics of formulas, we are able to resolve structural heterogeneities by defining paths among cubes that describe how to compute an indicator on the basis of others.

Queries are posed over the global model and rewritten over local cubes. To overcome the sparsity issue, we introduce a *completeness check* functionality, aimed to check the completeness of a result. In any case when a value is not complete (for instance if the query asks for a semester, but just three months are available), the completeness check determines alternative ways for its calculation, by exploiting two complementary expansion rules: besides the more traditional roll-up expansion, we introduce the indicator expansion. Through such a new operator, the value of an indicator at a given level can be calculated by other indicator values at the same level, through its formula. Moreover, through logic-based formula manipulation it is possible to derive non-explicitly defined formulas from others. This novel approach enables to (1) dynamically derive non-materialized data through expansion rules and (2) obtain an answer

---

[1] http://www.bivee.eu.

to a query if and only if either it is materialized or it is produced by complete aggregation, ensuring the quality of results as a by-product.

The rest of the work is structured as follows: in next Section we introduce a case study that is used along the paper. Section 3 is devoted to present the semantic multidimensional model while Sects. 4 and 5 respectively address the query rewriting mechanisms and the completeness check, together with some implementative details and computational aspects. In Sect. 6 we discuss some relevant related work in the Literature. Finally, Sect. 7 is devoted to draw conclusions and discuss future work.

2   Case Study

In this Section we present a case study that will be used as example in the paper. Let us consider two enterprises ACME1 and ACME2, each with a data mart and willing to join a Virtual Enteprise to cooperate for a certain project. In order to answer queries over the whole VE, their data should refer to the same schema. However, several structural heterogeneities exists. As for dimensions, in fact, ProductDim is missing in ACME2, and the others refer to different granularities (e.g., there are *Year* and *Semester* levels for TimeDim in ACME1, but only *Year* for ACME2). Finally, also the set of indicators are different: I1 only for ACME1, I2 and I3 for both. We also know that I1 is aggregated through SUM and is calculated by ACME1 as I2+I3. I2 and I3 have no formula; the first is aggregated through AVG and the second through SUM (Fig. 1).

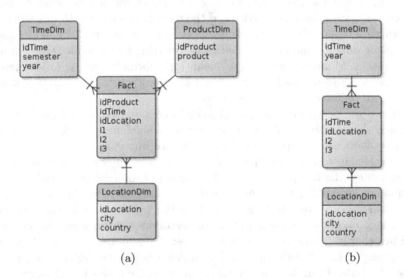

(a)   (b)

Fig. 1. Case study. The data marts for enterprises (a) ACME1 and (b) ACME2.

3   Semantic Multidimensional Model

The semantic multidimensional model is based on the formal representation of indicators and their formulas. We introduce here the basic notions of indicator, dimension and cube.

Indicator. An indicator is a way to measure a phenomenon and here it is defined as a pair $\langle aggr, f \rangle$, where $aggr$ is an aggregation function that represents how to summarize values of the indicator (e.g. sum, avg, min, max), while $f(ind_1, \ldots, ind_n)$ is a mathematical formula that describes the way the indicator is computed in terms of other indicators.

According to the classification adopted in data warehouse Literature, aggregation functions can be distributive, algebraic and holistic [1]. Indicator with a distributive aggregator can be directly computed on the basis of values at the next lower level of granularity, e.g. the Total Revenue for the 1st semester 2013 can be computed by summing values for the first six months of the 2013. Algebraic aggregators cannot be computed by means of values at next lower levels unless a set of other indicators are also provided, which transform the algebraic indicator in a distributive one; a classical example is given by the average aggregator (AVG), which must be computed on the basis of values at the lowest granularity level, whatever is the requested level. Indicators described by holistic functions (e.g. MEDIAN and MODE) can never be computed using values at next lower level. When no aggregation function is provided, indicators are computed by combining values of other indicators at the requested level.

We refer to [2,3] for further details about formulas and other properties of indicators.

Dimension. A Dimension is the coordinate/perspective along which the indicator is analysed (e.g., delivery time can be analysed with respect to means of transportation, the deliver and the date of delivery). Referring to the multidimensional model used in data warehouse [4], a dimension D is usually structured into a hierarchy of levels $L_1^D, \ldots, L_n^D$ where each level represents a different way of grouping members of the dimension (e.g., it is useful to group days by weeks, months and years). The domain of a level L_j^D is a set of members denoted by $\alpha(L_j^D)$, e.g. $\alpha(Country) = \{Portugal, Italy, Ireland, Greece, Spain, \ldots\}$.

Given $L_i^D \preceq L_j^D$, the transitive relation that maps a member of L_i^D to a member of L_j^D is defined as $partOf \subseteq \alpha(L_i^D) \times \alpha(L_j^D)$. This enables the possibility to move from a level to another with higher granularity through aggregation; this operation is called roll-up, the vice versa is the drill-down. To give some example, partOf("2013-01", "2013") means that the month "2013-01" is part of the year "2013", and partOf("Valencia", "Spain") that "Valencia" is part of "Spain". Although the fact that "Valencia is in Spain" is true in general, if an enterprise operating in Spain does not operate in Valencia, we should not consider (for this enterprise) values for the member "Valencia". Hence, we introduced a restriction $partOf_e$ of the $partOf$ relation, such that:

$$partOf(m_c, m_p) \wedge isValid(m_c, e) \rightarrow partOf_e(m_c, m_p),$$

where $isValid(m_c, e)$ is true if the member m_c makes sense for enterprise e. In this case each higher level member that is in $partOf$ relation with m_c is valid for e as well. More formally, $isValid(m_c, e) \wedge partOf(m_c, m_p) \rightarrow isValid(m_p, e)$.

The $partOf$ (hence $partOf_e$) relation is such that if $L_i^D \preceq L_j^D$, each member of L_i^D is in a $partOf$ relation with only one member of L_j^D; and each member of L_j^D is composed by at least one member of L_i^D. In other terms, the $partOf$ relation defines a partition of the members of level L_i^D.

As for the case study, let us assume for instance the following relations: $partOf_{A_1}(Barcelona, Spain)$, $partOf_{A_1}(Madrid, Spain)$, $partOf_{A_1}(Valencia, Spain)$ for ACME1 and only $partOf_{A_2}(Madrid, Spain)$ for ACME2.

Cube. According to the multidimensional model, each cell within a multidimensional structure contains aggregated data related to elements along each of its dimensions. We introduce this structure by the notion of *cubeElement*. A cubeElement is a storage element of the data mart for a given enterprise, and is defined as a tuple $ce = \langle ind \in I, m_1 \in \alpha(L_1^{D_1}), ..., m_n \in \alpha(L_j^{D_n}), e \in E, v \rangle$, where E is a set of enterprises and v is a value for ind.

A $cube(ind, e)$ is the logical grouping of all cubeElements provided by an enterprise e for an indicator ind, independently from the specific aggregation level adopted for the dimensions. We can assume the dimensional schema of a cube for ind (i.e., $ds(ind)$) to be the set of dimensions compatible with ind, e.g. $ds(I1) = \{TimeDim, ProductDim, LocationDim\}$. A cubeElement in $cube(ind, e)$ can be however defined over a subset of dimensions in the schema $ds(ind)$. In such a case, we assume that each missing dimension is aggregated at the highest level. Finally, the set of all cubes for a Virtual Enterprise is the global *data warehouse*.

4   Query Rewriting

Users query the global data warehouse for one or more indicators aggregated along the desired levels of the dimensions. Here we provide the definition of a multidimensional query and the query rewriting mechanism. For sake of simplicity, with no loss of generality the following definition refers to a single indicator.

Definition 1. A multidimensional query q over a Virtual Enterprise V is a tuple $\langle ind, W, K, V, \rho \rangle$, where:

- $ind \in I$ is the requested indicator,
- W is the set of levels $\{L_i^{D_1}, ..., L_j^{D_n}\}$ on which to aggregate,
- K is the collection of sets $K_i = \{m_{i1}, ..., m_{ik}\}$, $i = 1, ..., n$, of members on which to filter. K_i can be an empty set. In this case all members of the corresponding level are considered,
- $V = \{e_1, ..., e_n\} \subseteq E$

– ρ is *true* if the result is aggregated at the Virtual Enterprise level; otherwise the query will return a value for each enterprise in V.

The definition corresponds to the classical notion of aggregated OLAP query; in particular, W and *ind* are the elements of the target list, W is the desired roll-up level (or group-by component) while K allows slice and dice (suitable selections of the cube portion).

In our setting we face with a hybrid virtual-materialized data warehouse, since the aggregation levels of indicators are not the same for the different enterprises. In other terms, the query is posed on the global conceptual schema and needs to be rewritten in terms of the enterprise local schemas. Finally, the actual execution is realised over the enterprise logical schema[2]. We can rewrite a query for a given cube only if the dimensions requested in the query are a subset of the cube dimensions. Furthermore, not including a dimension D in the query means to request it at the highest level, that we conventionally denote by L_*^D.

Rule 1 *(Rewriting q Over a Cube).* Given a multidimensional query $q = \langle ind, W, K, V, \rho \rangle$, the multidimensional query q_c over the *cube(ind, e)*, with $e \in V$, is defined as $q_c = \langle ind, W', K', e \rangle$, where:

$$
\begin{cases}
(a) W' = W, K' = K & \text{if } W \text{ contains a level for each} \\
 & \text{dimension in } cube(ind, e) \\
(b) W' = W \cup \{L_*^{D_x}\}, K' = K \cup \{ALL\} & \text{if } \exists \text{ a dimension } D_x \in cube(ind, e) \\
 & \text{that has no corresponding level in W}
\end{cases}
$$

Otherwise q_c cannot be defined.

Example. Given the following query:

$\langle I1, \{Year, Country\}, \{\{2013, 2014\}, \{Spain\}\}, \{ACME_1, ACME_2\}, false \rangle$,

it is rewritten as:
$q_1 = \langle I1, \{Product, Year, Country\}, \{\{ALL\}, \{2013, 2014\}, \{Spain\} ACME_1 \rangle$
$q_2 = \langle I1, \{Year, Country\}, \{\{2013, 2014\}, \{Spain\}\}, ACME_2 \rangle$.

The result of the query q_c is defined as the set $R(q_c)$ of tuples $\langle a, m_1, ..., m_n, e \rangle$ such that a is a value for *ind*, every m_i belongs to a different level, $e \in V$ and a cubeElement $\langle ind, m_1, ..., m_n, e, a \rangle$ is either (1) materialized in the data mart or (2) it can derived[3]. The following rules define how to derive a cubeElement from other cubeElements.

Rule 2 *(Roll-up Expansion).* Given $m_i \in \alpha(L_x^D), L_x^D \succeq L_y^D$, if a set of cubeElements $\langle ind, m_1, ..., m_j, ..., m_n, e, a_j \rangle$ exists such that $m_j \in \alpha(L_y^D)$ and $partOf_e(m_j, m_i)$, then the cubeElement $\langle ind, m_1, ..., m_i, ..., m_n, e, a \rangle$ can be computed, where $a = agg(a_j)$ denotes the aggregation over all elements a_j of the cubeElements under the aggregation function of the indicator *ind*.

---

[2] We do not focus on this step as it depends on the specific technology used for storage.

[3] It is straightforward to see that the result $R(q_c)$ of the query q_c, derived by applying the Rule 1 to the query q, is a subset of $R(q)$.

This rule corresponds to the classical roll-up OLAP operation in data warehouse systems. We hasten to note that for distributive aggregation function L_y^D can be any level lower than L_x^D, while for algebraic functions L_y^D must be the bottom level to assure the exact result.

Rule 3 *(Indicator expansion).* Let $ind = f(ind_1, ..., ind_k)$ be the formula defining ind in terms of indicators $ind_1, ..., ind_k$. If $\exists ce_i = \langle ind, m_1, ..., m_n, e, a_i \rangle, i = 1, ..., k$ then the cubeElement $\langle ind, m_1, ..., m_n, e, f(a_1, ..., a_k) \rangle$ can be computed.

This rule introduces a novel operation that allows us to infer new formulas not explicitly given, which is made possible by exploiting their formal representation. It also has to be noted that if a formula for ind was not specified in the model, it could be derived through mathematical manipulation, as described in a previous work [5], e.g. from $I1 = I2 + I3$ a formula for $I2$ can be computed as $I1 - I3$.

5   Query Completeness

The notion of answer completeness allows to define rules for the retrieval of data together with the evaluation of the completeness level of a query.

Definition 2 *(Completeness).* Given a query $q = \langle ind, W, K, V, \rho \rangle$, the result set $R(q)$ is complete if $R(q_c)$ is complete for each q_c obtained from the application of Rule 1.
Given a query $q_c = \langle ind, W', K', e \rangle$ with $K' = \{K_1, ..., K_n\}$, the result set $R(q_c)$ is complete iif:

- Condition (1) $|R(q_c)| = \prod_{j=1}^{n} |K_j|$,
- Condition (2) $\forall$ tuple $\langle a, m_1, ..., m_n, e \rangle \in R(q_c)$,
  - $\exists$ a cubeElement $ce = \langle ind, m_1, ..., m_n, e, a \rangle$ or
  - ce can be computed by applying a complete roll-up expansion or the indicator expansion rules.

Checking completeness of a multidimensional query over a cube (q_c) can be understood as (condition 1) a tuple-by-tuple calculus of the result set, one for each possible combination of elements of sets $K_1, ..., K_n$. Each tuple calculus corresponds (condition 2) either to the retrieval of the corresponding cubeElement (if the tuple is materialized) or to the execution of a set of suitable multidimensional queries for that tuple (in this case we speak of a virtual tuple), that can turn out to be either complete or not. As a consequence, the definition of query completeness is recursive. Furthermore, Rules 2 and 3 give us rewriting rules to define the form of queries to compute virtual tuples. Traversing the path of completeness check gives also as side effect a way to completely rewrite the original query.

5.1   Completeness Procedure

Completeness check is a procedure aimed at verifying the completeness of the result set for a given query at enterprise level. The procedure takes as input a query q at enterprise level, and produces as output a list of complete items and the list of incomplete items. The procedure is composed of the following macro-processes:

(1) Evaluation of results for the query q
(2) For each incomplete item t of the result:
    (2.1) Use of Rules 2 or 3 to determine alternative ways to obtain the item
          through a set of new queries $q_1, \ldots, q_n$
    (2.2) Recursive execution for each q_i.

In the following, each step is discussed together with an example. Let us consider a query $\langle I1, \{Year, Country\}, \{\{2013, 2014\}, \{Spain\}\}, \{ACME1, ACME2\}\rangle$. For lack of space, we show the execution of the procedure only on the data mart for ACME1, after the Rule 1 is exploited. Table 1 shows the cubeElements materialized in the highly sparse data mart for ACME1, in a tabular form for convenience of representation.

Table 1. Case study: the content of the data mart for ACME1.

Indicator	ProductDim	TimeDim	LocationDim	value
$I1$	ALL	2013 (Year)	Spain (Country)	90
$I2$	ALL	2014-S1 (Semester)	Spain (Country)	70
$I2$	ALL	2014-S2 (Semester)	Spain (Country)	66
$I3$	ALL	2014 (Year)	Barcelona (City)	25
$I3$	ALL	2014 (Year)	Madrid (City)	30
$I3$	ALL	2014-S1 (Semester)	Valencia (City)	10
$I3$	ALL	2014-S2 (Semester)	Valencia (City)	12

Step 1. According to the set K of the query, a table T is defined, including one placeholder for each possible item of the final result set. The header of the table is given by the set of levels W specified in the query, while rows are the cartesian product generated on K. At the beginning, each item is searched in the data mart. All the items that cannot be found are (temporarily) marked as incomplete.

Example. Table T includes two items: $t_1 = \langle I1,ALL,2013,Spain\rangle$ and $t_2 = \langle I1,ALL,2014,Spain\rangle$. t_1 is immediately available as added to the result, while t_2 is not found.

Step 2. If the list of incomplete items is empty (base case), then the query q is complete, and the execution ends by producing as output the results. Otherwise for each item t in the list of incomplete items, Rules 2 and 3 are used to derive t from other cubeElements in the data warehouse. Several alternative derivations can be possible for t, by applying roll-up over different dimensions (Rule 2) or by decomposing the indicator through different formulas (Rule 3). Hence, for each possible derivation a new set of queries $q_1, \ldots, q_n$ is constructed and checked for completeness, in a recursive fashion, by calling the procedure again. After the recursive call ends, if every query q_i returned a value, the item t is calculated according to the chosen rule. Otherwise, given that the chosen derivation could not obtain the value for t, the derivation is discarded, and the next possible derivation is applied, until no derivation is left. In such a case, if no derivation is found to be complete, t is definitively marked as incomplete.

Example. t_2 is not found and the possible derivations are as follows. By executing Rule 2, drill-down can be used to expand either (a) Year to Semester or (b) Country to City levels. Product cannot be further expanded because it is the lowest level for ProductDim. By choosing[4] Rule 3, $I1$ is expanded into $I2$ and $I3$, given that its formula is $I2+I3$. Two new queries are then created:

- $q_1' = \langle I2, \{Product, Year, Country\}, \{\{ALL\}, \{2014\}, \{Spain\}\}, ACME1 \rangle$
- $q_2' = \langle I3, \{Product, Year, Country\}, \{\{ALL\}, \{2014\}, \{Spain\}\}, ACME1 \rangle$.

No cubeElement is available for q_1'. Let us suppose that Rule 2 is used in this case, drilling-down Year in Semester. As a result, three new queries are generated to retrieve I_2 for each semester of 2014 and for Spain. Results are available (i.e. 66 and 70), hence they are aggregated through the aggregation function of I_2 (AVG), and a result of 68 is obtained for q_1'.

As for q_2', given that no cubeElements is available as well, Rule 2 is exploited. In this case drilling-down Year in Semester produces no result, so the derivation Country in City is tried. For the three cities in the model (Barcelona, Madrid and Valencia)[5], cubeElements are retrieved for the first two (with values 25 and 30) and only Valencia is missing. Recursively, a new derivation is tried for this incomplete item. This time, Rule 2 is executed to drill-down Year to Semester. Finally, results are available for both semesters, and values are aggregated through the aggregation function of I_3 (SUM), obtaining $12+10=22$. q_2' is then computed as $25+30+22=77$. By backtracking of the recursive procedure, as last step, t_2 is computed as I_2+I_3, i.e. $68+77=145$.

5.2 Implementation and Computational Aspects

In order to define reasoning functionalities and support mathematical manipulation of PIs, the multidimensional model is represented in a logic-based language

---

[4] As the procedure explores the search space, if a solution exists, it is found whatever rule is chosen, although the order is critical w.r.t. execution time.

[5] As described in Sect. 3, in this drill-down for ACME1 we consider only the cities x such that the relations $partOf_{A_1}(x, Spain)$ hold in the data mart.

on which automatic inference can be exploited. To this end, we refer to Horn Logic Programming and in particular to Prolog, relying on XSB[6] (a logic programming and very efficient deductive database system) as reasoning engine.

While formulas are represented as facts, their manipulation as mathematical expressions is performed by a set of predicates. To this end, we have adapted and extended the Prolog Equation Solving System (PRESS) [6] that constitutes a first-order mathematical theory of algebra in LP, and allows for common algebraic manipulation of formulas. Among the offered functionalities, PRESS includes equation solving, equation simplification and rewriting, finding of inequalities and proving identities. Such predicates are useful to support more advanced reasoning functionalities including those needed to implement multidimensional query rewriting and completeness check.

This work can be framed in the general theory of query answering using views over incomplete data cubes. In the case of an implementation using only virtual views, its computational costs are dominated by the completeness check procedure. In the worst case, i.e. when no pre-aggregated data at intermediate levels exists, roll-up expansion has to traverse all the level hierarchy for every tuple of the result. This leads to a number of queries to be executed equal to $\frac{T(1-N^L)}{1-N}$, where T is the number of tuples in the database, N is the number of members of a level that are part of a member of the next higher level (e.g. three months are part of a quarter). To simplify the calculus, N is assumed to be constant for each level and member. Finally L is the number of levels in the hierarchy. Although in practice the worst case rarely occurs, it enlightens the value of materialization management.

In order to evaluate the costs related to execution times for completeness check, several steps ought to be considered. Here, we report the first results of the analysis of costs related to Rule 3, that is the novel type of rewriting proposed in this work. Conceptually, such an operation requires to find all possible rewritings for a formula and is independent on the size of data. We computed the execution time for all indicators in the model used for the BIVEE project [2], that includes 356 indicators where each of them has from 2 to 4 other indicators as operands (2.67 on average, only linear equations). The average execution time[7] in such a case is 735.60ms, ranging from a few ms to 2s.

6   Related Work

The sharing of data coming from distributed, autonomous and heterogeneous sources asks for methods to integrate them in a unified view, solving heterogeneity conflicts [7]. A common approach is to rely on a global schema to obtain an integrated and virtual view, either with a global-as-view (GAV) or a local-as-view (LAV) approach [8]. Query answering is one of the main tasks of a data integration system and involves a reformulation step, in which the query expressed

---

[6] http://xsb.sourceforge.net/.

[7] Experiments have been carried on a personal computer powered by an Intel Core i7-3630QM with 8 GB memory, running Linux Fedora 20.

by terms of the global schema is rewritten in terms of queries over the sources. According to the typology of the chosen approach (GAV or LAV), queries are answered differently, namely by unfolding or query rewriting based on views (see e.g. [9] in the context of distributed database systems, and [10,11] for data warehouse integration). Rewriting using views is used also for answering aggregate queries and as such can be related to the data warehouse field [11]. However, the multidimensional model has peculiarities that calls for specific studies. Heterogeneities hindering the integration of independent data warehouses may be classified on the basis of conflicts that occur at dimension or measure levels [12].

Recently, semantic representations of the multidimensional model have been proposed [13,14] mainly with the aim to reduce the gap between the high-level business view of indicators and the technical view of data cubes, to simplify and to automatise the main steps in design and analysis. In particular, in [15] a proprietary script language is exploited to define formulas, with the aim to support data cube design and analysis, while in [16] authors introduce an indicator ontology based on MathML, to define PI formulas in order to allow automatic linking of calculation definitions to specific data warehouse elements. Support for data cube design is considered in [15], while improvement of OLAP functionalities are presented in [17]. Both [13,14] tackle the problem of multidimensional modeling with semantics. The former relies on standard OWL-DL ontology and reason on it to check the multidimensional model and its summarizability, while the latter refers to Datalog inference to implement the abstract structure and semantics of multidimensional ontologies as rules and constraints. Although the complex nature of indicators is well-known, the compound nature of indicators is far less explored. Proposals in [11,15,16,18,19] include in the representations of indicators' properties some notion of formula in order to support the automatic generation of customised data marts, the calculation of indicators [15,18], or the interoperability of heterogeneous and autonomous data warehouses [11]. Anyway, in the above proposals, formula representation does not rely on logic-based languages, hence reasoning is limited to formula evaluation by ad-hoc modules.

Logic-based representations of dependencies among indicators are proposed in [19], although no manipulation of formulas is exploitable in this way.

7   Conclusion

In this paper we presented a semantics-based approach for dynamic calculation of performance indicators in the context of sparse distributed data marts. The methodology extends usual operators for query rewriting based on views, allowing to overcome structural heterogeneities among data mart schemas and making users aware of the completeness of the result of a query. Apart from the expressiveness of the reasoner, the approach does not set particular conditions on the types of formulas. At present our focus is on linear and polynomial formulas, since they represent the analytical definition of the 200 adopted indicators in the BIVEE framework. However, in general PRESS can deal with symbolic, trascendental and non-differential equations.

As extensions, while here we assume a unique aggregation function associated to each indicator, in future work we will consider different functions to aggregate data according to the dimension. For what concerns costs of computation, both a complete evaluation and the identification of methodological strategies to optimise completeness check exploitation (including pruning of search space and use of caching) will be discussed. Several strategies are possible for data materialization (e.g. lazy, eager or semi-eager), that is a classical issue in data warehouse implementation and will be discussed in a future work.

References

1. Gray, J., Chaudhuri, S., Bosworth, A., Layman, A., Reichart, D., Venkatrao, M., Pellow, F., Pirahesh, H.: Data cube: a relational aggregation operator generalizing group-by, cross-tab, and sub totals. Data Min. Knowl. Discov. **1**(1), 29–53 (1997)
2. Diamantini, C., Genga, L., Potena, D., Storti, E.: Collaborative building of an ontology of key performance indicators. In: Meersman, R., Panetto, H., Dillon, T., Missikoff, M., Liu, L., Pastor, O., Cuzzocrea, A., Sellis, T. (eds.) OTM 2014. LNCS, vol. 8841, pp. 148–165. Springer, Heidelberg (2014)
3. Diamantini, C., Potena, D., Storti, E.: SemPI: a semantic framework for the collaborative construction and maintenance of a shared dictionary of performance indicators. Future Generation Comput. Syst. (2015). http://dx.doi.org/10.1016/j.future.2015.04.011
4. Golfarelli, M., Rizzi, S.: Data Warehouse Design: Modern Principles and Methodologies, 1st edn. McGraw-Hill Inc, New York (2009)
5. Diamantini, C., Potena, D., Storti, E.: Extending drill-down through semantic reasoning on indicator formulas. In: Bellatreche, L., Mohania, M.K. (eds.) DaWaK 2014. LNCS, vol. 8646, pp. 57–68. Springer, Heidelberg (2014)
6. Sterling, L., Bundy, A., Byrd, L., O'Keefe, R., Silver, B.: Solving symbolic equations with press. J. Symb. Comput. **7**(1), 71–84 (1989)
7. Rahm, E., Bernstein, P.A.: A survey of approaches to automatic schema matching. VLDB J. **10**(4), 334–350 (2001)
8. Lenzerini, M.: Data integration: a theoretical perspective. In: Proceedings of the Twenty-first ACM SIGMOD-SIGACT-SIGART Symposium on Principles of Database Systems. PODS 2002, pp. 233–246. ACM, New York, NY, USA (2002)
9. Halevy, A.Y.: Answering queries using views: a survey. VLDB J. **10**(4), 270–294 (2001)
10. Cohen, S., Nutt, W., Sagiv, Y.: Rewriting queries with arbitrary aggregation functions using views. ACM Trans. Database Syst. **31**(2), 672–715 (2006)
11. Golfarelli, M., Mandreoli, F., Penzo, W., Rizzi, S., Turricchia, E.: Olap query reformulation in peer-to-peer data warehousing. Inf. Syst. **37**(5), 393–411 (2012)
12. Tseng, F.S., Chen, C.W.: Integrating heterogeneous data warehouses using xml technologies. J. Inf. Sci. **31**(3), 209–229 (2005)
13. Neumayr, B., Anderlik, S., Schrefl, M.: Towards Ontology-based OLAP: Datalog-based Reasoning over Multidimensional Ontologies. In: Proceedings of the Fifteenth International Workshop on Data Warehousing and OLAP, pp. 41–48 (2012)
14. Prat, N., Megdiche, I., Akoka, J.: Multidimensional models meet the semantic web: defining and reasoning on owl-dl ontologies for olap. In: Proceedings of the Fifteenth International Workshop on Data Warehousing and OLAP. DOLAP 2012, pp. 17–24. ACM, New York, NY, USA (2012)

15. Xie, G.T., Yang, Y., Liu, S., Qiu, Z., Pan, Y., Zhou, X.: EIAW: towards a business-friendly data warehouse using semantic web technologies. In: Aberer, K., Choi, K.-S., Noy, N., Allemang, D., Lee, K.-I., Nixon, L.J.B., Golbeck, J., Mika, P., Maynard, D., Mizoguchi, R., Schreiber, G., Cudré-Mauroux, P. (eds.) ASWC 2007 and ISWC 2007. LNCS, vol. 4825, pp. 857–870. Springer, Heidelberg (2007)
16. Kehlenbeck, M., Breitner, M.H.: Ontology-based exchange and immediate application of business calculation definitions for online analytical processing. In: Pedersen, T.B., Mohania, M.K., Tjoa, A.M. (eds.) DaWaK 2009. LNCS, vol. 5691, pp. 298–311. Springer, Heidelberg (2009)
17. Priebe, T., Pernul, G.: Ontology-based integration of OLAP and information retrieval. In: Proceedings of DEXA Workshops, pp. 610–614 (2003)
18. Horkoff, J., Barone, D., Jiang, L., Yu, E., Amyot, D., Borgida, A., Mylopoulos, J.: Strategic business modeling: representation and reasoning. Softw. Syst. Model. 13(3), 1015–1041 (2012)
19. Popova, V., Sharpanskykh, A.: Modeling organizational performance indicators. Inf. Syst. 35(4), 505–527 (2010)

Data Warehouses

Automatically Tailoring Semantics-Enabled Dimensions for Movement Data Warehouses

Juarez A.P. Sacenti[1]([⊠]), Fabio Salvini[2], Renato Fileto[1],
Alessandra Raffaetà[2], and Alessandro Roncato[2]

[1] PPGCC/INE/CTC, Federal University of Santa Catarina, Florianópolis, SC, Brazil
`juarez.sacenti@posgrad.ufsc.br`, `r.fileto@ufsc.br`
[2] DAIS, Università Ca' Foscari Venezia, Venezia, Italy
`salvini.fabio001@gmail.com`,
`{raffaeta,roncato}@unive.it`

Abstract. This paper proposes an automatic approach to build tailored dimensions for movement data warehouses based on views of existing hierarchies of objects (and their respective classes) used to semantically annotate movement segments. It selects the objects (classes) that annotate at least a given number of segments of a movement dataset to delineate hierarchy views for deriving tailored analysis dimensions for that movement dataset. Dimensions produced in this way can be quite smaller than the hierarchies from which they are extracted, leading to efficiency gains, among other potential benefits. Results of experiments with tweets semantically enriched with points of interest taken from linked open data collections show the viability of the proposed approach.

Keywords: Movement data · Data warehouses · Analysis dimensions · Semantic web · Linked Open Data (LOD) · Social media

1 Introduction

Nowadays, it is possible to accumulate and exploit large volumes of movement data, such as moving objects' trajectories gathered by using positioning technologies (e.g. GPS), and sequences of users' posts on social media [1,2]. On top of this, new methods have been developed to semantically enrich such data, for example with objects (instances) and concepts (classes of objects) that help to describe what goes on with the moving objects (e.g., places visited, transportation means employed, activities performed, goals of stops and moves) [3–6]. These developments unleash opportunities to build Movement Data Warehouses (MDWs), i.e., DWs of semantically enriched movement data [7–9], for a myriad of application fields, ranging from traffic and homeland security to geographically concentrated marketing and social behavior studies. However, to realize MDWs, besides accommodating large volumes of semantically annotated spatio-temporal data in the dimensional model [9], it is still necessary to develop appropriate ways to exploit semantic annotations for semantics-enabled movement data analysis.

© Springer International Publishing Switzerland 2015
S. Madria and T. Hara (Eds.): DaWaK 2015, LNCS 9263, pp. 205–216, 2015.
DOI: 10.1007/978-3-319-22729-0_16

This paper makes contributions towards building MDWs that support movement data analysis founded on their semantic annotations. It introduces an automatic approach to build tailored analysis dimensions from hierarchies of objects or classes used to semantically annotate movement segments, i.e., specific positions or sequences of positions occupied by moving objects, such as stops and moves [3]. For instance, a move can be semantically annotated with a concept that describes the transportation means employed (e.g., taxi, bus, tram) in a conceptual hierarchy or taxonomy. A stop can be semantically associated via some property (e.g. visits) with a Place of Interest (PoI). The PoIs are geographic objects that can be hierarchically organized according to their *part_of* relationships (such as the containment of a *shop* in a *shopping mall* that is in a particular *neighborhood* of a *city*). Notice that classes of objects such as PoIs can be regarded as concepts or categories (e.g., *city, neighborhood, shopping mall, shop, clothes shop, pet shop, barber shop*) and organized in another hierarchy determined by *is_a* relationships. Both, hierarchies of objects and hierarchies of their respective classes can be useful for information analysis in MDWs [9].

In this work, we consider semantic annotations of movement segments as associations with objects and concepts of Linked Open Data (LOD). Such associations are produced by methods such as those proposed in [4] and [6]. Hierarchies of LOD objects and hierarchies of their respective classes can be obtained by exploiting chains of partially ordering relationships, such as *part_of* relationships between objects and *is_a* relationships between concepts. They can also be built or complemented by relationships that are inferred from existing ones or data properties (e.g., geographic extensions). Then, their objects (classes) that annotate at least a given number of segments of a Movement Dataset (MoD) are selected and the hierarchy is traversed from them up to their least common ancestor, to produce tree-like hierarchy views which are the basis to build tailored analysis dimensions for that MoD.

Results of experiments with tweets taken via the Twitter API[1] and semantically enriched with PoIs of the DBpedia[2] and LinkedGeoData (LGD)[3] LOD collections testify the viability and some benefits of the proposed approach. The dimensions produced by our technique can be quite smaller than the hierarchies present in LOD from which they are extracted. Although we demonstrate our proposal on social media trails semantically enriched with PoIs of LOD collections, the proposed approach could be applied to other kinds of movement data enriched with other collections of objects and/or concepts, as it relies just on the number of hits of each object and concept, i.e., the number of times the object/concept is used to annotate segments of a particular MoD.

The rest of this paper is structured as follows. Section 2 provides basic definitions. Section 3 describes how to extract hierarchies from LOD collections, and how to tailor such hierarchies into MDWs' dimensions. Section 4 reports experiments that apply the proposed approach to build spatial dimensions of PoIs and

---

[1] https://dev.twitter.com/rest/public.

[2] http://dbpedia.org.

[3] http://linkedgeodata.org.

their categories taken from the LOD collections used to semantically annotate tweets. Section 5 discusses related work. Finally, Sect. 6 concludes the paper, by pointing out its contributions and enumerating future work.

2   Basic Definitions

This section introduces the definitions needed to understand our method for tailoring dimensions for MDWs. They describe the inputs of the proposed method, namely: movement data segments, semantically enriching resources (objects and their classes), associations of movement data with these resources to annotate the former, and resources hierarchies. A **Movement Dataset (MoD)** is a set of movement data segments like the one formally described in Definition 1.

Definition 1. *A **Movement Data Segment** (MDS) is a time-ordered sequence $\langle p_1, \ldots, p_n \rangle$ ($n \in \mathbb{N}^+$) of spatio-temporal positions of a moving object MO, where each position is a tuple $p_i = (x_i, y_i, t_i)$ with $1 \leq i \leq n$, where (x_i, y_i) are geographic coordinates, and t_i is a timestamp indicating the instant when MO was at those coordinates.*

An MDS can be any sequence of spatio-temporal coordinates occupied by a moving object, such as trajectory positions, geo-located user's posts on a social media, or geo-referenced logs of any kind of information system (e.g., Web logs). An MDS can be segmented into smaller MDSs at several abstraction levels (e.g., first in sub-trajectories that can refer to specific trips, and then in episodes such as stops and moves [3]). The resulting movement segments can be hierarchically organized according to containment relationships of their time spans, for information analysis in DWs as proposed in [9]. The movement segments can be enriched with semantic annotations to describe them according to several dimensions, such as those proposed in [5], namely space, time, characteristics of the moving objects, activities performed by these objects, their goal, transportation means employed, and the relevant environmental conditions during each movement segment time span. Such a semantic annotation can be an association with a resource (object or class) as stated by Definition 2.

Definition 2. *A **Semantically Enriching Resource** is a concept or instance of concept that can be used to semantically annotate an MDS.*

A semantically enriching resource (or simply resource) can be implemented as a reference (e.g., a Universal Resource Identifier - URI) to a class (concept) or object (instance) defined in an ontology, knowledge base, or even a LOD collection. In the latter a resource is denoted by a URI and relationships between resources are expressed by RDF triples.

For example, in Fig. 1 on the right some LOD resources are shown: the classes *restaurant* and *city*, two objects of the class restaurant, labeled *Ae Oche* and *Sushi Wok*, and an object of the class *city* named *Mestre*. These resources are used to annotate the tweets on the left side of the figure which are spatio-temporal positions. In fact they contain spatial coordinates expressed by longitude and latitude (in the lower right corner) and a temporal information specifying where

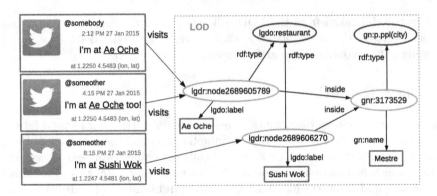

Fig. 1. Associations of tweets with PoIs and with their classes

and when tweets have been sent. The property that links tweets to resources is *visits*, to indicate that the tweets have been sent from the places (resources) where the person is.

LOD resources are convenient to semantically annotate movement data, as they have well-defined semantics and can be integrated more easily thanks to their adherence to semantic Web standards. Many information collections are publicly available in the Web by following Linked Data principles [10] (e.g., DBpedia, LGD). In addition, they have been continuously growing and kept up to date by active Web communities, which link their resources via a variety of properties, that include *owl:sameAs* (for objects duplicated in distinct collections), and *rdfs:equivalentClass* (for duplicated concepts). A direct association annotates an MDS with a property (e.g., visits) whose property value is a resource (e.g., the *restaurant Ae Oche*), as stated by Definition 3.

Definition 3. *A **direct association** is a triple of the form* $(mds, property, r)$*, where mds is an MDS, property is the association kind, and r is a resource.*

Many associations can annotate the same MDS. Annotating movement segments that are sequences of positions instead of individual positions grants compactness (e.g., a move can be annotated as a whole with a transportation means).

Indirect associations of MDS with other resources can be derived from the direct associations by following properties between resources of LOD collections, as stated by Definition 4.

Definition 4. *An **indirect association** between a movement data segment mds and a resource r is a tuple* $(mds, path, r)$ *where path is a chain of links constituted by the direct association* $(mds, property, r')$ *of mds with the resource* r'*, followed by a sequence of LOD properties that semantically links* r' *to r.*

In other words, an indirect association $(mds, path, r)$ indirectly links the *mds* to a resource r via a *path* composed by a direct association followed by some LOD property(ies). For example, the tweets on the left side of Fig. 1 are indirectly associated with the concept *restaurant* via direct *visits* associations followed by the *type* property between the respective resources and *restaurant*. They are also indirectly associated with the city of *Mestre* and its class (*gn:p.ppl (city)*).

Finally, a **Semantically-enriched Movement Dataset (SMoD)** is a set of movement data segments which are associated with resources for semantic annotation purposes. The use frequency of a resource in associations of an SMoD determines its number of hits with respect to that SMoD (Definition 5).

Definition 5. *Let S be a SMoD, $\mathcal{R}$ a set of resources, and $\mathcal{A}$ a set of direct and indirect associations between movement segments in S and resources in $\mathcal{R}$. The **number of hits** $h(r, \mathcal{A})$ of a resource $r \in \mathcal{R}$ with respect to A, is the number of distinct MDSs of S which are annotated with r via associations in $\mathcal{A}$.*

Of course, one could also count separately the number of direct and indirect hits of each resource, by considering only direct and indirect associations, respectively. In Fig. 1, while the number of (direct) hits of the *lgdr:node2689605789* is 2 hits, its class *lgdo:Restaurant* has 3 (indirect) hits. Also the city of Mestre *gnr:3173529* and its class (*gn:p.ppl (city)*) have 3 (indirect) hits each one.

3   Approach to Tailor Dimensions for MDWs

This section describes our proposal to automatically tailor dimensions from a given MoD. Figure 2 illustrates the steps of the Extract-Transform-Load (ETL) process that are required to build tailored dimensions, with those steps that are the focus of this work (2 and 3) in bold.

Step 1 (*Semantic enrichment*) takes as input a MoD and KBs, LOD, conventional databases (DBs), and/or spatio-temporal DBs (STDBs) containing information useful for making semantic annotations. As pointed out in [5], some important aspects to enrich MDSs are visited POIs, transportation means, environmental conditions, goals, and activities of the moving objects. This task is addressed in several works (e.g., [3,4,11]), but it is beyond the scope of this paper, whose focus is on steps 2 and 3.

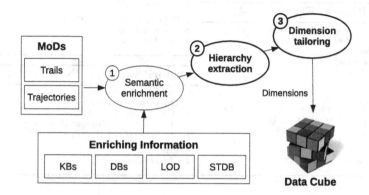

Fig. 2. The process for building semantic dimensions.

Step 2 is called *Hierarchy extraction* and it starts from the MoD which has been semantically enriched. For each semantic annotation, we explore the resources which are directly associated with MDSs in order to find properties that connect these resources to other resources by allowing to extract a hierarchy of concepts or of objects. Sometimes the extracted hierarchies are not so detailed, e.g., some levels are missing, so to improve them we search for other representations of a resource in a different KB by using the property *owl:sameAs* and then we integrate several properties. For example, for spatial objects, we can merge properties from GeoNames and GADM to build a spatial hierarchy connecting POIs to the continent they belong to, and having as intermediate levels districts, cities, states and countries. Remember that resources can be hierarchically organized according to a variety of partial order relationships expressed by properties (e.g., *isA*, *partOf* or *contains*). It enables information analysis of data annotated with them at different levels of detail.

Step 3 is called *Dimension tailoring* and is aimed at creating views of the extracted hierarchies that can serve as tailored analysis dimensions for the given MoD. In fact, hierarchies obtained from step 2 may contain non-relevant elements for the actual data analysis, that can be filtered away. Hierarchies produced in this way can be quite smaller than the ones from which they are extracted, leading to efficiency gains, among other potential benefits.

The proposed algorithm (Algorithm 1) takes as input a hierarchy $\mathcal{H}$ from step 2, a set of associations $\mathcal{A}$ which annotate MDS with resources in $\mathcal{H}$ and a given threshold σ. Then it proceeds as follows:

Algorithm 1. Tailor($\mathcal{A}$, $\mathcal{H}$, σ)

Input: Set of associations $\mathcal{A}$, resource
hierarchy $\mathcal{H}$, threshold σ
Output: dimension d
1 $d \leftarrow new\ dimension$;
2 $Filter(\mathcal{H}.root, \sigma, \mathcal{A})$;
3 $Merge(\mathcal{H}.root)$;
4 $d.root \leftarrow FindNewRoot(\mathcal{H}.root)$;
5 **return** d;

Algorithm 2. Filter(r, σ, $\mathcal{A}$)

Input: hierarchy root r, threshold σ,
set of associations $\mathcal{A}$
Output: none
1 **if** r *is a leaf* **then**
2 $\quad$ $r.hits \leftarrow h(r, \mathcal{A})$;
3 **else**
4 $\quad$ $r.hits \leftarrow 0$;
5 $\quad$ **for** *each child* $\in r.children()$ **do**
6 $\quad\quad$ $Filter(child, \sigma, \mathcal{A})$;
7 $\quad\quad$ $r.hits \leftarrow r.hits + child.hits$;
8 **if** $r.hits < \sigma$ **then**
9 $\quad$ $r.label \leftarrow$ "Others";

Algorithm 3. Merge(r)

Input: hierarchy root r
Output: none
1 $o \leftarrow new\ node$;
2 $o.label \leftarrow$ "Others";
3 $o.hits \leftarrow 0$;
4 **for** *each child* $\in r.children()$ **do**
5 $\quad$ **if** $child.label=$ "Others" **then**
6 $\quad\quad$ $o.addChildren(child.children())$;
7 $\quad\quad$ $r.removeChild(child)$;
8 $\quad\quad$ $o.hits \leftarrow o.hits + child.hits$;
9 $\quad$ **else**
10 $\quad\quad$ $Merge(child)$;
11 **if** $|o.children()| > 0$ **then**
12 $\quad$ $r.addChild(o)$;
13 $\quad$ $Merge(o)$;

Algorithm 4. FindNewRoot(r)

Input: hierarchy root r
Output: new hierarchy root
1 **if** $|r.children()| = 1$ **then**
2 $\quad$ $child \leftarrow$ the unique child of r;
3 $\quad$ **return** $FindNewRoot(child)$;
4 **else**
5 $\quad$ **return** r

1. via a post-order traversal of $\mathcal{H}$, the function *Filter* directly computes the number of hits (Definition 5) in the leaves of $\mathcal{H}$ whereas for an internal node it calculates such a number by summing up the hits in its children. Moreover, it checks whether a node has a number of hits greater or equal than σ. If this constraint is not satisfied the label of this node is replaced by "Others", which is a dummy value that states that this node is not relevant for the dimension (Algorithm 2).
2. After the execution of *Filter*, an internal node could have several children labeled by the dummy value "Others", so the function *Merge* is aimed at replacing all these nodes with a single child having such a label (Algorithm 3).
3. Finally, the least common ancestor is computed and the hierarchy view rooted at this node is returned as dimension (Algorithm 4).

4 Experiments

In this section, we present the experiments conducted on two datasets of tweets which have been semantically enriched by using two criteria: spatial proximity of the MDS to a resource and textual similarity between the name of the resource and the content of the tweet [6].

The first dataset, **SMoD1**, contains 1,530 tweets which have been directly associated with 258 spatial resources by considering a spatial proximity of 8 m and textual similarity of 94 %. These tweets are indirectly associated with 129 spatial resource categories via 1,328 *rdf:type* assertions.

The second dataset, **SMoD2**, is composed by 10,710 tweets directly associated with 1,501 spatial resources by using less restrictive conditions: a spatial proximity of 16 m and textual similarity of 90 %. Tweets are indirectly associated with 338 spatial resource categories via 7,422 *rdf:type* assertions.

Both SMoD1 and SMoD2 have their tweets successfully associated with a unique spatial resource. Each tweet has a geo-location inside a Minimum Bounding Rectangle (MBR) within the Brazilian territory (NE 5.264860 -28.839041, SW -33.750702 -73.985527) and was extracted between June 9, 2014 and July 14, 2014. The Twitter extraction includes a condition to filter only tweets that start with the string "I'm at" and end with a FourSquare URL. Spatial resources inside the same Brazilian territory MBR were extracted from DBpedia and LGD on July 21, 2014 and they were considered for the semantic enrichment step.

In this paper, we assume that the spatial resources used as annotations in SMoD1 and SMoD2 represent the current place where a user sends the tweet at the annotated timestamp. Hence, we apply our method in order to build a spatial object dimension and a spatial category dimension for both SMoDs.

The spatial objects hierarchy was extracted by searching for DBPedia and LGD resources used as annotations in SMoD1 and SMoD2, and connecting them to the respective GeoNames and GADM resources via *sameAs* properties, as illustrated in Fig. 3. Then, for the found resources, the hierarchies are completed by exploiting properties like *gadm:in_country* and *gn:parentfeature* to connect them with upper level resources. A portion of the resulting hierarchy is shown in Fig. 4. It is worth noticing that the nodes are organized in various levels of abstraction: *idresource*, *district*, *city*, *state*, *country* and *continent*.

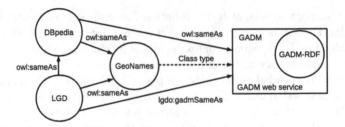

Fig. 3. LOD collections and properties used to extract hierarchies of spatial objects.

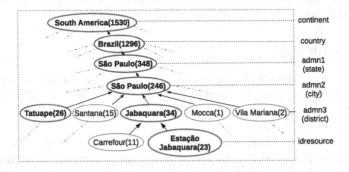

Fig. 4. A portion of a spatial objects hierarchy with the most popular nodes in bold.

Algorithm 1 was used to create views of hierarchies such as the one illustrated in Fig. 4, in which the number of hits per node appears in brackets. Assuming a threshold $\sigma = 20$ groups of nodes at each level (e.g., *Santana*, *Mocca* and *Vila Mariana* in the level *district*) are replaced with the dummy value "Others". This means that for the chosen threshold such resources are not so relevant since they annotate too few tweets. Hence they are filtered away. Figure 4 illustrates a spatial hierarchy, with a view composed of nodes having more than $\sigma = 20$ hits in bold. Such a view added with a single dummy node condensing each set of siblings that have a lower number of hits (e.g., *Santana*, *Mocca* and *Vila Mariana*) can be used as a tailored dimension.

Figure 5 shows the number of resources (nodes) in each level of the tailored objects dimension generated for different values of the minimum number of hits per resource (σ) in SMoD1 (left) and SMoD2 (right). Notice that those numbers of dimension nodes falls sharply as σ grows. In addition, even for $\sigma = 1$ there is a considerable reduction in the number of countries (7 that originated the tweets in SMoD1, instead of around 200 in the world), cities (around 100 generating those tweets, instead of more than 6000 in the region defined by an MBR encompassing Brazil that was used to filter those tweets), just to mention some reductions.

A hierarchy of PoI concepts indirectly associated with the tweets in SMoD1 was built by exploiting *rdfs:subClassOf* properties between such concepts. Figure 6 shows an extract of a hierarchy of 93 concepts built from SMoD1. In SMoD1, the most visited concepts are *Thing* (1509 hits), *Amenity* (981 hits), *Restaurant* (225 hits), *Station* (197 hits), *Shop* (156 hits), and *Cafe* (124 hits).

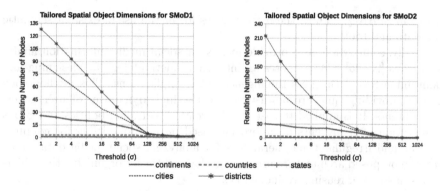

Fig. 5. Nodes per level of tailored spatial objects dimensions for increasing values of σ

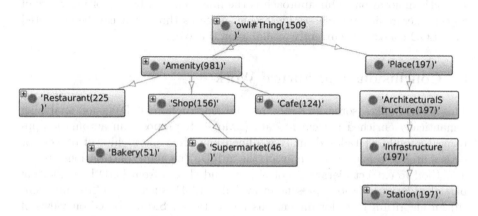

Fig. 6. LGD PoI categories with at least 15 indirect associations with tweets in SMoD1

5   Related Work

Several works propose models and approaches for analyzing movement data in DWs [7–9,12–14]. Among them, only [9] accommodates MoD semantically annotated with LOD in a multidimensional model that separates hierarchies of objects and hierarchies of categories of these objects, and also allows hierarchies of movement segments, movement patterns, and their respective categories as dimensions. However, it does not provide means to build dimensions with LOD.

Many other works exploit semantic Web technology for building and operating on DWs [15–22]. Some of these proposals use semantic Web data such as LOD to feed DWs. [19] proposes a semi-automatic method for extracting semantic data into a multidimensional database on-demand. In [18] the authors follow a similar approach, but restrict their case study to statistical LOD and use the RDF Data Cube vocabulary[4]. [21] discusses modeling issues to support OLAP

---

[4] http://www.w3.org/TR/vocab-data-cube.

queries in SPARQL. Thus, these approaches foster the use of existent semantic Web data and tools.

Finally, some works also provide means for social media information analysis in DWs [23–25]. [24] proposes an interactive methodology for designing and maintaining Social Business Intelligence (SBI) applications. This methodology aims at providing quick responses to face the high dynamism of user generated contents. It is applied to case studies related to Italian politics and goods consumption. [25] presents an approach based on meta-modeling for coupling dynamic and irregular topic hierarchies for analyzing social media data. However, the proposed architecture uses an ontology editor for organizing these hierarchies, instead of reusing available ones.

None of these works provide means to derive tailored dimensions for analyzing particular datasets, and more specifically movement datasets, from their semantic annotations. Our approach is the first to do so, by exploiting views of objects hierarchies and classes hierarchies, such as those that can be extracted from LOD collections currently available on the Web.

6 Conclusions and Future Work

This paper makes some advances in the construction of DWs for analyzing semantically enriched movement data (SMoD). It proposes an automatic approach for building tailored analysis dimensions from hierarchies of objects or classes used to semantically annotate the SMoD. The main contributions are: (i) a method to extract hierarchies of objects and classes from LOD by exploiting partial ordering relations present in available LOD, such as *part_of* and *is_a*; (ii) an algorithm to tailor dimensions for particular SMoD, based on views of the hierarchies of classes and objects used to semantically enrich the movement data; and (iii) a demonstration of the proposed method in a case study that generates spatial dimensions from hierarchies of PoIs and their classes used to semantically enrich social media user's trails composed of tweets.

The experience has taught us that it is usually easier to extract hierarchies of objects and classes from existing LOD collections than building them from scratch. Nevertheless, sometimes it is necessary to complement the semantic relations available in LOD collections by using other means (e.g., investigating containment relationships between geographic extensions of PoIs) to fulfill the absence of explicit relations between some pairs of resources. Experiments with our method to tailor dimensions in case studies with tweets semantically enriched with DBPedia and LGD resources showed considerable reductions in the size of the resulting dimensions compared to whole hierarchies extracted of LOD, even for low values of the hits count threshold (number of movement segments annotated per PoI or class of PoI).

This work, for the best of our knowledge, provides the first proposal for building DW dimensions from LOD used to semantically enrich movement data, and perhaps other kinds of data too. It constitutes an important step towards building DWs for semantics-enabled information analysis. Future work includes:

(i) deeper investigation of the implications of variations of the proposed method for tailoring dimensions; (ii) experiments with other movement datasets enriched with different LOD collections and knowledge bases; (iii) inspection of the features of other LOD collections to improve the hierarchy extraction; (iv) further evaluation of the benefits of the proposed method.

Acknowledgments. This work was supported by the European Union IRSES-SEEK (grant 295179), CNPq (grant 478634/2011-0), CAPES, and FEESC, the MIUR Project PON ADAPT (no. SCN00447), and by MOTUS (no. MS01 00015 - Industria2015). Special thanks to Cleto May and Douglas Klein for providing semantically enriched data for our experiments.

References

1. Parent, C., Spaccapietra, S., Renso, C., Andrienko, G.L., Andrienko, N.V., Bogorny, V., Damiani, M.L., Gkoulalas-Divanis, A., de Macêdo, J.A.F., Pelekis, N., Theodoridis, Y., Yan, Z.: Semantic trajectories modeling and analysis. ACM Comput. Surv. **45**, 1–32 (2013). Article 42
2. Pelekis, N., Theodoridis, Y.: Mobility Data Management and Exploration. Springer, New York (2014)
3. Yan, Z., Chakraborty, D., Parent, C., Spaccapietra, S., Aberer, K.: Semantic trajectories: mobility data computation and annotation. ACM TIST **4**, 1–38 (2013)
4. Fileto, R., Krüger, M., Pelekis, N., Theodoridis, Y., Renso, C.: Baquara: a holistic ontological framework for movement analysis using linked data. In: Ng, W., Storey, V.C., Trujillo, J.C. (eds.) ER 2013. LNCS, vol. 8217, pp. 342–355. Springer, Heidelberg (2013)
5. Bogorny, V., Renso, C., de Aquino, A.R., de Lucca Siqueira, F., Alvares, L.O.: CONSTAnT - a conceptual data model for semantic trajectories of moving objects. T. GIS **18**, 66–88 (2014)
6. May, C., Fileto, R.: Connecting textually annotated movement data with linked data. In: IX Regional School on Databases. ERBD, São Francisco do Sul, SC, Brazil (in Portuguese), SBC (2014)
7. Raffaetà, A., Leonardi, L., Marketos, G., Andrienko, G., Andrienko, N., Frentzos, E., Giatrakos, N., Orlando, S., Pelekis, N., Roncato, A., Silvestri, C.: Visual mobility analysis using T-warehouse. IJDWM **7**, 1–23 (2011)
8. Wagner, Ricardo, de Macedo, José Antonio Fernandes, Raffaetà, Alessandra, Renso, Chiara, Roncato, Alessandro, Trasarti, Roberto: Mob-warehouse: a semantic approach for mobility analysis with a trajectory data warehouse. In: Parsons, Jeffrey, Chiu, Dickson (eds.) ER Workshops 2013. LNCS, vol. 8697, pp. 127–136. Springer, Heidelberg (2014)
9. Fileto, R., Raffaetà, A., Roncato, A., Sacenti, J.A.P., May, C., Klein, D.: A semantic model for movement data warehouses. In: DOLAP 2014, pp. 47–56 (2014)
10. Ngomo, A.-C.N., Auer, S., Lehmann, J., Zaveri, A.: Introduction to linked data and its lifecycle on the web. In: Koubarakis, M., Stamou, G., Stoilos, G., Horrocks, I., Kolaitis, P., Lausen, G., Weikum, G. (eds.) Reasoning Web. LNCS, vol. 8714, pp. 1–99. Springer, Heidelberg (2014)

11. Rinzivillo, S., de Lucca Siqueira, F., Gabrielli, L., Renso, C., Bogorny, V.: Where have you been today? annotating trajectories with daytag. In: Nascimento, M.A., Sellis, T., Cheng, R., Sander, J., Zheng, Y., Kriegel, H.-P., Renz, M., Sengstock, C. (eds.) SSTD 2013. LNCS, vol. 8098, pp. 467–471. Springer, Heidelberg (2013)

12. Leal, B., de Macêdo, J.A.F., Times, V.C., Casanova, M.A., Vidal, V.M.P., de Carvalho, M.T.M.: From conceptual modeling to logical representation of trajectories in DBMS-OR and DW systems. JIDM **2**, 463–478 (2011)

13. Pelekis, N., Theodoridis, Y., Janssens, D.: On the management and analysis of our lifesteps. SIGKDD Explor. **15**, 23–32 (2013)

14. Leonardi, L., Orlando, S., Raffaetà, A., Roncato, A., Silvestri, C., Andrienko, G.L., Andrienko, N.V.: A general framework for trajectory data warehousing and visual OLAP. GeoInformatica **18**, 273–312 (2014)

15. Sell, D., Cabral, L., Motta, E., Domingue, J., dos Santos Pacheco, R.C.: Adding semantics to business intelligence. In: DEXA, pp. 543–547. IEEE Computer Society, Copenhagen (2005)

16. Filho, S.I.V., Fileto, R., Furtado, A.S., Guembarovski, R.H.: Towards Intelligent Analysis of Complex Networks in Spatial Data Warehouses. [26], pp. 134–145

17. Deggau, R., Fileto, R., Pereira, D., Merino, E.: Interacting with spatial data warehouses through semantic descriptions. [26], pp. 122–133

18. Kämpgen, B., Harth, A.: Transforming statistical linked data for use in OLAP systems. In: Proceedings of the 7th International Conference on Semantic Systems. I-Semantics 2011, pp. 33–40. ACM, New York (2011)

19. Nebot, V., Llavori, R.B.: Building data warehouses with semantic web data. Decis. Support Syst. **52**, 853–868 (2012)

20. Ibragimov, D., Hose, K., Pedersen, T.B., Zimányi, E.: Towards exploratory OLAP over linked open data – a case study. In: Castellanos, M., Dayal, U., Pedersen, T.B., Tatbul, N. (eds.) BIRTE 2013 and 2014. LNBIP, vol. 206, pp. 114–132. Springer, Heidelberg (2015)

21. Etcheverry, L., Vaisman, A., Zimányi, E.: Modeling and querying data warehouses on the semantic web using QB4OLAP. In: Bellatreche, L., Mohania, M.K. (eds.) DaWaK 2014. LNCS, vol. 8646, pp. 45–56. Springer, Heidelberg (2014)

22. Abelló, A., Romero, O., Pedersen, T.B., Llavori, R.B., Nebot, V., Cabo, M.J.A., Simitsis, A.: Using semantic web technologies for exploratory OLAP: a survey. IEEE Trans. Knowl. Data Eng. **27**, 571–588 (2015)

23. Berlanga, R., Aramburu, M.J., Llidó, D.M., García-Moya, L.: Towards a semantic data infrastructure for social business intelligence. In: Catania, B., Cerquitelli, T., Chiusano, S., Guerrini, G., Kämpf, M., Kemper, A., Novikov, B., Palpanas, T., Pokorny, J., Vakali, A. (eds.) New Trends in Databases and Information Systems. AISC, vol. 241, pp. 319–327. Springer, Heidelberg (2014)

24. Francia, M., Golfarelli, M., Rizzi, S.: A methodology for social BI. In: 18th International Database Engineering and Applications Symposium, (IDEAS), pp. 207–216. ACM, Porto (2014)

25. Gallinucci, E., Golfarelli, M., Rizzi, S.: Meta-stars: Dynamic, schemaless, and semantically-rich topic hierarchies in social BI. In: 18th International Conference on Extending Database Technology, EDBT 2015, Brussels, pp. 529–532. OpenProceedings.org (2015)

26. Bogorny, V., Vinhas, L., eds.: XI Brazilian Symposium on Geoinformatics, Campos do Jordão, São Paulo, Brazil, 28 November to 01 December 2010. In: Bogorny, V., Vinhas, L., eds.: GeoInfo, MCT/INPE (2010)

Real-Time Snapshot Maintenance
with Incremental ETL Pipelines
in Data Warehouses

Weiping Qu[(✉)], Vinanthi Basavaraj, Sahana Shankar, and Stefan Dessloch

Heterogeneous Information Systems Group,
University of Kaiserslautern, Kaiserslautern, Germany
{qu,k_vinanthi12,s_shankar,dessloch}@informatik.uni-kl.de

Abstract. Multi-version concurrency control method has nowadays been widely used in data warehouses to provide OLAP queries and ETL maintenance flows with concurrent access. A snapshot is taken on existing warehouse tables to answer a certain query independently of concurrent updates. In this work, we extend this snapshot with the deltas which reside at the source side of ETL flows. Before answering a query, relevant tables are first refreshed with the exact source deltas which are captured at the time this query arrives (so-called query-driven policy). Snapshot maintenance is done by an incremental recomputation pipeline which is flushed by a set of consecutive deltas belonging to a sequence of incoming queries. A workload scheduler is thereby used to achieve a serializable schedule of concurrent maintenance tasks and OLAP queries. Performance has been examined by using read-/update-heavy workloads.

1 Introduction

Nowadays companies are emphasizing the importance of data freshness of analytical results. One promising solution is executing both OLTP and OLAP workloads in an 1-tier *one-size-fits-all* database such as Hyper [8], where operational data and historical data reside in the same system. Another appealing approach used in a common 2-tier or 3-tier configuration is near real-time ETL [1] by which data changes from transactions in OLTP systems are extracted, transformed and loaded into the target warehouse in a small time window (five to fifteen minutes) rather than during off-peak hours. Deltas are captured by using Change-Data-Capture (CDC) methods (e.g. log-sniffing or timestamp [9]) and propagated using incremental recomputation techniques in micro-batches.

Data maintenance flows run concurrently with OLAP queries in near real-time ETL, in which an intermediate Data Processing Area (DPA, as counterpart of the data staging area in traditional ETL) is used to alleviate the possible overload of the sources and the warehouse. It is desirable for the DPA to relieve *traffic jams* at a high update-arrival rate and meanwhile at a very high query rate alleviate the burden of locking due to concurrent read/write accesses to shared data partitions. Alternatively, many databases like PostgresSQL [12] use

S. Madria and T. Hara (Eds.): DaWaK 2015, LNCS 9263, pp. 217–228, 2015.
DOI: 10.1007/978-3-319-22729-0_17

a Multi-Version Concurrency Control (MVCC) mechanism to solve concurrency issues. If serializable snapshot isolation is selected, a snapshot is taken at the beginning of a query and used during the entire query lifetime without interventions incurred by concurrent updates.However, for time-critical decision making, the snapshot should contain not only base data but also the latest deltas preceding the submission of the query [7]. The scope of our work is depicted in Fig. 1. Relevant warehouse tables are first updated by ETL flows using deltas with timestamps smaller or equal to the query submission time. A synchronization delay is thereby incurred by a maintenance flow execution before a query starts. We assume that a CDC process runs continuously and always pulls up-to-date changes without maintenance anomalies addressed in [5].Correct and complete sets of delta rows arrive in the DPA and are buffered as delta input streams. The submission of a query triggers a maintenance transaction to refresh relevant tables using a batch of newly buffered deltas (called query-driven refresh policy). A delta stream occurring in a specific time window is split into several batches to maintain snapshots for a number of incoming queries.

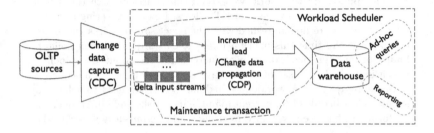

Fig. 1. Consistency scope in warehouse model

In our work, a snapshot is considered as *consistent* if it is refreshed by preceding delta batches before query arrival and is not interfered by fast committed succeeding batches (non-repeatable read/phantom read anomalies). While timestamps used to extend both delta rows and target data partitions could be a possible solution to ensure query consistency, this will result in high storage and processing overheads. A more promising alternative is to introduce a mechanism to schedule the execution sequence of maintenance transactions and OLAP queries. Moreover, sequential execution of ETL flows for maintaining snapshots can lead to high synchronization delay at a high query rate. Therefore, performance needs to be of concern as a main aspect as well.

In this work, we address the real-time snapshot maintenance problem in MVCC-supported data warehouse systems using near real-time ETL techniques. The objective of this work is to achieve high throughput at a high query rate and meanwhile ensure the serializability property among concurrent maintenance transactions on ETL tools and OLAP queries in warehouses. The contributions of this work are as follows: We define the consistency notion in our real-time ETL model based on which a workload scheduler is proposed for a serializable schedule of concurrent maintenance flows and queries that avoids using timestamp-based approach. An internal queue is used to ensure consistency with correct execution sequence. Based on the infrastructure introduced for near real-time ETL [1], we applied a pipelining technique using an open-source ETL tool called Pentaho Data Integration (Kettle) (shortly Kettle) [10]. Our approach is geared towards an incremental ETL pipeline flushed by multiple distinct delta batches for high query throughput. We define the delta batches as inputs for our incremental ETL pipeline with respect to query arrivals. Three levels of temporal consistency (open, closed and complete) are re-used here to coordinate the executions of pipeline stages. The experimental results show that our approach achieves nearly similar performance as in near real-time ETL while the query consistency is still guaranteed.

The remainder of this paper is constructed as follows. We discuss several related work in Sect. 2. Our consistency model is defined in Sect. 3. A workload scheduler is introduced in Sect. 4 to achieve the serializability property and an incremental recomputation pipeline based on Kettle is described in Sect. 5 for high throughput. We validate our approach with read-/update-heavy workloads and the experimental results are discussed in Sect. 6.

2 Related Work

Thomsen et al. addressed on-demand, fast data availability in so-called right-time DWs [4]. Rows are buffered at the ETL producer side and flushed to DW-side, in-memory tables with bulk-load insert speeds using different (e.g. immediate, deferred, periodic) polices. Timestamp-based approach was used to ensure accuracy of read data while in our work we used an internal queue to schedule the workloads for our consistency model. Besides, we also focused on improving throughput by extending an ETL tool.

Near real-time data warehousing was previously referred to as active data warehousing [2]. Generally, incremental ETL flows are executed concurrently with OLAP queries in a small time window. In [1], Vassiliadis et al. detailed a uniform architecture and infrastructure for near real-time ETL. Furthermore, in [3], performance optimization of incremental recomputations was addressed in near real-time data warehousing. In our experiments, we compare general near real-time ETL approach with our work which additionally guarantees the query consistency.

In [6,7], Golab et al. proposed temporal consistency in a real-time stream warehouse. In a certain time window, three levels of query consistency regarding a certain data partition in warehouse are defined, i.e. open, closed and complete, each which becomes gradually stronger. As defined, the status of a data partition is referred to open for a query if data exist or might exist in it. A partition at the level of closed means that the scope of updates to partition has been fixed even though they haven't arrived completely. The strongest level complete contains closed and meanwhile all expected data have arrived. We leverage these definitions of temporal consistency levels in our work.

3   Consistency Model

In this section, we introduce the notion of consistency which our work is building on using an example depicted in Fig. 2. A CDC process is continuously running and sending captured deltas from OLTP sources (e.g. transaction log) to the ETL maintenance flows which propagate updates to warehouse tables on which OLAP queries are executed. In our example, the CDC process has successfully extracted delta rows of three committed transactions T_1, T_2 and T_3 from the transaction log files and buffered them in the data processing area (DPA) of the ETL maintenance flows. The first query Q_1 occurs at warehouse side at time t_2. Before it is executed, its relevant warehouse tables are first updated by maintenance flows using available captured deltas of T_1 and T_2 which are committed before t_2. The execution of the second query Q_2 (at t_3) forces the warehouse table state to be upgraded with additional deltas committed by T_3. Note that, due to serializable snapshot isolation mechanism, the execution of Q_1 always uses the same snapshot that is taken from the warehouse tables refreshed with deltas of T_1 and T_2, and will not be affected by the new state that is demanded by Q_2. The third query Q_3 occurs at $t_{3,5}$ preceding the committing time of T_4. Therefore, no additional deltas needs to be propagated for answering Q_3 and it shares the same snapshot with Q_2.

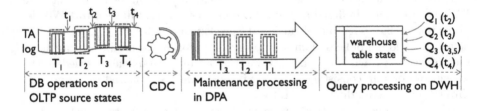

Fig. 2. Consistency model example

In our work, we assume that the CDC is always capable of delivering up-to-date changes to DPA for real-time analytics. However, this assumption normally does not hold in reality and maintenance anomalies might occur in this situation as addressed by Zhuge et al. [5]. In Fig. 2, there is a CDC delay between the recording time of T_4's updates in the transaction log and their occurrence time in the DPA of the ETL flow. The occurrence of the fourth query Q_4 arriving at t_4 requires a new warehouse state updated by the deltas of T_4 which are still not available in the DPA. We provide two realistic options here to compensate for current CDC implementations. The first option is to relax the query consistency of Q_4 and let it share the same snapshot with Q_2 and Q_3. The OLAP queries can tolerate small delays in updates and a "tolerance window" can be set (e.g., 30 s or 2 min) to allow scheduling the query without having to wait for all updates to arrive. This tolerance window could be set arbitrarily. Another option is to force maintenance processing to hang on until CDC has successfully delivered all required changes to DPA with known scope of input deltas for answering Q_4.

With these two options, we continue with introducing our workload scheduler and incremental ETL pipeline based on the scope of our work depicted in Fig. 1.

4   Workload Scheduler

In this section, we focus on our workload scheduler which is used to orchestrate the execution of ETL maintenance flows and OLAP queries. An OLAP query Q_i is first routed to the workload scheduler, which immediately triggers the creation of a new maintenance transaction M_i performing a single run of ETL maintenance flow to update the warehouse state for Q_i. The execution of Q_i stalls until M_i is completed with its commit action (denoted as $c(M_i)$). Query Q_i is later executed in a transaction as well in which the begin action (denoted as $b(Q_i)$) takes a snapshot of the new warehouse state derived by M_i. Therefore, the first integrity constraint enforced by our workload scheduler is $t(c(M_i)) < t(b(Q_i))$ which means that M_i should be committed before Q_i starts.

With arrivals of a sequence of queries $\{Q_i, Q_{i+1}, Q_{i+2}, ...\}$, a sequence of corresponding maintenance transactions $\{M_i, M_{i+1}, M_{i+2}, ...\}$ are constructed. Note that, once the $b(Q_i)$ successfully happens, the query Q_i does not block its successive maintenance transaction M_{i+1} for consistency control since the snapshot taken for Q_i is not interfered by M_{i+1}. Hence, $\{M_i, M_{i+1}, M_{i+2}, ...\}$ are running one after another while each $b(Q_i)$ is aligned with the end of its corresponding $c(M_i)$ into $\{c(M_i), b(Q_i), c(M_{i+1}), ...\}$. However, with the first constraint the serializability property is still not guaranteed since the commit action $c(M_{i+1})$ of a subsequent maintenance transaction might precede the begin action $b(Q_i)$ of its preceding query. For example, after M_i is committed, the following M_{i+1} might be executed and committed so fast that Q_i has not yet issued the begin action. The snapshot now taken for Q_i includes rows updated by deltas occurring later than Q_i's submission time, which incurs non-repeatable/phantom read anomalies. In order to avoid these issues, the second integrity constraint is $t(b(Q_i)) < t(c(M_{i+1}))$. This means that each maintenance transaction is not allowed to commit until its preceding query has successfully begun. Therefore, a serializable schedule can be achieved if the integrity constraint $t(c(M_i)) < t(b(Q_i)) < t(c(M_{i+1}))$ is not violated. The warehouse state is incrementally maintained by a series of consecutive maintenance transactions in response to the consistent snapshots required by incoming queries.

Figure 3 illustrates the implementation of the workload scheduler. An internal queue called ssq is introduced for a serializable schedule of maintenance and query transactions. Each element e in ssq represents the status of corresponding transaction and serves as a waiting point to suspend the execution of its transaction. We also introduced the three levels of query consistency (i.e. open, closed and complete) defined in [6] in our work to identify the status of maintenance transaction. At any time there is always one and only one open element stored at the end of ssq to indicate an open maintenance transaction (which is M_4 in this example). Once a query (e.g. Q_4) arrives at the workload scheduler①, the workload scheduler first changes the status of the last element in ssq from open

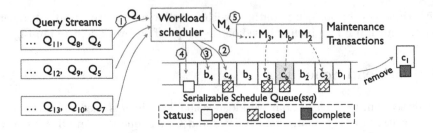

Fig. 3. Scheduling maintenance transactions and OLAP queries

to `closed`. This indicates that the scope of the input delta rows for `open` M_4 has been fixed and the commitment c_4 of M_4 should wait on the completion of this *ssq* element②. Furthermore, a new element b_4 is pushed into *ssq* which suspends the execution of Q_4 before its begin action③. Importantly, another new `open` element is created and put at the end of *ssq* to indicate the status of a subsequent maintenance transaction triggered by the following incoming query (e.g. Q_5)④. M_4 is triggered to started afterwards⑤. When M_4 is done and all the deltas have arrived at warehouse site, it marks its *ssq* element c_4 with `complete` and keeps waiting until c_4 is removed from *ssq*. Our workload scheduler always checks the status of the head element of *ssq*. Once its status is changed from `closed` to `complete`, it removes the head element and notifies corresponding suspended transaction to continue with subsequent actions. In this way, the commitment of M_4 would never precede the beginning of Q_3 which takes a consistent snapshot maintained by its preceding maintenance transactions $\{M_2, M_b, M_3\}$. Besides, Q_4 begins only after M_4 has been committed. Therefore, the constraints are satisfied and a serializable schedule is thereby achieved. It is worth to note that the workload scheduler will automatically trigger several maintenance transactions M_b to start during the idle time of data warehouses to reduce the synchronization delay for future queries. A threshold is set to indicate the maximum size of buffered deltas. Once it is exceeded, M_b is started without being driven by queries.

5   Incremental ETL Pipeline on Kettle

To achieve consistent snapshots for a sequence of queries $\{Q_i, Q_{i+1}, Q_{i+2}, ...\}$, we introduced a workload scheduler in previous section to ensure a correct order of the execution sequence of corresponding maintenance transactions $\{M_i, M_{i+1}, M_{i+2}, ...\}$. However, sequential execution of maintenance flows one at a time definitely incurs significant synchronization overhead for each query. The situation even gets worse at a very high query rate. In this section, we address the performance issue and propose an incremental ETL pipeline based on Kettle [10] for high query throughput.

Take a data maintenance flow from TPC-DS benchmark [11] as an example (see Fig. 4). An existing warehouse table called *store sales* is refreshed by

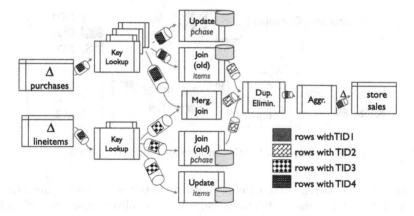

Fig. 4. Incremental ETL pipeline

an incremental ETL flow taking two delta streams *purchases* and *lineitems* as inputs. A set of key lookup steps[1] are applied on the deltas to lookup surrogate keys in warehouses. Subsequent steps (`Join (old) items`, `Merge join`, `Join (old) pchase` and `Dup. Elimin.`) form an incremental join variant to calculate deltas for previous join results. We followed a refined delta rule for natural join: $(\Delta T \bowtie S_{old}) \cup (T_{old} \bowtie \Delta S) \cup (\Delta T \bowtie \Delta S)$ introduced in [3]. By joining Δpurchases with a pre-materialized table, *items*, matching rows can be identified as deltas for *store sales*. The same holds for finding matching rows in another pre-materialized table *pchase* with new Δlineitems. After eliminating duplicates, pure deltas for incremental join can be derived plus the results of joining Δpurchases and Δlineitems. Meanwhile, in order to maintain mutual consistency in two pre-materialized tables and deliver correct results for subsequent incremental flow execution, *pchase* and *items* get updated directly with rows from respective join branches as well. After applying aggregations on the results of incremental join, final deltas for the target *store sales* table are derived.

With multiple queries $\{Q_1, Q_2, ..., Q_n\}$ occurring at the same time, *purchases* and *lineitems* streams are split into n partitions each of which is used as a specific input for the ETL flow to construct a new state s_i of *store sales* for Q_i. We refer to these partitions as *delta batches*. Before we introduce our incremental pipeline, the original implementation of Kettle is given as follows. In Kettle, each step of this incremental ETL flow is running as a single thread. As shown in Fig. 4, steps are connected with each other through in-memory data queues called *rowsets*. Rows are transferred by a provider step thread (using *put* method) through a single rowset to its consumer step thread (using *get* method). Once this incremental ETL flow gets started, all step threads start at the same time and gradually kill themselves once ones have finished with processing their batches.

Instead of killing step threads, all the steps in our incremental pipeline keep running and waiting for new delta batches. In contrast to continuous processing,

---

[1] ETL transformation operations are called *steps* in Kettle.

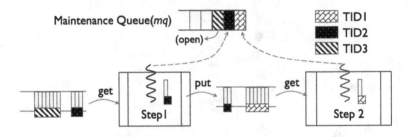

Fig. 5. Step coordination in incremental ETL pipeline

each step in our incremental pipeline must re-initialize its own state after a specific delta batch is finished. This is important to deliver consistent deltas for each query. For example, the result of sorting two delta batches together at a time could be different from sorting each batch separately. The same also holds for aggregation. Therefore, a step thread should be able to identify the scope of the rows in each delta batch, which is addressed by our incremental pipeline solution.

In Fig. 5, a two-stage pipeline is shown as an example. The incremental pipeline uses a local queue called maintenance queue mq to coordinate the execution of pipeline stages. The mq is a subset of the ssq described in Sect. 4 and only stores the status information of maintenance transactions. In addition, each element in mq stores an identifier TID_i belonging to a specific maintenance transaction M_i. Furthermore, each step thread maintains a local read point ($readPt$) and an queue index which points at a certain element in mq. Initially, all the incoming deltas were marked with the TID_1 of an **open** maintenance transaction. All step threads pointed at this **open** element using local indexes and waited for its status change from **open** to **closed**. With a query arrival, this element is marked as **closed** so that all the step threads are started with their the $readPt$s set to TID_1. At this time, the deltas with TID_1 are grouped to the first delta batch and become visible to step 1. After step 1 is finished with the first delta batch, it moves its queue index to the next entry in mq which contains TID_2. This entry is also closed by another new query arrival and step 1's $readPt$ is re-assigned TID_2 from the second entry. Figure 5 represents exactly a state where step 1 continues with processing the second delta batch after re-initialization while step 2 is still working on the first one. Once the second batch is done, the queue index of step 1 is moved onto a new **open** element in mq and the $readPt$ of the step 1 is not assigned TID_3 for further processing even though there have been new deltas coming into its input delta streams.

In this way, we provide the execution of a sequence of maintenance transactions $\{M_1, M_2, ..., M_n\}$ with pipelining parallelism. Figure 4 shows a state where the example incremental pipeline is flushed by multiple delta batches at the same time to delivers consistent deltas for each Q_i. Meanwhile, the execution of each maintenance transaction starts earlier than that in sequential flow execution mode, hence reducing the synchronization overhead.

Step Synchronization in Pipeline: We address another potential inconsistency issue here caused by concurrent reads and writes to the same pre-materialized tables with different delta batches. Take the same example shown in Fig. 4. After applying four lookups on the purchase deltas, intermediate results are used to update the pre-materialized table *pchase* and meanwhile joined with another pre-materialized table *items*. The same table *pchase* and *items* are meanwhile in turn read and updated, respectively by the lineitems deltas in the other join branch. A problem might occur in this case: due to diverse processing speeds, the upper join branch could perform faster than the lower one so that the upper one has begun to update the *pchase* table using the rows belonging to the 4th TID while the same table is still being joined with the rows with the 3rd TID in the lower join branch, thus leading to anomalies and incorrect delta results for the 3rd incoming query. Without an appropriate synchronization control of these two join branches, the shared materialization could be read and written by arbitrary deltas batches at the same time.

We solve this obstacle by first identifying an explicit dependency loop across the steps in two join branches, i.e. `Update pchase` is dependent on `Join (old)` `pchase` and `Update items` depends on `Join (old) items`. With a known dependency loop in a pipeline, all four above mentioned steps are bundled together and share a synchronization aid called *CyclicBarrier*. Any acquisition of a new TID from *mq* by a step in this bundle would not precede the rest of associated steps. This forces all four steps to start to work on the same batch at the same time, thus leading to consistent delta output.

6 Experimental Results

We examine the performance in this section with read-/update-heavy workloads running on three kinds of configuration settings.

Test Setup: We used the TPC-DS benchmark [11] in our experiments. Our test-bed is composed of a target warehouse table *store sales* (SF 10) stored in a Postgresql (version 9.4) [12] instance which was fine-tuned, set to serializable isolation level and ran on a remote machine (2 Quad-Core Intel Xeon Processor E5335, 4×2.00 GHz, 8 GB RAM, 1 TB SATA-II disk), a maintenance flow (as depicted in Fig. 4) running locally (Intel Core i7-4600U Processor, 2×2.10 GHz, 12 GB RAM, 500 GB SATA-II disk) on Kettle (version 4.4.3) together with our workload scheduler and a set of query streams each of which issues queries towards remote *store sales* once at a time. The maintenance flow is continuously fed by deltas streams from a CDC thread running on the same node. The impact of two realistic CDC options (see Sect. 3) was out of scope and not examined.

We first defined three configuration settings as follows. **Near Real-time (NRT):** simulates a general near real-time ETL scenario where only one maintenance job was performed concurrently with query streams in a small time window. In this case, there is no synchronization of maintenance flow and queries. Any query can be immediately executed once it arrives and the consistency is not guaranteed. **PipeKettle:** uses our workload scheduler to schedule the execution sequence of a set of maintenance transactions and their corresponding queries. The consistency is thereby ensured for each query. Furthermore, maintenance transactions are executed using our incremental ETL pipeline. **Sequential execution (SEQ):** is similar to **PipeKettle** while the maintenance transactions are executed sequentially using a flow instance once at a time.

Orthogonal to these three settings, we simulated two kinds of read-/update-heavy workloads in the following. **Read-heavy workload**: uses one update stream (SF 10) consisting of *purchases* (♯: 10 K) and *lineitems* (♯: 120 K) to refresh the target warehouse table using the maintenance flow and meanwhile issues totally 210 queries from 21 streams, each of which has different permutations of 10 distinct queries (generated from 10 TPC-DS ad-hoc query templates, e.g. q[88]). For PipeKettle and SEQ, each delta batch consists of 48 new purchases and 570 new lineitems in average. **Update-heavy workloads**: uses two update streams (♯: 20 K and 240 K) while the number of query streams is reduced to seven (totally 70 queries). Before executing a query in PipeKettle and SEQ, the number of deltas to be processed is 6-times larger than that in read-heavy workloads.

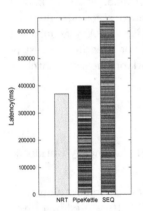
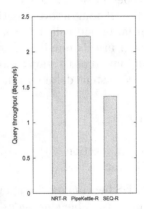
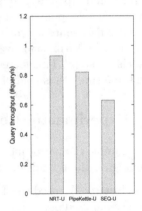

Fig. 6. Performance comparison without queries

Fig. 7. Query throughput in read-heavy workload

Fig. 8. Query throughput in update-heavy workload

Test Results: Fig. 6 illustrates a primary comparison among NRT, PipeKettle and SEQ in terms of flow execution latency without query interventions. As the baseline, it took 370 s for NRT to processing one update stream. The update stream was later split into 210 parts as deltas batches for PipeKettle and SEQ. It can be seen that the overall execution latency of processing 210 delta batches in PipeKettle is 399 s which is nearly close to the baseline due to pipelining parallelism. However, the same number of delta batches is processed longer in SEQ (∼650 s, which is significantly higher than the others).

Figures 7 and 8 show the query throughputs measured in three settings using both read-/update-heavy workloads. Since the delta batch size is small in read-heavy workload, the synchronization delay for answer each query is also small.

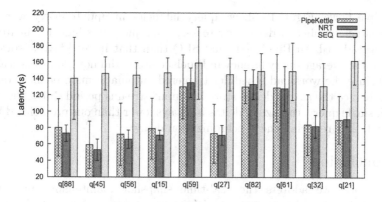

Fig. 9. Average latencies of 10 ad-hoc query types in read-heavy workload

Therefore, the query throughput achieved by PipeKettle (2.22 queries/s) is very close to the one in baseline NRT (2.30) and much higher than the sequential execution mode (1.37). We prove that our incremental pipeline is able to achieve high query throughput at a very high query rate. However, in update-heavy workload, the delta input size becomes larger and the synchronization delay grows increasingly, thus decreasing the query throughput in PipeKettle. Since our PipeKettle automatically triggered maintenance transactions to reduce the number of deltas buffered in the delta streams, the throughput (0.82) is still acceptable as compared to NRT(0.93) and SEQ (0.63).

The execution latencies of 10 distinct queries recorded in read-heavy workload is depicted in Fig. 9. Even with synchronization delay incurred by snapshot maintenance in PipeKettle, the average query latency over 10 distinct queries is approaching the baseline NRT whereas NRT does not ensure the serializability property. SEQ is still not able to cope with read-heavy workload in terms of query latency, since a query execution might be delayed by sequential execution

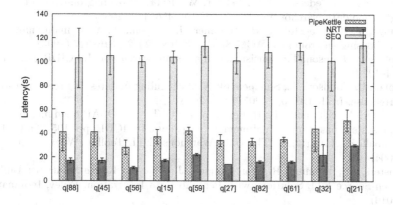

Fig. 10. Average latencies of 10 ad-hoc query types in update-heavy workload

of multiple flows. Figure 10 shows query latencies in update-heavy workload. With a larger number of deltas to process, each query has higher synchronization overhead in both PipeKettle and SEQ than that in read-heavy workload. However, the average query latency in PipeKettle still did not grow drastically as in SEQ since the workload scheduler triggered automatic maintenance transactions to reduce the size of deltas stored in input streams periodically. Therefore, for each single query, the size of deltas is always lower than our pre-defined batch threshold, thus reducing the synchronization delay.

7   Conclusion

In this work, we address the real-time snapshot maintenance in MVCC-supported data warehouse systems. Warehouse tables are refreshed by delta streams in a query-driven manner. Based on the consistency model defined in this paper, our workload scheduler is able to achieve a serializable schedule of concurrent maintenance flows and OLAP queries. We extended current Kettle implementation to an incremental ETL pipeline which can achieve high query throughput. The experimental results show that our approach achieves average performance which is very close to traditional near real-time ETL while the query consistency is still guaranteed.

References

1. Vassiliadis, P., Simitsis, A.: Near real time ETL. In: Kozielski, S., Wrembel, R. (eds.) New Trends in Data Warehousing and Data Analysis, pp. 1–31. Springer, Heidelberg (2009)
2. Karakasidis, A., Vassiliadis, P., Pitoura, E.: ETL queues for active data warehousing. In: Proceedings of the 2nd International Workshop on Information Quality in Information Systems, pp. 28–39. ACM (2005)
3. Behrend, A., Jörg, T.: Optimized incremental ETL jobs for maintaining data warehouses. In: Proceedings of the Fourteenth International Database Engineering and Applications Symposium, pp. 216–224. ACM (2010)
4. Thomsen, C., Pedersen, T.B., Lehner, W.: RiTE: Providing on-demand data for right-time data warehousing. In: ICDE, pp. 456–465 (2008)
5. Zhuge, Y., Garcia-Molina, H., Hammer, J., Widom, J.: View maintenance in a warehousing environment. ACM SIGMOD Rec. 24(2), 316–327 (1995)
6. Golab, L., Johnson, T.: Consistency in a stream warehouse. In: CIDR, vol. 11, pp. 114–122 (2011)
7. Golab, L., Johnson, T., Shkapenyuk, V.: Scheduling updates in a real-time stream warehouse. In: ICDE, pp. 1207–1210 (2009)
8. Kemper, A., Neumann, T.: HyPer: a hybrid OLTP and OLAP main memory database system based on virtual memory snapshots. In: ICDE, pp. 195–206 (2011)
9. Kimball, R., Caserta, J.: The Data Warehouse ETL Toolkit. Wiley, Indianapolis (2004)
10. Casters, M., Bouman, R., Van Dongen, J.: Pentaho Kettle Solutions: Building Open Source ETL Solutions with Pentaho Data Integration. Wiley, Indianapolis (2010)
11. http://www.tpc.org/tpcds/
12. http://www.postgresql.org/

Eco-Processing of OLAP Complex Queries

Amine Roukh[1]([✉]) and Ladjel Bellatreche[2]

[1] University of Mostaganem, Mostaganem, Algeria
`roukh.amine@univ-mosta.dz`
[2] LIAS/ISAE-ENSMA, Poitiers University, Futuroscope, Poitiers, France
`bellatreche@ensma.fr`

Abstract. With the Era of *Big Data* and the spectacular development
of High-Performance Computing, organizations and countries spend considerable efforts and money to control/reduce the energy consumption.
In data-centric applications, DBMS are *one of the major energy consumers* when executing complex queries. As a consequence, integrating
the energy aspects in the advanced database design becomes an economic
necessity. To predict this energy, the development of mathematical cost
models is one of the avenues worth exploring. In this paper, we propose a
cost model for estimating the energy required to execute a workload. This
estimation is obtained by the means of statistical regression techniques
that consider three types of parameters related to the query execution
strategies, the used deployment platform and the characteristics of the
data warehouses. To evaluate the quality of our cost model, we conduct
two types of experiments: one using our mathematical cost model and
another using a real DBMS with dataset of TPC-H and TPC-DS benchmarks. The obtained results show the quality of our cost model.

1 Introduction

Electrical power is a valuable resource and its consumption is important in
computing platforms such as devices and data centers that are highly energy-
intensive. This is due to the high power requirements of IT equipment and cool-
ing infrastructure needed to support these equipment. The report issued by the
US Environmental Protection Agency in 2007 showed that data centers servers
consumed 1.5 % of the nation's electricity, up to 61 billion kilowatt-hours, worth
4.5 billion Dollars in 2006. This consumption would nearly double from 2005 to
2010 to roughly 100 billion kWh of energy at an annual cost of $7.4 billion [19].

Note that the database is a major component of the data centers, since it
consumes the largest part of the power [14]. Traditionally, designing advanced
database applications (statistical, scientific, etc.) mainly focuses on speeding up
the query processing cost to satisfy the needs of end users and decision makers.
Usually, this design ignores the energy dimension. Therefore, several research
efforts have been triggered to put the energy in the heart of DBMS development
[18]. These efforts consider aspects related to hardware and software. Concerning
hardware, new devices with low energy-consuming have been developed such as
SSD, *GPU*, etc. and technologies for energy-proportional that offers to any device

S. Madria and T. Hara (Eds.): DaWaK 2015, LNCS 9263, pp. 229–242, 2015.
DOI: 10.1007/978-3-319-22729-0_18

the ability to reduce its power consumption to the actual workload (e.g., *Energy-proportional hardware* (DVFS)) [17]. From the software perspective, several new energy-driven algorithms for implementing relational operations have been proposed. Some DBMS editors propose solutions to integrate energy. Following the same directions, several well-established councils offer researchers benchmarks integrating energy (ex. TPC-Energy benchmark [1,14]).

To predict the energy consumption, the development of *accurate cost models* is necessary [6,22]. Note that the energy estimation in the context of the DBMS is a complex task, since it has to consider all hardware components of the servers hosting the target DBMS. Two tendencies exist to perform this estimation: (i) the *modular estimation* for each component of the server(s), and the (ii) *global estimation* that considers the server(s) as a whole. The first tendency requires a precise knowledge of all components that are usually heterogeneous. In our work, we adopt the second tendency to avoid this heterogeneity. Note in the database context, there exists a long experience in developing cost models. The recent query optimizers use cost-based approaches to identify the optimal query plans. These cost models use parameters issued from several database components: (a) *the database/data warehouse* (e.g., the size of tables, the length of tuples, etc.), (b) *the workload* (e.g., type of queries, the selectivity factors of joins and selections predicates, sizes of intermediate results, etc.), (c) *the hardware* (e.g., size of buffer, page size of the disk, etc.) and (d) *the deployment platform* (e.g., RAM, disk size, number of nodes, etc.). Three main cost metrics exist: (a) *Inputs-outputs* (IO) metric that mainly estimates the number of disk pages that have to be loaded in the main memory when executing a workload, (b) *CPU metric* dedicated to the amount of CPU work needed to execute a workload, and (c) *communication metrics* between DBMS devises. Contrary to query processing techniques that usually consider IO metrics [12], energy processing techniques have to integrate the three cost dimensions: IO, CPU, and communication. One of the main difficulties to define energy cost models is the identification of parameters related to the used materials, the workload, and the execution strategies.

Our contributions in this paper are: (i) we experimentally identify the key factors (query pipeline, size of databases, complexity of the queries, behaviour of CPU and IO when executing *simple and complex queries*, etc.) that may impact consumption power estimation, (ii) we develop a *multivariate regression model* that predicts the power consumption for a given query in terms of its I/O and CPU costs and (iii) we show via our experimental results obtained by the use of traditional and advanced datasets and workloads (TPC-H and TPC-DS) that confirm our claims and show the effectiveness of our predictive models (our predictions are on the average within 4 % of the actual values and 9 % as maximum).

The remainder of this paper is organized as follows. Section 2 reviews the most important works on power consumption in the database context. In Sect. 3, our methodology in developing energy-cost model is given and describes all required components. Section 4 highlights our theoretical and real experiments that we conduct to evaluate the quality of our cost model. Section 5 concludes the paper by summarizing the important results and point out some open issues.

2   Related Work

By exploring the most important studies of the state-of-art, we identify the large panoply of research studies on power and energy-efficient resource management in computing systems. As we said before, these works can be divided into two approaches: *Hardware-Driven Approaches* ($\mathcal{HDA}$) and *Software-Driven Approaches* ($\mathcal{SDA}$). Due to the lack of space, we mainly focus on $\mathcal{SDA}$.

Solutions in $\mathcal{HDA}$ goes from naive to advanced. In the naive category, we can cite the technique that allows disabling electronic component during periods of inactivity. In advanced solutions, we can cite for instance the use of the *Dynamic Voltage and Frequency Scaling* technique to dynamically change the performance of component for better energy efficiency [3]. From database perspective, commercial DBMS establish serious collaborations with hardware companies to satisfy the key requirement related to green computing goals. *Oracle Exadata Database Machine* [7] is a fruitful collaboration between Oracle and Intel.

$\mathcal{SDA}$ to incorporate energy in databases has been developed in 90's. To increase the battery life of mobile computers, the authors in [2] proposed to modify the traditional objective function of query optimizer by considering energy dimension when selecting query plans. The work in [11] proposed an *Improved Query Energy-Efficiency* by introducing explicit delays mechanism, which uses query aggregation to leverage common components of queries of a workload. In [20], the authors dealt with energy-based query optimization. Consequently, they propose a power cost model connected to PostgreSQL's cost model. A static power profile for each basic database operation is defined. These profiles are incorporated into the DBMS. The power cost (like CPU power cost to access tuple, power cost for reading/writing one page, and so on) of a plan can be then calculated from the basic *SQL* operations via different access methods offered by the DBMS. The main drawbacks of this work are: **(i)** its ignorance of query pipelines and **(ii)** the use of a simple linear regression technique to identify parameters related to energy. However, the relationship between energy and system parameters is not linear. In the same direction, the work of [10] shows that processing a query as fast as possible does not always turn out to be the most energy-efficient way to operate a DBMS. Based on this observation, the authors proposed a framework for an energy-aware database query processing. It augments query plans produced by traditional query optimizer with an energy consumption prediction for some specific database operators like select, project and join. The proposed model for quantifying power cost of a query is quite simple and does not consider complex queries. The authors of [9] propose deep research in modelling the peak power of database operations, where the peak power is the maximum power consumed during query execution. This work proposes to consider the query pipelines. The main limitation of this work is the necessity of training the model for new unseen pipelines. More recently, [22] extends their prior work [20] to improve model accuracy under system dynamics and the variations of workload characteristics, by developing an on-line model estimation scheme that dynamically corrects the static model parameters based on feedback control mechanism.

3   Our Methodology

Before detailing our methodology, some concepts and definitions are needed.

The energy is defined as the ability to do work. It is usually measured in joules. The power represents the rate of doing work, or energy per unit of time. It is measured in watts. Formally, energy and power can be defined as: $P = \frac{W}{T}$; $E = P \times T$, where P, T, W and E represent respectively, a power, a period of time, the total work performed in that period of time, and the energy. The power consumption of a given system can be divided into two parts:

1. *Baseline Power*: The power consumption when the machine is idle. This includes the power consumption of the fans, CPU, memory, I/O and other motherboard components in their idle state.
2. *Active Power*: The power consumption due to the execution of the workload. The active power component is determined by the kind of workload that executes on the machine, and the way it utilizes CPU, memory and I/O components.

Two concepts of power that have to be considered during the evaluation of power utilization in DBMS exist representing the average power consumed during the query execution and peak power representing the maximum power. In this paper, we consider the *average power*.

The architecture of our power cost model is illustrated in Fig. 1. The components of our architecture are: (i) query execution strategy, (ii) CPU and IO parameters identification and (iii) final query energy costs.

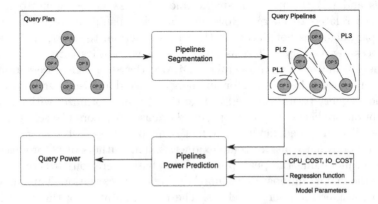

Fig. 1. Overview over power cost model

3.1   What Are the Key Parameters?

Due to the complexity of identifying the key parameters that have to be considered to estimate energy when executing queries, we adopt an experimental strategy in which we test the execution of several simple and complex queries in different size datasets and then we observe parameters related to query execution strategies, databases and hardware (IO and CPU).

Pipeline Parameters. Note that the most actual DBMS uses the pipeline strategy to execute queries. As a consequence, its incorporation in our cost model is mandatory. Before studying its effects, let us introduce some definitions.

Definition 1. *A pipeline consists of a set of concurrently running operators [12].*

Definition 2. *An operator is blocking if it cannot produce any output tuple without reading at least one of its inputs (e.g., sort operator).*

In order to study its impact on energy processing, let us consider the execution of the query **Q9** of the TPC-H benchmark[1]. The datasets and the system setup we used in our experiments can be found in Sect. 4. The active power consumption during the execution of this query is given in Fig. 2. It starts at 10 s and finishes at 143 s. Three different segments (pipelines) are identified (Fig. 2b). There are 7 pipelines, and a partial order of the execution of these pipelines is enforced by their terminal *blocking* operators (e.g., PL6 cannot begin until PL5 is complete). Although the query has 7 different pipelines, only 4 pipelines are important in our discussion (the others end faster). The determination of power-to-pipelines is performed thank to the DBMS Real-Time SQL Monitoring module. It offers real-time statistics at each step of the execution plan with the elapsed time. Based on this, we segment the power log to their respective pipelines.

If we compare the power changing points with the pipeline switching points (in Fig. 2), we observe the following: *when a query switches from one pipeline to another, its power consumption also changes.* During the execution of a pipeline, the power consumption usually tends to be approximately constant. This observation contradicts the assumption made by the Xu et al. [21] saying that in the period with only one query running in the system, the power cost is stable (steady state). This is because their estimation is done at query level. To have more accurate energy cost models, the estimation has to be performed at the fine-grained level that includes the pipelines.

Impact of the Database Size on IO and CPU. Usually, we admit the following fact: *"bigger size, more energy".* To check this fact, we conduct experiments by considering queries of the TPC-DS benchmark[2] with two data sets, one with 10 GB database scale factor and the second with 100 GB in terms of active power consumption. Our experiments clearly show that queries running on small data sizes *sometimes* are more energy-consuming than those running

---

[1] http://www.tpc.org/tpch/.

[2] http://www.tpc.org/tpcds/.

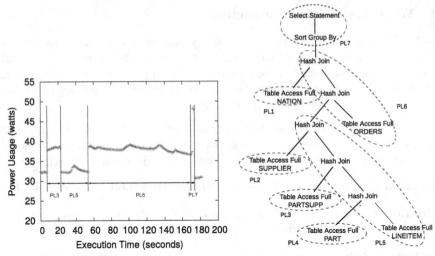

(a) Pipelines segmented power consumption (b) Pipelines segmented execution plan

Fig. 2. Power consumption and execution plan of TPC-H benchmark query $Q9$ with corresponding pipeline annotation.

on big size databases. The same observation has been identified by considering datasets and workload of TPC-H benchmark.

Our explanation for this observation is: for some queries, reading large files needs more I/O tasks. When intermediate results cannot fit into memory, it will be written to disk and read later. Such overhead is translated in more CPU waits. In other terms, the query spends its time reading/writing data more than processing its tuples. Thus, queries dominated by I/O cost lead to less power consumption. On the other side, when files are small, reading data finishes quickly and in its remaining time, the query is dominated by CPU processing. This can be translated in high power consumption.

This observation negates the premise made by Xu et al. [22] that the marginal energy consumption (active energy) of a query is positively related to its size. In fact, we found many queries that have different power consumption regardless of the data sizes.

Inspired by this observation, we design and implement a cost-based power model. Basically, our methodology first decomposes the query execution plan into a set of power independent pipelines delimited by blocking/semi-blocking operators and then estimates power consumption related to CPU and I/O for each pipeline.

As in previous work [4], the pipelines are created in an inductive manner, starting from the leaf operators of the plan. Whenever we encounter a blocking operator, the current pipeline ends, and a new pipeline starts. As a result, the original execution plan can be viewed as a *tree of pipelines* (cf. Fig. 2b). Based on our analysis, we found that the DBMS executes pipelines in a sequential order.

3.2   CPU and IO Costs for Pipelines

In a traditional DBMS, query execution cost is treated as a linear combination of three components: CPU cost, I/O cost and communication cost [22]. We follow the same logic to propose an energy cost models for pipelines integrating CPU and IO.

Our strategy for pipeline modelling is to leverage the cost models that are built into the database systems for query optimization. To process tuples, each operator in a pipeline needs to perform CPU and/or I/O tasks. The cost of these tasks represents the "cost of the pipeline", which is the active power to be consumed in order to finish the tasks. In this paper, we focus on a single server setup. Thus, the communication cost can be ignored.

More formally, for a given workload W of n complex queries $\{Q_1, Q_2, \ldots, Q_n\}$. The power cost model for overall queries is defined as follows:

$$Power(W) = \sum_{i=1}^{n} Power(Q_i) \tag{1}$$

As we said before, the $Power(Q_i)$ strongly depends on the number of pipelines. To illustrate this, let p_i be the number of pipelines of the query Q_i: $\{PL_1, PL_2, \ldots, PL_p\}$. As a consequence, the $Power(Q_i)$ is given by the following equation:

$$Power(Q_i) = \frac{\sum_{j=1}^{p} Power(PL_j) * Time(PL_j)}{Time(Q_i)} \tag{2}$$

$Time(PL_j)$ and $Time(Q_i)$ represent respectively, the execution time of the pipeline j and the query Q_i. These two factors are completely ignored by Xu et al. [22]. They allow identifying CPU or IO dominated pipelines for a given query. In our estimation, we pick these factors from the DBMS statistics module. However, to test the quality of our model, we will use the real execution time.

Note that the cost of a pipeline j has also to be decomposed since it may involve several algebraic operations. Let n_j be the number of these operations $(\{OP_1, OP_2, \ldots, OP_n\})$. The cost of a pipeline j $(Power(PL_j))$ is given by the following equation:

$$Power(PL_j) = \beta_{cpu} \times \sum_{k=1}^{n_j} CPU\_COST_k + \beta_{io} \times \sum_{k=1}^{n_j} IO\_COST_k \tag{3}$$

where β_{cpu} and β_{io} are the model parameters (i.e., unit power costs) for the pipelines.

These costs are then converted to time by using system-specific parameters such as CPU speed and IO transfer speed. In our model, we take IO and CPU estimations before converting it to time. Note that the system parameters are calculated at the creation of the database. Usually, they may be unreliable. As a consequence, they have to be calibrated to ensure the accuracy of the computation, as recommended by the DBMS vendor.

The *IO_COST* is the predicted number of *I/O* required for executing a specific operator. The *CPU_COST* is the predicted number of *CPU Cycle* and *buffer cache get* that DBMS needs to run a specific operator. Our cost model does not rely on the type of SQL operators and their implementations. As a consequence, it is well adapted to complex queries like those of the *TPC-DS benchmark*.

Now, it remains to estimate the parameters of β_{cpu} and β_{io} of the Eq. 3. This estimation will be described in the next section.

3.3 Estimation of IO and CPU Parameters

To do this estimation, we propose the use of the regression technique. More precisely, the power cost of a pipeline PL_j may be computed as follows:

$$Power(PL_j) = \beta_0 + \beta_1 \times IO\_COST + \beta_2 \times CPU\_COST + \epsilon \qquad (4)$$

ϵ represents the noise of measurement. The β parameters are regression coefficients that will be estimated while learning the model from training data. Thus, the linear regression models are solved by estimating the model parameters β. This is typically done by finding the least squares solution [13].

Firstly, we naively borrow the linear regression with the formula in 4, as done in [9,10,22]. Unfortunately, this model does not work well, especially when the data size changes. This is due to the no linearity of the relationships between the data size and the power (see Sect. 3.1) (*"bigger size, more energy"* slogan). Therefore, we decide to use multiple polynomial regression model with degree m which is well suitable when a *non-linear* relationship between the independents variables and the corresponding dependent variables occurs. We perform experiments to fix the degree of our multiple polynomial regression model. The degree equals 2 ($m = 2$) gives the best results. Our polynomial is:

$$Power(PL_j) = \beta_0 + \beta_1(IO\_COST) + \beta_2(CPU\_COST) +$$
$$\beta_3(IO\_COST^2) + \beta_4(CPU\_COST^2) + \beta_5(IO\_COST \times CPU\_COST) + \epsilon \qquad (5)$$

Training. As mentioned above, the β parameters are estimated while learning the model from training data. We then perform series of observations in which queries are well-chosen, and the power values consumed by the system are collected using measuring equipment while running these queries. In the same time, for each training instance, we segment the query operators in a set of pipelines and calculate their costs.

To generate training instances, we create our custom query workload based on TPC-H datasets (at scale factor 10 and 100). The workload containing 110 SQL queries divided into two main categories: (i) queries with operations that exhaust the system processor (CPU intensive queries) and (ii) queries with exhaustive storage subsystem resource operations (I/O intensive queries). Note that the considered queries include: queries with single table scan, queries with multiple joins with different predicates. They also contain sorting/grouping conditions and simple and advanced aggregation functions as in [9].

We note that for some queries, especially those containing aggregation and analytical functions, our DBMS does not compute their CPU cost. Note that these two types of functions are power intensive. To overcome this constraint, we propose to *manually* calculate the CPU Cycles required to execute aggregation functions, using our machine by the means of basic instructions such as ADD, SUB, MUL, DIV, and COMPARE. The obtained estimations are then multiplied by the number of tuples [8].

After collecting power consumption training queries, we apply the regression equation (5) to find our model parameters. Once we get them, an estimation of new queries is obtained without the use of our measurement equipment.

4   Experimental Evaluation

To evaluate the effectiveness of our proposal, we conduct several experiments. Before describing them in details, we first present our experimental machine to compute the energy and the used datasets.

The used Machine, Software, and Hardware. We use a similar set-up used in the state-of-arts [9,16,20] (Fig. 3). Our machine is equipped with a "Watts UP? Pro ES" power meter with one second as a maximum resolution. The device is directly placed between the power supply and the database workstation under test to measure the workstation's overall power consumption (Fig. 3). The power values are logged and processed in a separate monitor machine.

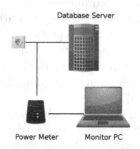

Fig. 3. The experiment setup

We used a Dell PowerEdge R210 II workstation having an Intel Xeon E3-1230 V2 3.30 GHz processor, 10 GB of DDR3 memory and a 2 × 500 GB hard drive. Note that techniques like dynamic voltage-frequency were not applied in our experiments. Our workstation machine is installed with the latest version of Oracle 11gR2 DBMS under Ubuntu Server 14.04 LTS with kernel 3.13, to minimize spurious influences. We also disable unnecessary background tasks, clear the system and Oracle cache for each query execution.

Datasets. We use TPC-H and TPC-DS benchmarks datasets and queries. The queries are executed in an isolated way.

The Obtained Results. Recall that our analytical model is composed of two parts: the power parameters that can be learned through a "training" procedure, and pipeline parameters (e.g., IO and CPU costs) that can be read from available database statistics.

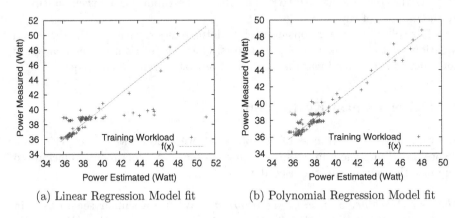

(a) Linear Regression Model fit   (b) Polynomial Regression Model fit

Fig. 4. Training workload power consumption and regressions fit.

For precision reasons, each experiment is performed several times.

In the training phase, at each query execution, we collect statistics of pipelines and power usage. These collected data are used by the *R language software*[3] to find the best values of our model parameters.

The results of the training phase, against the fitted values from linear and polynomial models (cf. Eqs. 4 and 5) are plotted in Fig. 4. As we can see from Fig. 4a, there are significant differences between the estimated power and the real power consumption for several queries. On the other hand, the predicted and actual power consumption approximate the diagonal lines closely when using polynomial regression (Fig. 4b).

In the testing phase, given the estimated power cost offered by the pipeline model (E), we compare it with the real system active power consumption (M). To quantify the model accuracy, we used the following error ratio metric: $Error = \frac{|M-E|}{M}$.

To evaluate our proposal in large datasets, we run all 22 queries of the TPC-H benchmark with two scale factors: 10 GB and 100 GB. Note that most of the queries contain more than 4 pipelines. The results are shown in Table 1.

As we can easily figure out that the average error is typically small (1.6 % in 10 GB datasets and 2.1 % in 100 GB), and the maximum error is usually below than 5 %. The larger errors in the estimation made by our estimation model (the case of the queries *Q2* and *Q13*) are due to the cardinality estimation

---

[3] http://www.r-project.org/.

Table 1. Estimation errors in TPC-H benchmark queries with different database sizes.

Query	10 GB error	100 GB error	Query	10 GB error	100 GB error
Q1	1.2	0.9	Q12	1.9	0.05
Q2	10.1	8.9	Q13	12.8	6.1
Q3	1.0	0.1	Q14	0.4	0.6
Q4	0.5	0.3	Q15	2.3	1.0
Q5	1.2	0.6	Q16	0.4	1.7
Q6	1.3	0.1	Q17	0.2	1.4
Q7	1.1	0.7	Q18	1.9	3.7
Q8	0.5	0.1	Q19	0.6	1.0
Q9	1.9	0.8	Q20	1.8	2.1
Q10	0.6	1.2	Q21	0.9	0.4
Q11	4.8	0.3	Q22	0.007	2.1

errors given by the query optimizer. By focusing on the query Q2 in the case of 100 GB dataset, we identify that its third pipeline is the longest one. The number of its estimated rows is equal to 180 692 while the real number is 1 362 975. Thus, the values of I/O and CPU cost are significantly wrong. Since our study treats the DBMS as "black box", therefore, we cannot check the model based on true cardinalities which are obtained after the execution of the query.

To verify the portability of our model, we experimentally test it against TPC-DS benchmark, which is more complex than TPC-H benchmark. This complexity is due to its schema diversity, dataset distribution and decision support nature of its query workload. We use a dataset with 100 GB. Motivated by [15], we selected 16 queries out of 99 queries template, wherein the queries are characterized by resource utilization pattern, such as CPU intensive queries or I/O intensive queries. The average length of pipelines is 7. Figure 5 shows the obtained results.

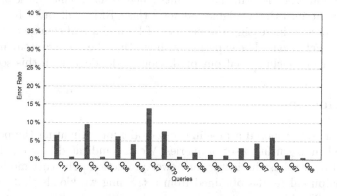

Fig. 5. Power estimation errors in TPC-DS benchmark queries.

We can observe that the prediction error is less than 10 %, except the query $Q47$. By analysing it, we found that it uses analytic function `rank()`. DBMS is not able to give its cost. We argue that this is the source of this error. To verify this, we remove the portion of the code containing the `rank()` function. This updated query is denoted by $Q47p$. As we can see in Fig. 5, the estimation error comes down to 7.5 %.

This situation (computation of aggregations and analytical functions costs) has to be considered by DBMS vendors to facilitate the development of energy-aware databases, as previously highlighted in [5,6].

Table 2. Estimation errors in TPC-H benchmark queries with 2 GB RAM.

Query	10 GB error	Query	10 GB error
$Q1$	1.4	$Q12$	0.8
$Q2$	11.9	$Q13$	4.2
$Q3$	1.4	$Q14$	1.7
$Q4$	0.7	$Q15$	1.9
$Q5$	2.4	$Q16$	2.4
$Q6$	1.5	$Q17$	0.8
$Q7$	2.7	$Q18$	6.5
$Q8$	1.5	$Q19$	0.0
$Q9$	2.2	$Q20$	0.3
$Q10$	2.7	$Q21$	2.3
$Q11$	5.6	$Q22$	0.07

Finally, we conduct experiments to evaluate the impact of the size of the main memory on our prediction models. We then set up the DBMS memory to 2 GB and we run TPC-H workload. The obtained results are shown in Table 2. We observe that when intermediate results (such as hash build phase or external sort) cannot fit into the available memory, the DBMS will save them to disk. This contributes in increasing the number of IO and a real change of the power consumption behaviour. The experiment shows that the average estimation error is 2.5 %. We can conclude that our prediction model deals with this scenario.

5   Conclusion

In this paper, we proposed a pipeline cost model for estimating the power consumption when executing a set of queries. We also indicated how pipeline modelling could be a robust indicator for power prediction. The cost model is built based on empirical results obtained from a training workloads that were carefully created. The hardware parameters are obtained by multivariate polynomial regression algorithms. Furthermore, we performed tests on our framework with real databases, by running a set of queries, and comparing their power costs with those predicted by our prediction model. Our results show that the model can predict power with a low error.

Currently, we are working on two directions: (i) studying the effect of multi-query optimization on energy reduction and (ii) incorporating energy in the query optimization.

References

1. http://www.tpc.org/tpc_energy/
2. Alonso, R., Ganguly, S.: Energy efficient query optimization. Matsushita Info Tech Lab, Citeseer (1992)
3. Beloglazov, A., Buyya, R., Lee, Y.C., Zomaya, A., et al.: A taxonomy and survey of energy-efficient data centers and cloud computing systems. Adv. Comput. **82**(2), 47–111 (2011)
4. Chaudhuri, S., Narasayya, V., Ramamurthy, R.: Estimating progress of execution for sql queries. In: SIGMOD, pp. 803–814. ACM (2004)
5. Graefe, G.: Database servers tailored to improve energy efficiency. In: Proceedings of the 2008 EDBT Workshop on Software Engineering for Tailor-Made Data Management, pp. 24–28. ACM (2008)
6. Harizopoulos, S., Shah, M., Meza, J., Ranganathan, P.: Energy efficiency: the new holy grail of data management systems research. arXiv preprint (2009). arXiv:0909.1784
7. Intel and Oracle. Oracle exadata on intel® xeon® processors: Extreme performance for enterprise computing. White paper (2011)
8. Intel Corporation. Intel® 64 and IA-32 Architectures Optimization Reference Manual, September 2014
9. Kunjir, M., Birwa, P.K., Haritsa, J.R.: Peak power plays in database engines. In: EDBT, pp. 444–455. ACM (2012)
10. Lang, W., Kandhan, R., Patel, J.M.: Rethinking query processing for energy efficiency: slowing down to win the race. IEEE Data Eng. Bull. **34**(1), 12–23 (2011)
11. Lang, W., Patel, J.: Towards eco-friendly database management systems. arXiv preprint (2009). arXiv:0909.1767
12. Li, J., Nehme, R., Naughton, J.: Gslpi: a cost-based query progress indicator. In: ICDE, pp. 678–689. IEEE (2012)
13. McCullough, J.C., Agarwal, Y., Chandrashekar, J., Kuppuswamy, S., Snoeren, A.C., Gupta, R.K.: Evaluating the effectiveness of model-based power characterization. In: USENIX Annual Technical Conference (2011)
14. Poess, M., Nambiar, R.O.: Energy cost, the key challenge of today's data centers: a power consumption analysis of tpc-c results. PVLDB **1**(2), 1229–1240 (2008)
15. Poess, M., Nambiar, R.O., Walrath, D.: Why you should run tpc-ds: a workload analysis. In: VLDB, pp. 1138–1149. VLDB Endowment (2007)
16. Rodriguez-Martinez, M., Valdivia, H., Seguel, J., Greer, M.: Estimating power/energy consumption in database servers. Procedia Comput. Sci. **6**, 112–117 (2011)
17. Schall, D., Härder, T.: Towards an energy-proportional storage system using a cluster of wimpy nodes. In: BTW, pp. 311–325 (2013)
18. Tu, Y.-C., Wang, X., Zeng, B., Xu, Z.: A system for energy-efficient data management. ACM SIGMOD Rec. **43**(1), 21–26 (2014)

19. Wang, J., Feng, L., Xue, W., Song, Z.: A survey on energy-efficient data management. ACM SIGMOD Rec. **40**(2), 17–23 (2011)
20. Xu, Z., Tu, Y.-C., Wang, X.: Exploring power-performance tradeoffs in database systems. In: ICDE, pp. 485–496 (2010)
21. Xu, Z., Tu, Y.-C., Wang, X.: Power modeling in database management systems. Technical report CSE/12-094, University of South Florida (2012)
22. Xu, Z., Tu, Y.-C., Wang, X.: Dynamic energy estimation of query plans in database systems. In: ICDCS, pp. 83–92. IEEE (2013)

Materializing Baseline Views for Deviation Detection Exploratory OLAP

Pedro Furtado[1], Sergi Nadal[3], Veronika Peralta[2], Mahfoud Djedaini[2], Nicolas Labroche[2], and Patrick Marcel[2(✉)]

[1] University of Coimbra, Coimbra, Portugal
pnf@dei.uc.pt
[2] University of Tours, Tours, France
{veronika.peralta,mahfoud.djedaini,nicolas.labroche,
patrick.marcel}@univ-tours.fr
[3] Universitat Politècnica de Catalunya, Barcelona, Spain
snadal@essi.upc.edu

Abstract. Alert-raising and deviation detection in OLAP and exploratory search concerns calling the user's attention to variations and non-uniform data distributions, or directing the user to the most interesting exploration of the data. In this paper, we are interested in the ability of a data warehouse to monitor continuously new data, and to update accordingly a particular type of materialized views recording statistics, called baselines. It should be possible to detect deviations at various levels of aggregation, and baselines should be fully integrated into the database. We propose Multi-level Baseline Materialized Views (BMV), including the mechanisms to build, refresh and detect deviations. We also propose an incremental approach and formula for refreshing baselines efficiently. An experimental setup proves the concept and shows its efficiency.

Keywords: OLAP · Deviation detection · Materialized views

1 Introduction

Data warehouses are used to support analysis over large amounts of data and decision making based on the results of those analysis. Online analytic processing (OLAP) tools enable analysis of multidimensional data interactively using multiple perspectives, roll-up, drill-down, slicing and dicing. That multitude of analysis perspectives contributes to information overload. While a huge amount of information is available and browsable in many different ways, it is commonly agreed that mechanisms to ease the job of the analyst are of key importance.

Deviation detection is a useful mechanism to identify variations in data distributions that can be relevant to explore, and alert users to look at those uncommon patterns. In order to do this, it is necessary to build, refresh and monitor baselines. Baselines are statistical artifacts that maintain information about the

This work was done while Pedro Furtado was visiting University of Tours.

S. Madria and T. Hara (Eds.): DaWaK 2015, LNCS 9263, pp. 243–254, 2015.
DOI: 10.1007/978-3-319-22729-0_19

normal distribution of some data. When new values are outside that normality users are alerted, or the variations are used to guide navigation. In this work we restrict deviation detection to identifying if values are outside a usual band, on future work we intend to explore more complex models.

Baselines should be integrated into databases as "first-class citizens". They have many similarities to materialized views in terms of their life-cycle. In this paper we propose Multi-level Baseline Materialized Views (BMV). The approach assumes baselines can be created and monitored at different levels of aggregation, such as daily and monthly totals. BMVs are integrated into the data warehouse, managed and refreshed as a kind of materialized view, at multiple levels of aggregation. The proposed architecture should be efficient, therefore we design the physical mechanisms for baseline monitoring and refreshing taking efficiency into account. The resulting architecture is evaluated using a proof of concept and the dataset of the Star Schema Benchmark (SSB).

The rest of this paper is organized as follows. Section 2 discusses related work. Section 3 motivates Multi-level Baseline Materialized Views with an example and possible uses. Section 4 defines the concept formally. Section 5 proposes the mechanisms for managing BMVs (creation, refreshing and detection). Section 6 presents experimental results to validate our proposal and Sect. 7 concludes.

2 Related Work

For a long time now, researchers have been interested in finding out interesting patterns in datasets [4,6,11,12], and interestingness measures have been studied as well [5]. Some previous works have dealt with the more specific issue of deviation detection [2,4,11,12].

In [2], the authors adapt attribute focusing algorithms to multidimensional hierarchical data. Attribute focusing discovers interesting attribute values, in the sense that the observed frequency of those values deviate from their expected frequency. In order to make attribute focusing useful for OLAP, they have adapted the method to hierarchical dimensions often found in data cubes.

In [11], the authors introduce an operator that summarizes reasons for drops or increases observed at an aggregated level. They developed an information theoretic formulation and designed a dynamic programming algorithm to find the reasons. This work is then also evolved in [12], where the authors propose automation through the *iDiff* operator that in a single step returns summarized reasons for drops or increases observed at an aggregated level. Comparing to our proposal, this approach does not materialize baselines, therefore it is very computationally intensive during navigation, since it requires searching for deviations on the fly. In [4], a more general framework is proposed for measuring differences in data characteristics based in quantifying the deviation between two datasets in terms of the models they induce. The framework covers a wide variety of models, including frequent itemsets, decision tree classifiers, and clusters. The authors show how statistical techniques can be applied to the deviation measure to assess whether the difference between two models is significant. In [3] a set of approaches was designed for summarizing multidimensional cubes, including quantization, multidimensional histograms and an efficient approach for computation and maintenance of confidence intervals.

Our previous work on the subject [8], for raising alerts in real-time data warehouses, introduced a first definition of baselines, an approach to compute them and to detect deviation in real-time data warehouses. We now evolve the concept in some directions. We generalize the concept of baseline to detect deviations at any level of aggregation. Multi-level Baseline Materialized Views materialize baselines, and are fully integrated in the data warehouse life-cycle. The mechanisms for building, refreshing and monitoring deviations are also different and achieve integration into the data warehouse architecture. BMVs share many of the mechanisms of materialized views, but with specific approaches for refreshing the baseline. We also propose a mechanism for holding and computing data normality intervals incrementally within the baseline.

3   Using Baseline Materialized Views

In this section we make use of examples to better motivate and clarify the concept of BMV, its relevance in the framework of exploratory search and query recommendation, and how it could be used in such context.

A Baseline Materialized View defines two levels of aggregation: (i) a detailed level (DL), at which deviations are detected, and (ii) a more aggregated level (AL), at which statistical information is summarized to the user. For each cell of the AL, a set of cells of the DL allow to study deviations.

As an example, consider a data warehouse that records each individual sale of products. A sales manager is interested in knowing when the sales of some products are above or below its typical sales values. He creates a baseline that alerts when the weekly sales of a product are abnormal when compared to the average weekly sales of the product category along the year. The baseline monitors weekly sales of products (DL: *product* × *week*) and provides statistics by year and category (AL: *category* × *year*). If the user wanted the baseline for a specific subset of the cube, he could add the appropriate filters (e.g. country="Canada").

A straightforward use of BMV is alerting. This means that the baseline will compare the current week with the remaining weeks of the same year or of the previous year (if there are not enough weeks in the current year yet to compare to). In order for this to work, (*product* × *week*) must be a hierarchical detail of (*category* × *year*). Weekly product sales will be automatically integrated into a BMV after each week ends (because of the week field in the definition of the baseline). The system writes into an alert log when the value to be integrated is out of a typical interval, which we define as $[AVG(sales)-CI, AVG(sales)+CI]$, where CI is a confidence interval.

In addition, the concept of BMVs is extremely useful in exploratory OLAP. In the case of a user exploring cubes from some perspective, deviation detection with BMVs can be used to efficiently direct him to drill-down from the current perspective directly into some interesting facts that the BMV maintains.

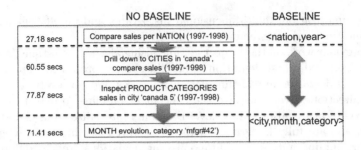

Fig. 1. Drill-down scenario. Time on the left refers to experiments in Sect. 6.

As another example, consider the drill-down scenario (shown in Fig. 1). A user initially compares sales per nation for *1997* and *1998*. He then focuses on *Canada*, comparing sales per city. Finding that *Canada 5* city has an abnormal volume of sales, he drills-down to product category, and finds out that category *MFGR#42* has an enormous volume of sales. Analyzing the monthly evolution of sales of this category, he discovers that it has had an abnormal amount of orders in *February 1998*. The length of this simple navigation is 4 queries. This compares with 2 queries if using a baseline with AL: *nation × year* and DL: *city × month × category*. When the user is navigating at the more aggregated level, the baseline immediately indicates the abnormal increase in sales of category *MFGR#42*. But, even more important, while the user navigating through the data may never find the interesting data, the interesting data is suggested to him when using baselines. In the experimental section we revisit this example, and show the improvement also in execution time.

BMVs are compatible with, and complement ordinary materialized views (MV) in two ways. First, while an ordinary MV provides one-level aggregation of finest granularity data (i.e. primary fact events), a BMV analyzes data at some level of detail (DL), not necessarily the finest granularity, and describes the distribution of those data at a more aggregated level (AL). Second, the most relevant innovation in BMVs is that they are associated with automatic deviation checks at each BMV refresh instant, generating a log of detected deviations and/or alerts.

4   Formal Framework

In this section we define baseline materialized views and illustrate them using simple examples. To keep the formalism simple, we consider read-only cubes under a ROLAP perspective, described by a star schema. We consider the classic definition of hierarchies, where a hierarchy h_i is a set $Lev(h_i) = \{l_0, \ldots, l_d\}$ of levels together with a *roll-up* total order $\succeq_{h_i}$ of $Lev(h_i)$.

A *multidimensional schema* (or briefly, a *schema*) is a triple $\mathcal{M} = \langle A, H, M \rangle$ where:

- $A = \{l_1, \ldots, l_m\}$ is a finite set of *levels*, with pairwise disjoint domains,
- $H = \{h_1, \ldots, h_n\}$ is a finite set of *hierarchies*, such that each level l_i of A belongs to at least one hierarchy of H.
- $M = \{m_1, \ldots, m_p\}$ is a finite set of *measure* attributes.

Given a schema $\mathcal{M} = \langle A, H, M \rangle$, let $Dom(H) = Lev(h_1) \times \ldots \times Lev(h_n)$; each $g \in Dom(H)$ being a *group-by set* of $\mathcal{M}$. We use the classical partial order on group-by sets, defined by: $g \succeq g'$, if $g = \langle l_1, \ldots, l_n \rangle$, $g' = \langle l'_1, \ldots, l'_n \rangle$ and $\forall i \in [1, n]$ it is $l_i \succeq_{h_i} l'_i$. We note g_0 the most specific (finest) group by set.

Definition 1 Baseline. *A baseline over schema* $\mathcal{M} = \langle A, H, M \rangle$ *is a 5-tuple* $B = \langle G, G_d, P, Me, Stats \rangle$ *where:*

1. $G \in Dom(H)$ *is the baseline group-by set;*
2. $G_d \in Dom(H)$, *with* $G \succeq G_d$ *(G is a rollup of G_d), is the baseline deviation detection group-by set;*
3. $P = \{p_1, \ldots, p_n\}$ *is a set of Boolean predicates, whose conjunction defines the selection predicate for q; they are of the form $l = v$, or $l \in V$, with l a level, v a value, V a set of values. Conventionally, if no selection on h_i is made in B (all values being selected) nothing appears in P and we write $P = \emptyset$ if no selection is made at all;*
4. $Me \subseteq M$ *is a subset of measures;*
5. $Stats = \{f_1, \ldots, f_k\}$ *over Me is a set of aggregation functions, one per measures in Me. In this paper, these functions compute the average and confidence interval for each measure in Me.*

A baseline can be viewed as a query, that specifies how to aggregate (using functions in *Stats*) a set of facts (defined by group by set G_d and predicates in P) at a given aggregation level (group by set G). A baseline materialized view (BMV) is a baseline materialized as a stored data set and refreshed automatically using the procedure described in Sect. 5.

Example 1. Assume a baseline by Month and Brand, describing the statistics (in terms of a confidence interval) for the average sales by Day and Product. This baseline is:
$\langle \{month, brand\}, \{day, product\}, \{nation = \text{``France''}\}, \{sales\}, \{f_{CI}\} \rangle$[1],

where f_{CI} computes the confidence interval of a set of numerical values. The baseline will update the sales total, the average sales and the confidence interval when products are added to the baseline, by addition to the (month $\times$ brand) cell to which the product and current month belong. The baseline will also allow detecting deviations, by verifying if the new values being inserted are within or outside of the confidence interval. In some cases this verification needs to be done with the previous period. For instance, when a new month starts, there is no sale yet for that month, therefore new sales should be compared with the values of the previous month, until there are enough values in the current month to compare with it.

---

[1] For the sake of readability, baseline expressions are simplified in the examples: the top levels are dropped in the group-by sets.

5   Baseline Materialized Views Life-Cycle

The BMV life-cycle is composed by a set of mechanisms that includes building the BMV initially, refreshing it, detecting deviations and selecting the best BMVs to materialize.

5.1   Updating Baseline Materialized Views

In a data warehouse, materialized views (MVs) are refreshed periodically, either as soon as "new data arrives", or in deferred mode. Conceptually, the update is based on a merge operation, whereby the MV is merged with new data.

The approach for refreshing BMVs in a star schema of a ROLAP data warehouse shares a lot of similarities with refreshing MVs. An MV would have the following elements: $MV = \langle G, P, Me, agg \rangle$. In this section we describe the general approach to refresh these structures and how to merge the measures (Me). The operation that merges the statistics ($Stats$) of BMVs is described in Sect. 5.2.

The refresh is done by first running the query associated with the deviation detection group set (G_d in BMV) or with the group-set (G in MV) against the new data items. The result is an aggregation of the data items with granularity G_d or G respectively. Now, new values are merged with those already in the BMV/MV using additive formulas. Examples of such formulas to merge measures of MVs or BMVs are:

$$sum_{\text{new}} = sum_{\text{old}} + SUM(\text{new items}) \tag{1}$$

$$count_{\text{new}} = count_{\text{old}} + COUNT(\text{new items}) \tag{2}$$

$$min_{\text{new}} = MIN(min_{\text{old}}, MIN(\text{new items})) \tag{3}$$

$$max_{\text{new}} = MAX(max_{\text{old}}, MAX(\text{new items})) \tag{4}$$

$$avg_{\text{new}} = \frac{sum_{\text{old}} + SUM(\text{new items})}{count_{\text{old}} + COUNT(\text{new items})} \tag{5}$$

Note that this is similar to the standard approach used for incremental maintenance of data cubes and summary tables in a Warehouse [9].

Example 2. Consider the following examples of, respectively, BMV1, BMV2 and BMV3:
$\langle \{product, year\}, \{product, day\}, \{nation = France\}, \{sales\}, \{sum(sales)\} \rangle$
$\langle \{product, year\}, \{product, week\}, \{nation = France\}, \{sales\}, \{sum(sales)\} \rangle$
$\langle \{brand, year\}, \{product, week\}, \emptyset, \{sales, \{sum(sales)\} \rangle$
Consider a new data item to be added with the following fields:

product	brand	date	nation	sales
P101	Philips	01-01-2015	France	...

It is integrated into BMV_1 at the end of the day, as it needs to be aggregated into *(product,day)*. It is integrated into BMV_2 and BMV_3 at the end of the week, as it is pre-aggregated into *(product, week)* to represent sales of each product for a week. The measure *sum(sales)* for product *P101* will be updated by summing all sales of that product in the period.

5.2 Statistics Structure and Merging Operation

The process of refreshing BMVs is incremental, with the addition of new data to the existing data. New data needs to refresh the statistics describing it.

Consider the structure of BMVs, $BMV = \langle G, Gd, P, Me, Stats \rangle$. G_d is the deviation detection group-by set. The BMV is refreshed when there is complete data at that level. For instance, if we are considering daily sales per product, the deviation detection group-by set is daily sales per product and we can refresh the baseline after each day has passed (or later, if deferred). Now we describe how to refresh the *Stats* element. Note that the statistics to be considered must have an incremental property to guarantee the usability and the efficiency of the approach.

According to the *Central Limit Theorem* it is possible to estimate a confidence interval around the mean of a population with the hypothesis of an underlying *Normal* distribution if the number of samples is 30 or more. The mean μ and the standard deviation σ of a population $X = \{x_i\}_{i=1}^{n}$ can be computed incrementally as follows if we keep in memory the tuple $Stats = (n, \sum_{i=1}^{n} x_i, \sum_{i=1}^{n} x_i^2)$:

$$\mu = \frac{1}{n} \sum_{i=1}^{n} x_i, \qquad \sigma = \left(\sum_{i=1}^{n} x_i^2 - \frac{\left(\sum_{i=1}^{n} x_i \right)^2}{n} \right)^{\frac{1}{2}} \tag{6}$$

The confidence interval CI is then defined as follows:

$$CI = \mu \pm Z_{(1-\rho)} \cdot \frac{\sigma}{\sqrt{n}} \tag{7}$$

where $Z_{(1-\rho)}$ is a parameter that is read from the *Normal* distribution table and that indicates how many standard deviations covers the interval when the confidence is in the interval is $1 - \rho$, or in other words, when the probability that the true mean is outside the confidence interval is ρ.

Finally, we define the interval I as being the sum of the confidence interval of the estimator of the mean CI, that reflects the uncertainty in the estimator, and a value that can be expressed as either a fraction f of the mean itself or a constant value c or both (defined by hand). Both f and c should be positive real values or zero.

Example 3. The following is the information in a BMV, from which the average and the confidence intervals are promptly computed using (5), (6) and (7). The confidence interval is configured for $1 - \rho = 0.90$ which corresponds to a parameter $Z_{0.90} = 1.645$.

$\langle \{month, brand\}, \{day, product\}, \emptyset, \{sum(sales)\},$
$\quad \{((count(sales), sum(sales), sum(sales * sales)), 1.645\}\rangle$

5.3 Materialization Algorithm

Before introducing the algorithm, we motivate it showing the unfeasibility of materializing the complete set of BMVs. Consider a roll-up lattice, where each node n_i represents a group-by set G, hence the number of BMVs that can be generated from the node, is the total number of available nodes from n_i to the top node n_0. That represents all possible perspectives G that can be observed from G_d Fig. 2 depicts a simple lattice structure with 3 dimensions, where each node shows the number of possible BMVs. Thus, the number of BMVs it can contain is 22. On top of that, that number will be increased by the possible combinations of selection predicates P at each level n_i.

In [8] we introduced the concept of monitoring query set MQ, that contains a set of queries q which reflect the user's interest for deviation detection. Here we extend the definition of MQ in order to incorporate the elements of a BMV, therefore $MQ = \{m_1, \dots, m_n\}$ where $m_i = \langle G, G_d, P, Me, Stats \rangle$. MQ can be obtained by mining the query log, with techniques such as [1,12], tailored to the user's interest, or following a cost-based approach based on empirical measures as the well-known greedy algorithm [7].

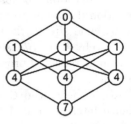

Fig. 2. A 3-dimensional lattice

Algorithm 1 depicts the materialization phase, valid for both generation and refreshment of BMVs. For each monitoring query, the related BMV is materialized using the formulas presented in Sects. 5.1 and 5.2. We assume the existence of a *CDC* (Change Data Capture) table containing the new incoming items. Operation `isMaterialized` checks if a set of BMVs contains the BMV in the second parameter. On the other hand, `mergeStatistics` given a set of BMVs, a BMV and a sample of items applies Eqs. (1) to (5) from the sample to the BMV and includes it in the set. Likewise, `refreshConfidenceInterval` applies Eqs. 6 and 7.

Algorithm 1. Materializing BMVs

Input MQ: monitoring query set, BMV: set of baseline materialized views, CDC
    table containing new items
Output BMV refreshed set of baseline materialized views
1: **for** $q = \langle G, G_d, P, Me, Stats \rangle \in MQ$ **do**
2:     $sample = q(CDC)$ ▷ Run query against CDC table
3:     **if** `isMaterialized`(BMV, q) **then**
4:         `mergeStatistics`$(BMV, q, sample)$
5:     `refreshConfidenceInterval`$(BMV, q, sample)$
6: **return** BMV

5.4   Detection and Alerting

Detection refers to the system detecting deviations in data. The baselines contain reference statistics, so detection is based on verifying whether a new value is within the interval. For instance,

```
if (newSalesValue > avgSales + IC + f · IC + c or
    newSalesValue < avgSales − IC − f · IC − c) {
      dump alert to log
}
```

where the information dumped to the log has the structure $\langle product, day,$ $month, year, new\ sales, average\ sales, confidence\ interval\ limits \rangle$.

Deviations are tested immediately prior to integrating the new data into the baseline. If a deviation is detected, information is dumped into a deviation log. Only cases with more than a pre-specified minimum number of samples in the baseline will be dumped. Then there are different possible courses of action. It is possible to raise alerts for the user to look at the data, but the log can also be used to direct the user to the most interesting data during navigation. Besides the deviation log, there can be a "new cases" log, which records new cases that are not yet in the baseline when they appear (e.g. a new product).

At the start of a new period, there are no values registered yet in the BMV for that period, so it is not possible to compare new values with the same period. If there is at least a minimum number of samples, the value is compared within the period, otherwise it is compared against the previous period instead of the current one.

6   Proof of Concept

We illustrate the interest of our approach on the Star Schema Benchmark [10], recording sales of products (parts) over time (the date dimension), customers and suppliers. We modified the generated data to add skewness/deviations (an abnormal amount of sales of specific products or brands in specific time intervals). SSB generates sales data from 1992 to 1999. The sales of product 1 and brand MFGR#241 where increased to 5 % of total sales in the first week of 1998. 1 GB of data was generated (SF=1).

The set of BMVs created is listed in Table 2. We apply the approach to detect weekly abnormalities in the first week of 1998 (parameters f and c were set to 0), show sales evolution and detection results, and evaluate the following efficiency metrics: (a) time to build the BMV; (b) time to refresh the BMV; (c) time to detect the deviations (generate the alert log). The time to build the BMVs and MVs (materialized views) is compared. All the experiments were run on a Toshiba Qosmio machine, on a multicore Intel i7 CPU q720 (1.6 GHz), 8 GB of RAM, 1 TB 7400 RPM disk . The DBMS was MySQL version 5.5.41 bundled with MAMP out-of-the-box with no tuning except for the creation of indexes.

Figure 3 shows the evolution of sales (sum of quantity) in 52 weeks of 1997 and the first week of 1998 of two particular brands (MFGR#111 and MFGR#421).

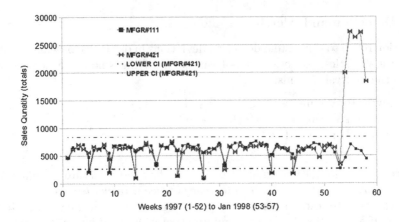

Fig. 3. Charting brand by week 1997 and Jan 1998

Table 1. Content of alert log for jan 1998

Brand	Week	Year	Sales	Num of Sales	Avg Sales	Upper Limit	Lower Limit
MFGR#412	1	1998	19,945	62	5457.47	8313.62	2601.32

The injected increase in sales can be seen in the evolution of sales of brand MFGR#421, with a steep rise in the first weeks of 1998. We can also see the deviation interval limits for the same brand (MFGR#421). Figure 1 shows the output log with deviations detected for the first week of 1998. It correctly identifies brand MFGR#421 as having sales of 19,945, a value well above the mean (5,457) and outside the interval limits also identified (2,601 to 8,313). This was the only row in the log, indeed correct since we only injected variations in MFGR#421.

Table 2 depicts the times taken to build, refresh and detect deviations in BMVs, and the size of the BMVs. The build time is incurred only once and depends on the size of the dataset. It was between 60 and 620 seconds, depending on the amount of data and aggregation level. Figure 4 details the build times for product BMVs, and compare with the time taken to build MVs of the same aggregation level. The same test was done for the brand BMVs but is not reported due to lack of space. The total times to build a BMV, an MV following the same pre-aggregation and an MV directly from the base data were 620, 556 and 258 seconds respectively. Creation of an MV is significantly faster if directly from the base data, but times to build BMVs are still acceptable and offer the advantage of having deviation detection (Table 1).

The refresh times and time taken to generate the deviation detection logs are shown in Table 2. Refreshing was very quick for BMVs with few rows and took between 25 and 125 seconds for huge BMVs (updating sales statistics for 200,000 products). The time taken to test and generate deviation logs was insignificant in all cases.

Table 2. Time to operate baseline materialized Views

BMV	name	initial build time	size(rows)	refresh time	time to generate alert log
PY-PD	prod-year, prod-day	367	1.2M	29.36	1.07
PM-PD	prod-month, prod-day	620	4.45 M	124.41	0.96
BY-BD	brand-year,brand-day	80	384	0.11	0.12
BM-BD	brand-month,brand-day	79.85	4608	0.09	0.46
PY-PW	prod-year,prod-week	484	1.2M	24.9	3.6
BY-BW	brand-year,brand-week	59.35	384	0.1	0.09

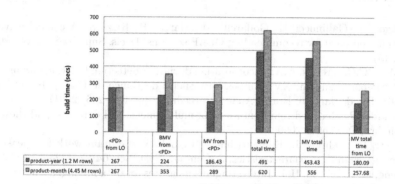

	\<PD\> from LO	BMV from \<PD\>	MV from \<PD\>	BMV total time	MV total time	MV total time from LO
product-year (1.2 M rows)	267	224	186.43	491	453.43	180.09
product-month (4.45 M rows)	267	353	289	620	556	257.68

Fig. 4. Detailed timing to build products BMVs

Finally, we tested the drill-down scenario detailed in Sect. 3. Without using BMVs, the user had to submit **4 different queries**, for a cost of **237 secs**. When using right BVMs, which provided him assistance, he only had to submit **2 queries** with a cost of **98.5 secs**.

7 Conclusions and Future Work

In this paper we have proposed Multi-Level Baseline Materialized Views (BMV) as a way of integrating deviation detection functionality into the data warehouse, to efficiently detect abnormal values. We have defined baseline materialized views, described their structure and the mechanisms for building, refreshing, monitoring and alerting on various levels of aggregation based on those baselines. In order to prove that the approach works, we created an experimental setup with the Star Schema Benchmark and tested it over sales data. The results of the experiments show that the approach works well and does not introduce significant overheads. It provides an effective way to integrate baselines, deviation detection and alerting into databases.

An interesting avenue for future work is to relax the uniform distribution assumption underlying the approach, and to devise baseline materialization algorithms by looking at the navigation sessions of a user, to decide automatically to build a BMV based on her knowledge of the data. In other words, there should be a BMV selection algorithm for choosing the best set to materialize. Additionally, more complex methods can be used to detect anomalies by comparing data distributions, such as the Kullback Leibler divergence. We would like to explore real-time capabilities further and to apply these concepts to help in exploratory OLAP, since the abnormalities detected in data are crucial clues to the most interesting paths to follow during analysis and exploration. Finally, we also need to deal with the information overload that results from too many alerts.

References

1. Aligon, J., Gallinucci, E., Golfarelli, M., Marcel, P., Rizzi, S.: A collaborative filtering approach for recommending OLAP sessions. Decis. Support Syst. **69**, 20–30 (2015)
2. Fabris, C.C., Freitas, A.A.: Incorporating deviation-detection functionality into the OLAP paradigm. In: XVI Simpósio Brasileiro de Banco de Dados, 1–3 Outubro 2001, Rio de Janeiro, Brasil, Anais/Proceedings, pp. 274–285 (2001)
3. Furtado, P.: Reduced representations of multidimensional datasets, phd thesis, u. coimbra, December 2000
4. Ganti, V., Gehrke, J., Ramakrishnan, R., Loh, W.: A framework for measuring differences in data characteristics. J. Comput. Syst. Sci. **64**(3), 542–578 (2002)
5. Geng, L., Hamilton, H.J.: Interestingness measures for data mining: a survey. ACM Comput. Surv. (CSUR) **38**(3), 9 (2006)
6. Han, J., Cheng, H., Xin, D., Yan, X.: Frequent pattern mining: current status and future directions. Data Min. Knowl. Disc. **15**(1), 55–86 (2007)
7. Harinarayan, V., Rajaraman, A., Ullman, J.D.: Implementing data cubes efficiently. SIGMOD Rec. **25**(2), 205–216 (1996)
8. Lòpez, M., Nadal, S., Djedaini, M., Marcel, P., Peralta, V., Furtado, P.: An approach for raising alert in real-time data warehouses. Journes francophones sur les Entrepts de Donnes et lAnalyse en ligne Bruxelles, Belgique, 2–3 avril 2015, 2015(1), 55–86 (2015)
9. Mumick, I.S., Quass, D., Mumick, B.S.: Maintenance of data cubes and summary tables in a warehouse. In: SIGMOD 1997, Proceedings ACM SIGMOD International Conference on Management of Data, May 13–15, 1997, Tucson, Arizona, USA, pp. 100–111 (1997)
10. O'Neil, P., O'Neil, E., Chen, X., Revilak, S.: The star schema benchmark and augmented fact table indexing. In: Nambiar, R., Poess, M. (eds.) TPCTC 2009. LNCS, vol. 5895, pp. 237–252. Springer, Heidelberg (2009)
11. Sarawagi, S.: Explaining differences in multidimensional aggregates. In: VLDB 1999, Proceedings of 25th International Conference on Very Large Data Bases, September 7–10, 1999, Edinburgh, Scotland, UK, pp. 42–53 (1999)
12. Sarawagi, S.: idiff: Informative summarization of differences in multidimensional aggregates. Data Min. Knowl. Discov. **5**(4), 255–276 (2001)

Stream Processing

Binary Shapelet Transform for Multiclass Time Series Classification

Aaron Bostrom$^{(\boxtimes)}$ and Anthony Bagnall

University of East Anglia, Norwich, NR47TJ, UK
a.bostrom@uea.ac.uh

Abstract. Shapelets have recently been proposed as a new primitive for time series classification. Shapelets are subseries of series that best split the data into its classes. In the original research, shapelets were found recursively within a decision tree through enumeration of the search space. Subsequent research indicated that using shapelets as the basis for transforming datasets leads to more accurate classifiers.

Both these approaches evaluate how well a shapelet splits all the classes. However, often a shapelet is most useful in distinguishing between members of the class of the series it was drawn from against all others. To assess this conjecture, we evaluate a one vs all encoding scheme. This technique simplifies the quality assessment calculations, speeds up the execution through facilitating more frequent early abandon and increases accuracy for multi-class problems. We also propose an alternative shapelet evaluation scheme which we demonstrate significantly speeds up the full search.

1 Introduction

Time series classification (TSC) is a subset of the general classification problem, the primary difference being that the ordering of attributes within each instance is important. For a set of n time series, $\mathbf{T} = \{T_1, T_2, ..., T_n\}$, each time series has m ordered real-valued observations $T_i = \{t_{i1}, t_{i2}, ..., t_{im}\}$ and a class value c_i. The aim of TSC is to determine a function that relates the set of time series to the class values.

One recently proposed technique for TSC is to use shapelets [17]. Shapelets are subseries of the series $\mathbf{T}$ that best split the data into its classes. Shapelets can be used to detect discriminatory phase independent features that cannot be found with whole series measures such as dynamic time warping. Shapelet based classification involves measuring the similarity between a shapelet and each series, then using this similarity as a discriminatory feature for classification. The original shapelet-based classifier [17] embeds the shapelet discovery algorithm in a decision tree, and uses information gain to assess the quality of candidates. A shapelet is found at each node of the tree through an enumerative search. More recently, we proposed using shapelets as a transformation [9]. The shapelet transform involves a single-scan algorithm that finds the best k shapelets in a set of n time series. We use this algorithm to produce a transformed dataset,

© Springer International Publishing Switzerland 2015
S. Madria and T. Hara (Eds.): DaWaK 2015, LNCS 9263, pp. 257–269, 2015.
DOI: 10.1007/978-3-319-22729-0_20

where each of the k features is the distance between the series and one shapelet. Hence, the value of the ith attribute of the jth record is the distance between the jth record and the ith shapelet. The primary advantages of this approach are that we can use the transformed data in conjunction with any classifier, and that we do not have to search sequentially for shapelets at each node. However, it still requires an enumerative search throughout the space of possible shapelets and the full search is $O(n^2m^4)$. Improvements for the full search technique were proposed in [12,17] and heuristics techniques to find approximations of the full search were described in [4,6,15].

Our focus is only on improving the exhaustive search. One of the problems of the shapelet search is the quality measures assess how well the shapelet splits all the classes. For multi-class problems measuring how well a shapelet splits all the classes may confound the fact that it actually represents a single class. Consider, for example, a shapelet in a data set of heartbeat measurements of patients with a range of medical conditions. It is more intuitive to imagine that a shapelet might represent a particular condition such as arrhythmia rather than discriminating between multiple conditions equally well. We redefine the transformation so that we find shapelets assessed on their ability to distinguish one class from all, rather than measures that separate all classes. This improves accuracy on multi-class problems and allows us to take greater advantage of the early abandon described in [17].

A further problem with the shapelet transform is that it may pick an excessive number of shapelets representing a single class. By definition, a good shapelet will appear in many series. The best way we have found to deal with this is to generate a large number of shapelets then cluster them [9]. However, there is still a risk that one class is generally easier to classify and hence has a disproportionate number of shapelets in the transform. The binary shapelet allows us to overcome this problem by balancing the number of shapelets we find for each class.

Finally, we describe an alternative way of enumerating the shapelet search that facilitates greater frequency of early abandon of the distance calculation.

2   Shapelet Based Classification

The shapelet transform algorithm described in [9] is summarised in Algorithm 2. Initially, for each time series, all candidates of length min to max are generated (i.e. extracted and normalised in the method $generateCandidates$). Then the distance between each shapelet and the other $n-1$ series are calculated to form the order list D_S. Distance between a shapelet S length l and a series T is given by

$$sDist(S,T) = \min_{w \in W_l}(dist(S,w)) \tag{1}$$

where W_l is the set of all l length subseries in T and $dist$ is the Euclidean distance between the equal length series S and w. The order list is used to determine the quality of the shapelet in the $assessCandidate$ method. Quality can be assessed by information gain [17] or alternative measures such as the F, moods median or rank order statistic [11]. Once all the shapelets for a series are evaluated they are sorted and the lowest quality overlapping shapelets are removed. The remaining candidates are then added to the shapelet set. By default, we set $k = 10n$ with the caveat that we do not accept shapelets that have zero information gain.

Algorithm 1. FullShapeletSelection($\mathbf{T}$, min, max, k)

Input: A list of time series $\mathbf{T}$, min and max length shapelet to search for and k,the maximum number of shapelets to find)
Output: A list of k shapelets
1: $kShapelets \leftarrow \emptyset$
2: **for all** T_i in $\mathbf{T}$ **do**
3:     $shapelets \leftarrow \emptyset$
4:     **for** $l \leftarrow min$ to max **do**
5:         $W_{i,l} \leftarrow generateCandidates(T_i, l)$
6:         **for all** subseries S in $W_{i,l}$ **do**
7:             $D_S \leftarrow findDistances(S, \mathbf{T})$
8:             $quality \leftarrow assessCandidate(S, D_S)$
9:             $shapelets.add(S, quality)$
10:     $sortByQuality(shapelets)$
11:     $removeSelfSimilar(shapelets)$
12:     $kShapelets \leftarrow merge(k, kShapelets, shapelets)$
13: **return** $kShapelets$

Once the best k shapelets have been found, the transform is performed with Algorithm 2. A more detailed description can be found in [8].

Extensions to the basic shapelet finding algorithm can be categorised into techniques to speed up the average case complexity of the exact technique and those that use heuristic search. The approximate techniques include reducing the dimensionality of the candidates and using a hash table to filter [15], searching the space of shapelet values (rather than taking the values from the train set series) [6] and randomly sampling the candidate shapelets [4]. Our focus is on improving the accuracy and speed of the full search. Two forms of early abandon described in [17] can improve the average case complexity. Firstly, the Euclidean distance calculations within the $sDist$ (Eq. 1) can be terminated early if they exceed the best found so far. Secondly, the shapelet evaluation can be abandoned early if $assessCandidate$ is updated as the $sDist$ are found and the best possible outcome for the candidate is worse than the current top candidates.

A speedup method involving trading memory for speed is proposed in [12]. For each pair of series T_i, T_j, cumulative sum, squared sum, and cross products of T_i and T_j are precalculated. With these statistics, the distance between subseries can be calculated in constant time, making the shapelet-discovery algorithm $O(n^2m^3)$. However, precalculating of the cross products between all series prior to shapelet discovery requires $O(n^2m^2)$ memory, which is infeasible for most problems. Instead, [12] propose calculating these statistics prior to the start of the scan of each series, reducing the requirement to $O(nm^2)$ memory, but increasing the time overhead. Further refinements applicable to shapelets were described in [14], most relevant of which was a reordering of the sequence of calculations within the *dist* function to increase the likelihood of early abandon. The key observation is that because all series are normalised, the largest absolute values in the candidate series are more likely to contribute large values in the distance function. Hence, if the distances between positions with larger candidate values are evaluated first, then it is more likely the distance can be abandoned early. This can be easily implemented by creating an enumeration through the normalised candidate at the beginning, and adds very little overhead. We use this technique in all experiments.

Algorithm 2. *FullShapeletTransform*(Shapelets **S,T**)

1: $\mathbf{T}' \leftarrow \emptyset$
2: **for all** T in **T do**
3:     $T' \leftarrow <>$
4:     **for all** shapelets S in **S do**
5:         $dist \leftarrow sDist(S, T)$
6:         $T' \leftarrow append(T', dist)$
7:         $T' \leftarrow append(T', T.class)$
8:     $\mathbf{T}' \leftarrow \mathbf{T}' \cup T'$
9: **return T'**

---

3   Classification Technique

Once the transform is complete we can use any classifier on the problem. To reduce classifier induced variance we use a heterogenous ensemble of eight classifiers. The classifiers used are the WEKA [7] implementations of k Nearest Neighbour (where k is set through cross validation), Naive Bayes, C4.5 decision tree [13], Support Vector Machines [3] with linear and quadratic basis function kernels, Random Forest [2] (with 100 trees), Rotation Forest [16] (with 10 trees) and a Bayesian network. Each classifier is assigned a weight based on the cross validation training accuracy, and new data are classified with a weighted vote. The set of classifiers were chosen to balance simple and complex classifiers that use probabilistic, tree based and kernel based models. With the exception of k-NN, we do not optimise parameter settings for these classifiers via cross validation. More details are given in [8].

4   Shapelet Transform Refinements

4.1   Binary Shapelets

The standard shapelet assessment method measures how well the shapelet splits up all the classes. There are three potential problems with this approach when classifying multi-class problems. The problems apply to all possible quality measures, but we use information gain to demonstrate the point. Firstly, useful information about a single class may be lost. For example, suppose we have a four class problem and a shapelet produces the order line presented in Fig. 1, where each colour represents a different class.

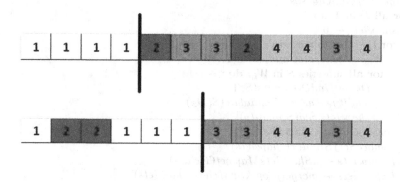

Fig. 1. An example order line split for two shapelets. The top shapelet discriminates between class 1 and the rest perfectly, yet has lower information gain than the orderline shown below it.

The first shapelet groups all of class 1 very well, but cannot distinguish between classes 2, 3 and 4 and hence has a lower information gain than the split produced by the second shapelet in Fig. 1 which separates class 1 and 2 from class 3 and 4. The more classes there are, the more likely it is that the quantification of the ability of a shapelet to separate out a single class will be overwhelmed by the mix of other class values. We can mitigate against this potential problem by defining a binary shapelet as one that is assessed by how well it splits the class of the series it originated from all the other classes. The second problem with searching all shapelets with multi-class assessment arises if one class is much easier to classify than the others. In this case it is likely that more shapelets will be found for easy class than other classes. Although our principle is to find a large number of shapelets (ten times the number of training cases) and let the classifier deal with redundant features, there is still a risk that a large number of similar shapelets for one class will crowd out useful shapelets for another class. If we use binary shapelets we can simply allocate a maximum number of shapelets to each class. We adopt the simple approach of allocating a maximum of k/c shapelets to each class, where c is the number of classes. Finally, the shapelet early abandon described in [17] is not useful for multi-class problems. Given a partial orderline and a split point, the early abandon works by upper bounding the information gain by assigning the unassigned series to the side of the split that would give the maximum gain. However, the only way to do this with multi-class problems is to try all permutations. The time this takes quickly

---

Algorithm 3. BinaryShapeletSelection(**T**, min, max, k)

Input: A list of time series **T**, min and max length shapelet to search for and k,the
    maximum number of shapelets to find)
Output: A list of k Shapelets
 1: $numClasses \leftarrow getClassDistribution(T)$
 2: $kShapeletsMap \leftarrow \emptyset$
 3: $prop \leftarrow k/numClasses$
 4: **for all** T_i in **T do**
 5:    $shapelets \leftarrow \emptyset$
 6:    **for** $l \leftarrow min$ to max **do**
 7:      $W_{i,l} \leftarrow generateCandidates(T_i, l)$
 8:      **for all** subseries S in $W_{i,l}$ **do**
 9:        $D_S \leftarrow findDistances(S, \mathbf{T})$
10:        $quality \leftarrow assessCandidate(S, D_S)$
11:        $shapelets.add(S, quality)$
12:    $sortByQuality(shapelets)$
13:    $removeSelfSimilar(shapelets)$
14:    $kShapelets \leftarrow kShapeletsMap.get(T.class)$
15:    $kShapelets \leftarrow merge(prop, kShapelets, shapelets)$
16:    $kShapeletsMap.add(kShapelets, T.class)$
17: **return** $kShapeletsMap.asList()$

---

rises to offset the possible benefits from the early abandon. If we restrict our attention to just binary shapelets then we can take maximum advantage of the early abandon. The binary shapelet selection is described by Algorithm 3.

4.2 Changing the Shapelet Evaluation Order

Shapelets are phase independent. However, for many problems the localised features are at most only weakly independent in phase, i.e. the best matches will appear close to the location of the candidate. Finding a good match early in $sDist$ increases the likelihood of an early abandon for each $dist$ calculation. Hence, we redefine the order of iteration of the $dist$ calculations within $sDist$ so that we start with the index the shapelet was found at and move consecutively left and right from that point. Figure 2 demonstrates the potential benefit of this approach. The scan from the beginning is unable to early abandon on any of the subseries before the best match. The scan originating at the candidate's location finds the best match faster an hence can early abandon on all the distance calculations at the beginning of the series. Hence, if the location of the best shapelet is weakly phase dependent, we would expect to observe an improvement in the time complexity. The revised function $sDist$, which is a subroutine of $findDistances$ (line 9 in Algorithm 3), is described in Algorithm 4.

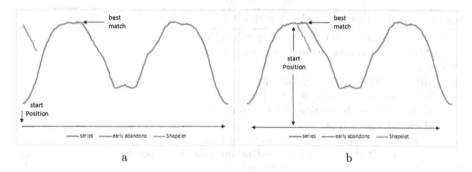

a   b

Fig. 2. An example of Euclidean distance early abandon where the $sDist$ scan starts from the beginning (a) and from the place of origin of the candidate shapelet (b). For the scan from the beginning, there are no early abandons until the scan has passed the best match. Because the best match is close to the location of the candidate shapelet, starting from the shapelets original location allows for a greater number of early abandons.

5 Results

We demonstrate the utility of our approach through experiments using 31 benchmark multi-class datasets from UCR [10] and UEA repositories [1]. In common with the vast majority of research in this field, we present results on the standard train/test split. The min and max size of the shapelet transform are found through a train set cross validation described in [9]. As a sanity check, we have also evaluated the binary shapelets on two class problems to demonstrate there is negligible difference. On 16 two class problems, the full transform was better on 7, the binary transform better on 7 and they were tied on 2. All the results and the code to generate them are available from [1].

5.1 Accuracy Improvement on Multi-class Problems

Table 1 gives the results for the full shapelet transform, the binary shapelet transform and standard benchmark classifiers on problems with 3–50 classes. Overall, the binary shapelet transform is better on 19 data sets, the full transform better on 10 and on two they are equal. The difference between the full and the binary shapelet transform is significant at the 5 % level using a paired T test and at the 10 % level using the binomial and signed rank test. Although neither the binary nor full shapelet transform is significantly more accurate than 1-nearest neighbour with dynamic time warp distance measure (1-NN DTW) and 1-NN DTW with window size set through cross validation (1-NN DTWCV), in a larger study, we have shown that the shapelet transform is significantly better than both 1-NN DTW and 1-NN DTWCV [8]. The main point of including the nearest neighbour results is to highlight the wide variation in accuracy. For example, the shapelet approach is clearly superior on the FacesUCR, fish and CBF problems, but much worse on MiddlePhalanxTW, CricketX and FaceAll. This demonstrates the importance of finding the correct transformation space for a given problem.

Algorithm 4. $sDist($ shapelet S,series $T_i)$

1: $subSeq \leftarrow getSubSeq(T_i, S.startPos, S.length)$
2: $bestDist \leftarrow euclideanDistance(subSeq, S)$
3: $i \leftarrow 1$
4: **while** $leftExists \, || \, rightExists$ **do**
5:    $leftExists \leftarrow S.startPos - i \geq 0$
6:    $rightExists \leftarrow S.startPos + i < T_i.length$
7:    **if** $rightExists$ **then**
8:       $subSeq \leftarrow getSubSeq(T_i, S.startPos + i, S.length)$
9:       $currentDist \leftarrow earlyAbandonDistance(subSeq, S, bestDist)$
10:       **if** $currentDist > bestDist$ **then**
11:          $bestDist \leftarrow currentDist$
12:    **if** $leftExists$ **then**
13:       $subSeq \leftarrow getSubSeq(T_i, S.startPos - i, S.length)$
14:       $currentDist \leftarrow earlyAbandonDistance(subSeq, S, bestDist)$
15:       **if** $currentDist > bestDist$ **then**
16:          $bestDist \leftarrow currentDist$
17:    $i \leftarrow i + 1$
18: **return** $bestDist$

Figure 3 shows the plot of the difference in accuracy of the full and binary shapelet transform plotted against the number of classes. There is a clear trend of increasing accuracy for the binary transform as the number of classes continues. This is confirmed in Table 2, which presents the same data grouped into bins of ranges of number of classes.

5.2   Accuracy Comparison to Other Shapelet Methods

Our goal is to improve the enumerative shapelet transform. Nevertheless, it is informative to compare performance against alternative shapelet based techniques. Table 3 presents the results on 22 multiclass problems for logical shapelets [12], fast shapelets [15] and learning shapelets [5]. We present two sets of results for learning shaplets: those presented on the website (LST1) and those we have recreated using the code available on the website (LST2). Differences between the two may be down to random variation. Overall, there is no significant difference between the classifiers (as tested with Friedman's rank sum test). A more extensive study is required to determine whether any one of the approaches outperforms the others in terms of accuracy.

5.3   Average Case Time Complexity Improvements

One of the benefits of using the binary transform is that it is easier to use the shapelet early abandon described in [12]. Early abandon is less useful when finding the best k shapelets than it is for finding the single best, but when it can be employed it can give real benefit. Figure 4 shows that on certain datasets, using the binary shapelet discovery means millions of fewer $sDist$ evaluations.

Table 1. Test errors for 5 classifiers on 31 multi class data sets. The binary shapelet and full shapelet both use the weighted ensemble described in Sect. 3. The three nearest neighbour classifiers use Euclidean distance (1NN-ED), dynamic time warping (1NN-DTW) and dynamic time warping with window size set through cross validation (1NN-DTWCV).

dataSet	#classes	Binary Shapelet	Full Shapelet	1NN-ED	1NN-DTW	1-NNDTWCV
fiftywords	50	**0.262**	0.281	0.63	0.623	0.584
Adiac	37	0.322	0.435	0.369	0.31	**0.235**
WordSynonyms	25	0.346	0.403	0.253	**0.096**	0.132
SwedishLeaf	15	**0.062**	0.093	0.526	0.539	0.539
FaceAll	14	0.246	0.263	0.101	**0.05**	0.062
FacesUCR	14	**0.085**	0.087	0.467	0.5	0.5
CricketX	12	0.221	0.218	0.086	0.066	**0.065**
CricketY	12	0.246	0.236	0.217	0.166	**0.166**
MedicalImages	10	**0.337**	0.396	0.35	0.352	0.351
MALLAT	8	0.087	0.06	0.148	**0.003**	0.006
fish	7	**0.006**	0.023	0.267	0.267	0.233
Lightning7	7	0.301	0.26	**0.183**	0.429	0.217
Plane	7	0	0	0.25	0.179	0.214
DistalPhalanxTW	6	**0.367**	**0.367**	0.474	0.506	0.506
MiddlePhalanxTW	6	0.409	0.461	**0.127**	0.137	**0.127**
ProximalPhalanxTW	6	0.195	0.229	0.087	0.093	**0.027**
Symbols	6	**0.073**	0.114	0.121	0.165	0.134
SyntheticControl	6	**0.003**	0.017	0.127	0.137	0.127
Beef	5	**0.1**	0.167	0.382	0.351	0.251
Haptics	5	0.497	0.523	0.216	0.17	**0.114**
Car	4	0.1	0.267	0.231	0.095	**0.092**
DiatomSizeReduction	4	**0.118**	0.124	0.425	0.274	0.288
FaceFour	4	0.091	0.057	0.12	**0.007**	0.017
OliveOil	4	**0.067**	0.1	0.4	0.35	0.35
Trace	4	0	0.02	0.325	0.105	0.101
ArrowHead	3	0.257	**0.229**	0.389	0.396	0.391
CBF	3	0.018	**0.003**	0.426	0.223	0.236
ChlorineConcentration	3	0.311	0.3	0.356	0.208	**0.197**
DistalPhAgeGroup	3	0.252	0.259	0.038	0	0
MiddlePhAgeGroup	3	0.396	0.37	0.252	0.245	**0.201**
ProximalPhAgeGroup	3	0.156	**0.146**	0.252	0.245	0.201
total wins		13	5	2	4	10
Average Rank		2.613	3.06	3.66	3.08	2.58

Table 2. Number of data sets the binary shapelet beats the full shapelet split by number of classes.

Number of classes	Binary Better	Full Better
10 and above	7	2
6 to 9	5	2
3 to 5	7	6
All	19	10

We assess the improvement from using Algorithm 4 by counting the number of point wise distance calculations required from using the standard approach and the alternative enumeration. For the datasets used in our accuracy experiments, changing the order of enumeration reduces the number of calculations in the distance function by 76 % on average. The improvement ranges from negligible (e.g. Lightning7 requires 99.3 % of the calculations) to substantial (e.g. Adiac operations count is 63 % of the standard approach). This highlights that the best shapelets may or may not be phase independent, but nothing is lost from changing the evaluation order and often substantial improvements are achieved. Full results, all the code and the datasets used can be downloaded from [1].

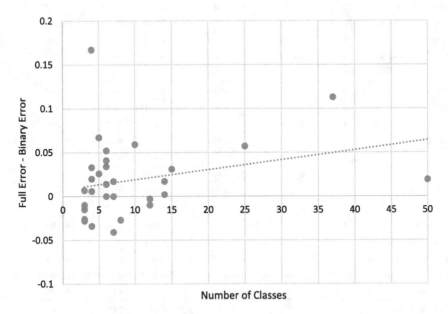

Fig. 3. Number of classes plotted against the difference in error between the full shapelets and the binary shapelets. A positive number indicates the binary shapelets are better. The dotted line is the least squares regression line.

Table 3. Test errors for 6 shapelet based classifiers on 22 multiclass data sets.

Problem	Binary	Full	Logical	Fast	LST1	LST2
Adiac	**0.322**	0.435	0.414	0.514	0.437	0.737
Beef	**0.1**	0.167	0.433	0.447	0.24	0.433
CBF	0.018	**0.003**	0.114	0.053	0.006	0.006
ChlorineConcentration	0.311	**0.3**	0.382	0.417	0.349	0.406
CricketX	0.221	0.218		0.528	**0.209**	0.328
CricketY	0.246	**0.236**		0.52	0.249	0.346
DiatomSizeReduction	0.118	0.127	0.199	0.117	0.033	**0**
FaceAll	0.246	0.262	0.341	0.411	**0.218**	0.24
FaceFour	0.091	0.057	0.511	0.09	0.048	**0.034**
FacesUCR	0.085	0.086	0.338	0.328	**0.059**	0.092
fiftywords	0.262	0.281		0.489	**0.232**	0.25
fish	**0.006**	0.023	0.223	0.197	0.066	0.091
Haptics	**0.497**	0.519		0.624	0.532	0.5286
Lightning7	0.301	0.26	0.452	0.403	**0.197**	0.233
MALLAT	0.087	0.06	0.344	**0.033**	0.046	0.038
MedicalImages	0.337	0.396	0.413	0.433	**0.271**	0.33
OliveOil	**0.067**	0.1	0.167	0.213	0.56	0.6
SwedishLeaf	**0.062**	0.093	0.187	0.269	0.087	0.102
Symbols	0.073	0.114	0.357	0.068	**0.036**	0.086
SyntheticControl	**0.003**	0.017	0.53	0.081	0.007	0.023
Trace	**0**	0.02	**0**	0.002	**0**	**0**
WordSynonyms	0.346	0.401		0.563	**0.34**	0.409
total wins	8	3	1	1	9	3
Average Rank	2.48	3	4.97	4.73	2.205	3.38

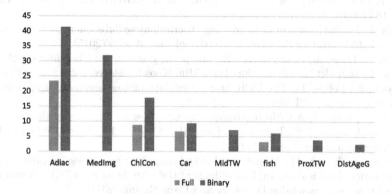

Fig. 4. Number of $sDist$ measurements that were not required because of early abandon (in millions) for both full and binary shapelet discovery on seven datasets.

6   Conclusion

Shapelets are useful for classifying time series where the between class variability can be detected by relatively short, phase independent subseries. They offer an alternative representation that is particularly appealing for problems with long series with recurring patterns. The downside to using shapelets is the time complexity. The heuristic techniques described in recent research [6,15] offer potential speed up (often at the cost of extra memory) but are essentially different algorithms that are only really analogous to shapelets described in the original research [17]. Our interest is in optimizing the original shapelet finding algorithm within the context of the shapelet transform. We describe incremental improvements to the shapelet transform specifically for multi-class problems. Searching for shapelets assessed on how well they find a single class is more intuitive, faster and becomes more accurate than the alternative as the number of classes increases.

References

1. Bagnall, A.: UEA time series classification website. http://www.uea.ac.uk/computing/tsc
2. Breiman, L.: Random forests. Mach. Learn. **45**(1), 5–32 (2001)
3. Cortes, C., Vapnik, V.: Support-vector networks. Mach. Learn. **20**(3), 273–297 (1995)
4. Gordon, D., Hendler, D., Rokach, L.: Fast randomized model generation for shapelet-based time series classification (2012). arXiv preprint arXiv:1209.5038
5. Grabocka, J.: Learning time series shapelets. http://fs.ismll.de/publicspace/LearningShapelets/
6. Grabocka, J., Schilling, N., Wistuba, M., Schmidt-Thieme, L.: Invariant time-series classification. In: Proceedings of the 20th ACM SIGKDD International Conference on Knowledge Discovery and Data Mining (2014)
7. Hall, M., Frank, E., Holmes, G., Pfahringer, B., Reutemann, P., Witten, I.H.: The WEKA data mining software: an update. ACM SIGKDD Explor. Newsl. **11**(1), 10–18 (2009)
8. Hills, J.: Mining Time-series Data using Discriminative Subsequences. Ph.D. thesis, School of Computing Sciences, University of East Anglia (2015)
9. Hills, J., Lines, J., Baranauskas, E., Mapp, J., Bagnall, A.: Classification of time series by shapelet transformation. Data Min. Knowl. Discovery **28**, 851–881 (2014)
10. Keogh, E., Folias, T.: The UCR time series data mining archive. http://www.cs.ucr.edu/eamonn/TSDMA/
11. Lines, J., Bagnall, A.: Alternative quality measures for time series shapelets. In: Yin, H., Costa, J.A.F., Barreto, G. (eds.) IDEAL 2012. LNCS, vol. 7435, pp. 475–483. Springer, Heidelberg (2012)
12. Mueen, A., Keogh, E., Young, N.: Logical-shapelets: an expressive primitive for time series classification. In: Proceedings of the 17th ACM SIGKDD International Conference on Knowledge Discovery and Data Mining (2011)
13. Quinlan, J.R.: C4. 5: Programs for Machine Learning. Morgan kaufmann, Los Altos (1993)

14. Rakthanmanon, T., Campana, B., Mueen, A., Batista, G., Westover, B., Zhu, Q., Zakaria, J., Keogh, E.: Searching and mining trillions of time series subsequences under dynamic time warping. In: Proceedings of the 18th ACM SIGKDD International Conference on Knowledge Discovery and Data Mining (2012)
15. Rakthanmanon, T., Keogh, E.: Fast shapelets: a scalable algorithm for discovering time series shapelets. In: Proceedings of the 13th SIAM International Conference on Data Mining (2013)
16. Rodriguez, J.J., Kuncheva, L.I., Alonso, C.J.: Rotation forest: a new classifier ensemble method. IEEE Trans. Pattern Anal. Mach. Intell. **28**, 1619–1630 (2006)
17. Ye, L., Keogh, E.: Time series shapelets: a novel technique that allows accurate, interpretable and fast classification. Data Min. Knowl. Discov. **22**(1), 149–182 (2011)

StreamXM: An Adaptive Partitional Clustering Solution for Evolving Data Streams

Robert Anderson and Yun Sing Koh[⊠]

Department of Computer Science, University of Auckland, Auckland, New Zealand
rand079@aucklanduni.ac.nz, ykoh@cs.auckland.ac.nz

Abstract. A challenge imposed by continuously arriving data streams is to analyze them and to modify the models that explain them as new data arrives. We propose StreamXM, a stream clustering technique that does not require an arbitrary selection of number of clusters, repeated and expensive heuristics or in-depth prior knowledge of the data to create an informed clustering that relates to the data. It allows a clustering that can adapt its number of classes to those present in the underlying distribution. In this paper, we propose two different variants of StreamXM and compare them against a current, state-of-the-art technique, StreamKM. We evaluate our proposed techniques using both synthetic and real world datasets. From our results, we show StreamXM and StreamKM run in similar time and with similar accuracy when running with similar numbers of clusters. We show our algorithms can provide superior stream clustering if true clusters are not known or if emerging or disappearing concepts will exist within the data stream.

Keywords: StreamXM · StreamKM · X-means · Clustering · Streaming · Adaptive · Data-streams · Unsupervised learning

1 Introduction

Data stream mining has arisen as a key technique in data analysis. As detailed by Gaben et al. [1], data streams may continually generate new information, with no upper bound on the information produced by the stream. With data streams, clustering has a broad range of real-world applications. For instance, it has been found to be effective for differentiating Twitter posts by bots and real people [2], detecting fraudulent credit-card transactions [3] and detecting malicious traffic over networks [4]. In all of these cases, running the analysis in real-time provides extra value: bots can be screened from Twitter feeds, fraudulent credit card transactions can be denied and malicious traffic can be rejected, with hosts banned. In all of these cases, clustering data-streams allows potential value that traditional approaches to analysis would not.

Currently, there are few partitional clustering techniques for data stream mining that can adapt to unseen concepts as they emerge in a data stream. Partitional clustering creates a simple and effective way to describe datasets

© Springer International Publishing Switzerland 2015
S. Madria and T. Hara (Eds.): DaWaK 2015, LNCS 9263, pp. 270–282, 2015.
DOI: 10.1007/978-3-319-22729-0_21

that can provide an interpretable explanation of how data is changing over the life of a dataset, making it a useful approach to clustering for our problem. However, most partitional clustering approaches in a data stream require a fixed number of clusters to be predefined that cannot change over the lifespan of the stream. This will not be an effective technique for analysing data which changes over time. In this research, we proposed a partitional clustering technique that can adapt to underlying concepts without being constrained by a fixed level of clustering.

We introduce the novel idea of combining an approximation-based data-stream clustering approach with a clustering approach that uses the Bayes Information Criterion, a penalised-likelihood measure, to create an effective clustering for a given window of data. We propose two variants of our technique: StreamXM, and StreamXM with Lloyds, which relies upon Lloyds algorithm. Both these techniques are evolving partition-based clustering techniques for data streams. By adapting previous work in the area StreamKM's coreset construction [5] and X-means analyses upon these coresets, we combine the benefit of an efficient and cheap algorithm with the flexibility of another.

Two major contributions of our research include: (1) proposing implementations that can achieve similar quality of the less-flexible StreamKM's clustering on static data while retaining levels of memory usage and runtime that are acceptable in a streaming environment. (2) providing a partitional clustering technique that can adapt to appearing or disappearing concepts in underlying data, thus achieving similar or better quality and performance of clustering than less-flexible techniques without requiring a specific level of clustering to use.

In our next section, we discuss the research work in the area. In Sect. 3 we discuss our proposed algorithm. In Sect. 4, we briefly discuss our experimental setup and the datasets to be used for our comparative analysis. In the following section, we explore the significance of our results. Finally, we conclude the paper in Sect. 6.

2   Related Work

In recent years there has been a broad range of research in the area of clustering data streams [6–8]. A highly-regarded and well-known clustering technique that utilises hierarchical clustering is BIRCH [9]. This algorithm takes n instances and a desired number of clusters, and calculates the clustering feature of the set, comprising of the linear sum and square sum of instances. This method is shown to run quickly, but does not minimise sum-of-squared error well when compared to other current data stream clustering methods [5].

Guha et al. [10] proposed a method based upon a k-median approach, entitled STREAM. A k-median approach is similar to k-means, except that the value selected for a cluster centre must be the median value for instances within the cluster grouping. This clusterer reduces the sum-of-squared error when compared to a clusterer like BIRCH, but runs in superlinear time.

StreamKM [5] proposed a clustering algorithm that would allow k-means++ to achieve speeds required for data-stream clustering. They used a merge-and-reduce technique to reduce a large set of data points to a representative coreset tree. This allows fast seeding for k-means++ and by using a coreset size significantly smaller than the total dataset size, they could justify running complex operations such as k-means++ upon it. They explore the adaptive coreset, which provides a weighted multiset as a basis for the seeding. Using sampling based upon merge-and-reduce, they could maintain a sample of a fixed size that could represent all data seen, with a greater weighting placed on recent data. Use of this technique has been supported by recent applications for real-world data. For example, Miller et al. [2] use a variant of StreamKM for a proposed Twitter spammer detector that performed very accurately.

3   The StreamXM Algorithms

Here we detail our proposed algorithms, (1) StreamXM with Lloyds and (2) StreamXM, to allow adaptive clustering of unseen data.

Both of our proposed algorithms use X-Means [11], an approach that takes a minimum and maximum clustering size, and measures quality of clusterings in that range by the Bayes Information Criterion. This is a penalised-likelihood approach that weighs the explanatory value of more clusters against the increased complexity of the model. It considers the relative improvement in BIC by splitting each cluster. To measure this, it runs local k-means within each cluster, if it were to be split into two new clusters, and calculates whether these child clusters better explain the data than their parent. This allows an informed approach to choosing a sensible clustering without prior knowledge. Both of our technique use X-means in a slightly different manner.

3.1   StreamXM with Lloyds

The first implementation of StreamXM that we propose combines the strength of StreamKM's implementation of the Lloyd's algorithm with the adaptability of X-Means. The Lloyd's algorithm [12] is a local search that seeks to iteratively minimise the sum of squares after initialising with random centres, until reaching a stable level. This is common to many implementations of k-means. However, traditional k-means implementations often only guarantee finding local optima. It utilises D^2-weighting, encouraging centres to maximise their distance from each other on initialisation, and shows that it leads to significant speed improvements over traditional k-means clustering. We initially run X-means to determine the correct number of cluster and the centroids. Using the information derived from X-means, we then run k-means on static data for the minimum number of clusters set. Essentially, in our first iteration, it uses X-Means to determine the clustering level k for a coreset, and then uses the powerful implementation of k-means++ used by StreamKM to cluster the coreset.

The technique takes a stream of data instances and uses the merge-and-reduce technique to construct a coreset tree as it receives them, with the aim of keeping the entire data stream summarised within $\lceil log_2(n/m) + 2 \rceil$ buckets. A full bucket represents $2^{i-1}m$ instances sent to the algorithm, with i varying by the total number of instances inserted in the coreset tree [5]. Once the width is reached, an X-means clustering of the coreset is performed. It should be noted that in our implementation, and as per the initial X-means algorithm, the weight of each coreset point is disregarded. The coreset points that exist should provide a sufficient approximation of the source data without weights, especially once the data stream has been running for some time.

The algorithm we propose can be seen as working as follows. First, it compresses a set interval of streamed instances into a coreset, similarly to StreamKM. It then runs X-Means on the interval, resulting in a clustering of a set number of clusters. It checks the sum of squared errors between the cluster centres their points, and tries running k-means++, using Lloyds algorithm, five times (as found as a good combination of speed and quality by Ackermann et al. [5]), seeking a clustering with the same number of centres as the X-Means run, but with a reduced sum of squared error. It compares the X-Means clustering and five k-means++ clusterings and chooses that with the lowest sum of squared error. The process then repeats for each further interval.

Algorithm 1. StreamXM with Lloyds

1: **procedure** STREAMXM WITH LLOYDS
2:     **for each data point**:
3:     Insert data point into coreset tree
4:     **if** $pointCounter \% width == 0$ **then**
5:         $xmeansClustering \leftarrow$ X-means($coreset, minClusters, maxClusters$)
6:         $numCentres \leftarrow xmeansClustering.numClusters()$
7:         $curCost \leftarrow xmeansClustering.sumOfSquares()$
8:         $clusterCentres \leftarrow xmeansClustering.centres()$
9:         **repeat 5 times**:
10:           $kmeansPlusPlusClustering \leftarrow$ k-means++($coreset, numCentres$)
11:           **if** $kmeansPlusPlusClustering.cost() < curCost$ **then**
12:             $curCost \leftarrow kmeansPlusPlusClustering.cost()$
13:             $clusterCentres \leftarrow kmeansPlusPlusClustering.centres()$
14:     **return** $clusterCentres$

X-means will return the optimal clustering it can find for the coreset given different levels of k between **minClusters** and **maxClusters** [11]. It performs an initial k-means split with $k = 2$ upon the data. For the resulting two clusters, it considers splitting each cluster through running k-means with $k = 2$ on individual clusters. It calculates the relative Bayes Information Criterion (BIC) change for the optimal single-cluster split, and considers how well this model explains the coreset data when compared to the simpler model. The BIC of Pelleg and Moore's model penalises the log-likelihood of each model against its complexity, when deciding which to use.

Once this is completed, StreamXM repeats k-means++ on the data five times (as it is the number of iterations that the authors of StreamKM found worked best for their algorithm), with a k value decided by X-means. It will select the optimal clustering based upon sum of squared distance. This ensures that our algorithm provides the adaptivity of repeated X-means runs while gaining the benefit of repeated efficient k-means runs upon the coreset. This algorithm shares StreamKM's parameters of coreset size and width, and X-Means's parameters of minimum and maximum clusters to use.

We can infer the worst case time complexity of the algorithm through considering the algorithms it is based upon. We have inherited the coreset operations from StreamKM, as well as the repeated runs of k-means++. In [5], the authors show that by performing its complex operations (such as k-means++ and merge-and-reduce steps upon the coreset tree) upon the coreset of fixed size m rather than the total dataset, StreamKM avoids super-linear complexity. They show the algorithm performs in $\mathcal{O}(dnm)$ time, with d referring to the dimensionality of the data.

The BIC calculation can be performed in $\mathcal{O}(k)$ time and could be run a maximum $1 + k(k + 1)$ times per X-means, where k represents **maxClusters**. Traditional k-means runs in $\mathcal{O}(knd)$ time, but X-means only ever runs k-means for $k = 2$ for any run, though runs k-means $\frac{k(k+1)}{2}$ times. The implementation uses kd-trees, which avoid recalculating distance for subsets of points joined to a centroid through a process called 'blacklisting' described in [13], which leads to an average time-complexity that is often sub-linear [11]. X-means is only being run on the coreset, and not n. None of the factors that affect running time of X-means change as n changes, so we can state that the total additional time to run X-means scales linearly with n. We must select a coreset size m, a dimensionality d and a suitably restrained **maxClusters** value so that the increased time fits within the constraints of our application. Model memory requirements of StreamKM are shown to be $\Theta(dm \log(n/m))$ in [5], and our model introduces no further memory requirements beyond the coreset tree stored in StreamKM.

3.2   StreamXM

The second implementation of StreamXM we propose abandons k-means++ and relies on clusterings delivered by X-Means. Our technique takes an interval of data and constructs a coreset to represent it with a reduced number of points. It then runs X-Means upon the data, finding a level of k that is appropriate for the interval, as well as coordinates of k cluster centres. It then repeats X-Means on the same set of data $xmIter - 1$ times. Each time it reruns X-Means, it considers whether the new clustering it finds has a reduced sum-of-squared errors when compared to the previous clusterings it found. It selects the clustering with the lowest sum-of-squared errors as the chosen clustering for that interval. This algorithm is run on every interval of data until the stream ends. The pseudocode is shown in Algorithm 2.

Algorithm 2. StreamXM

1: **procedure** STREAMXM
2:    **for each data point**:
3:       Insert data point into coreset tree
4:       **if** $pointCounter \% width == 0$ **then**
5:          $xmeansClustering \leftarrow$ X-means$(coreset, minClusters, maxClusters)$
6:          $numCentres \leftarrow xmeansClustering.numClusters()$
7:          $curCost \leftarrow xmeansClustering.sumOfSquares()$
8:          $clusterCentres \leftarrow xmeansClustering.centres()$
9:          **repeat** $xmIter - 1$ **times**:
10:         $newXMeansClustering \leftarrow$ X-means$(coreset, minClusters, maxClusters)$
11:         **if** $newXMeansClustering.cost() < curCost$ **then**
12:            $curCost \leftarrow newXMeansClustering.cost()$
13:            $clusterCentres \leftarrow newXMeansClustering.centres()$
14:    **return** $clusterCentres$

This algorithm has the same parameters as StreamXM with Lloyds. It also has $xmIter$, the number of times X-Means is to be run on each interval of data. This allows StreamXM multiple chances to find a better clustering for the coreset, and decides on the best through a minimised sum-of-squares. Therefore, an increased value of $xmIter$ will tend to lead to a higher average clustering, and will also take more time. As described in the Experimental Setup section, we test to find the ideal value of $xmIter$, which allows a quality clustering without severely impacting runtime. The memory requirements of this algorithm will be the same as StreamXM with Lloyds.

4 Experimental Setup

To test our implementation, we used the MOA API (available from http://moa. cms.waikato.ac.nz), which provided an implementation of StreamKM to compare against. All experiments were run on an Intel Core i7-3770S running at 3.10GHz with 16.0GB of RAM on a 64-bit Windows 7.

Across our experiments, we used several evaluation measures to explore our algorithms' accuracy, speed and memory use. For every set of tests run, the experiment was repeated 30 times per set of factors, so as to receive a suitably representative mean and standard deviation for the underlying distribution.

To compare the quality of our clusterings we used two measures. First, we measured the sum of squared errors (SSQ) between cluster centres and actual instances belonging to each cluster centre across the experiment. This was calculated for each instance according to the clustering at the end of the interval that the instance was present within. As per Eq. 1, the SSQ is the sum across all clusters $i = 1$ to $i = K$ of all instances in each cluster x of the squared Euclidean distance between the instance and the cluster centre m_i. The sum of SSQ across all intervals gave the total SSQ for an experiment.

$$SSQ_{interval} = \sum_{i=1}^{K} \sum_{x \in C_i} dist^2(m_i, x) \tag{1}$$

Second, we measured purity of resulting clusterings. We calculated purity as per Eq. 2. ω represents the set of clusters in each interval-clustering while $\mathbb{C}$ represents the set of classes. We take the sum of the maximum class j within each cluster k across all intervals. Divided by the total number of instances N, we retrieve the purity for an experiment [14].

$$Purity(\Omega, \mathbb{C}) = \frac{1}{N} \sum_k \max_j |\omega_k \cap c_j| \tag{2}$$

4.1   Datasets

Our goal in testing our algorithms was: to show that they can perform comparably well in a static environment with StreamKM; to show that they can adapt to an underlying change in the number of clusters; and to demonstrate their value when used for analysis of real world data.

For thes synthetic dataset, we decided to use **BIRCHCluster**, a cluster generator included in the WEKA API (http://www.cs.waikato.ac.nz/ml/weka/). The advantage of this tool is that it creates a specified number of instances that fall into clusters in a spherical pattern, with a set dimensionality, with a set radius. BIRCH is primarily known as a clustering method [9], but can also generate clustering problems to test other techniques upon. Using **BIRCHCluster**, we created thirty datasets of differing size (250,000, 500,000 and 1 million instances), clusterings (5, 10, 20 and 50 underlying clusters). To generate an evolving dataset, we first generated the stream largest number of custers needed. To replicate dissapearing clusters, we would remove x number of clusters from the stream pass a point t. Vice versa if we simulated appearing clusters we would remove x number of clusters from the beginning of the stream until a point t. In our experiment t was set as half the entire stream size.

We ran both algorithms on the KDD Cup '99 Network Intrusion dataset. For consistency in our paper, we have split each dataset into 10 intervals for evaluation.

5   Results

In this section, we summarise results found from our experiments. First, we compare StreamXM's to StreamKM across in a static (non-evolving) stream. Second, we compare StreamXM's to StreamKM in an evolving stream. In the experiments, memory use are in bytes and runtime in milliseconds. For every set of tests run, the experiment was repeated 30 times per set of factors.

5.1   Performance and Quality of StreamXM on Static Datasets

These experiments compare our techniques StreamXM (**sxm**) and StreamXM with Lloyds (**sxmlloyds**) against StreamKM run with a value of k: half of the true clustering (**skmhalf**); the same as the true clustering (**skmfull**); and the same as the mean clustering used by StreamXM (**skm**). We have two main aims with these experiments. The first is to show that the quality of clusterings delivered by StreamXM and StreamXM with Lloyds is at least comparable with those delivered by StreamKM across datasets with varied clustering, dimensionality and stream size. The second is to show that the performance of StreamXM and StreamXM with Lloyds is suitable for a stream environment in which the runtime and memory have at worst a linear relationship with the varied factors in each experiment.

We are showing that our StreamXM variants work comparably to StreamKM on static datasets, as data streams when data is not evolving over time. All of these experiments used the parameters decided upon was: $xmIter = 3$ for StreamXM,a coreset of size 1000 and the BIC as the information criterion for StreamXM and StreamXM with Lloyds. $xmIter = 3$ was chosen as extensive experimentation showed that it was a reasonable setting. As these are static datasets, StreamXM variants will gain no advantage from their adaptive nature as there are no changes within the stream.

Here it is important to note that **skmfull** will always produce a close to perfect results for both Purity and SSQ measures as it was given perfect knowledge of the number of clusters available. In a real world data stream, determining the correct number of cluster before the stream has arrived is an impossible task. However for completeness we have included **skmfull** in our experiments.

Performance and Quality of StreamXM with Varying Clusters. Here, our experiments tested purity, sum-of-squared errors, memory usage and runtime in problems with 5, 10, 20 and 50 underlying (true) clusters. Here, we sought to explore how an increasing number of clusters affects the quality and performance of our techniques against StreamKM, and to ensure that increasing the underlying clustering of a dataset does not increase the time taken to cluster it in a superlinear fashion. We ran experiments with datasets of 5 dimensions and 250,000 instances.

Table 1 shows the performance of StreamXM and StreamXM with Lloyds in comparison with StreamKM. When we ran **skmfull** it was always be the best performer in a static environment because it is set to run with the true number of clusters in the underlying data stream. In reality it is highly improbable for the clusterer to be perfectly set for an unseen data stream, thus not reported in the table, due to space restrictions. In Table 2 we show the runtime and memory performance of our techniques. Although there is a slight overhead in our technique, we show in the next set of experiments that this is not an exponential problem as the stream size increases. Memory use across all techniques is consistently similar, and does not increase significantly with increases in the true clustering level. We can surmise that increased levels of k used in a model do not contribute to an increase in model memory requirements.

Table 1. Quality measures and clustering for algorithms run with varied true clustering

Model	Purity± 95%CI	SSQ×10³±95%CI	Mean k± 95%CI
True clustering = 5 and minClusters = 3			
skm	0.979 ± 0.020	2.277 ± 2.057	4.867 ± 0.124
skmhalf	0.681 ± 0.014	39.209 ± 3.857	3.000 ± 0.000
sxm	0.964 ± 0.011	4.798 ± 1.598	4.815 ± 0.085
sxmlloyds	0.898 ± 0.021	10.662 ± 2.951	4.347 ± 0.118
True clustering = 10 and minClusters = 5			
skm	0.857 ± 0.016	6.030 ± 0.906	8.100 ± 0.172
skmhalf	0.583 ± 0.010	28.323 ± 2.202	5.000 ± 0.000
sxm	0.834 ± 0.013	12.245 ± 1.446	8.037 ± 0.135
sxmlloyds	0.795 ± 0.012	10.589 ± 1.212	7.371 ± 0.129
True clustering = 20 and minClusters = 10			
skm	0.802 ± 0.013	5.575 ± 0.635	14.967 ± 0.239
skmhalf	0.580 ± 0.007	17.431 ± 1.063	10.000 ± 0.000
sxm	0.769 ± 0.011	11.176 ± 0.945	14.942 ± 0.192
sxmlloyds	0.765 ± 0.011	7.402 ± 0.704	14.145 ± 0.198
True clustering = 50 and minClusters = 25			
skm	0.750 ± 0.006	4.702 ± 0.282	34.467 ± 0.278
skmhalf	0.578 ± 0.005	10.807 ± 0.393	25.000 ± 0.000
sxm	0.684 ± 0.009	10.325 ± 0.559	34.459 ± 0.303
sxmlloyds	0.728 ± 0.007	5.536 ± 0.311	33.202 ± 0.274

Table 2. Performance measures for algorithms run with varied true clustering

Model	Runtime±95%CI	Memory×10⁶± 95%CI
True clustering = 5 and minClusters = 3		
skm	2912 ± 103	1.211 ± 0.019
skmhalf	2865 ± 99	1.207 ± 0.017
sxm	3115 ± 111	1.208 ± 0.018
sxmlloyds	3004 ± 108	1.209 ± 0.018
True clustering = 10 and minClusters = 5		
skm	2900 ± 91	1.211 ± 0.023
skmhalf	2822 ± 85	1.210 ± 0.022
sxm	3161 ± 102	1.210 ± 0.023
sxmlloyds	3016 ± 95	1.211 ± 0.024
True clustering = 20 and minClusters = 10		
skm	3083 ± 86	1.204 ± 0.021
skmhalf	2976 ± 82	1.205 ± 0.020
sxm	3626 ± 96	1.202 ± 0.020
sxmlloyds	3309 ± 91	1.204 ± 0.020
True clustering = 50 and minClusters = 25		
skm	3688 ± 70	1.222 ± 0.017
skmhalf	3528 ± 65	1.218 ± 0.018
sxm	5046 ± 89	1.214 ± 0.018
sxmlloyds	4466 ± 72	1.219 ± 0.017

Table 3. Quality measures and clustering for algorithms run with varied stream size

Model	Purity± 95%CI	SSQ ×10³± 95%CI	Mean k± 95%CI
Stream size = approx.250,000			
skm	0.857 ± 0.016	6.030 ± 0.906	8.100 ± 0.172
skmhalf	0.583 ± 0.009	28.323 ± 2.202	5.000 ± 0.000
sxm	0.834 ± 0.013	12.246 ± 1.446	8.037 ± 0.135
sxmlloyds	0.795 ± 0.012	10.590 ± 1.212	7.371 ± 0.129
Stream size = approx.500,000			
skm	0.855 ± 0.016	12.547 ± 1.859	8.067 ± 0.161
skmhalf	0.582 ± 0.010	56.572 ± 4.519	5.000 ± 0.000
sxm	0.833 ± 0.013	24.127 ± 2.773	8.022 ± 0.117
sxmlloyds	0.799 ± 0.012	20.618 ± 2.450	7.431 ± 0.130
Stream size = approx.1,000,000			
skm	0.848 ± 0.013	25.781 ± 3.425	8.000 ± 0.133
skmhalf	0.582 ± 0.009	112.763 ± 8.920	5.000 ± 0.000
sxm	0.833 ± 0.012	48.202 ± 5.206	8.019 ± 0.110
sxmlloyds	0.796 ± 0.011	42.015 ± 4.669	7.389 ± 0.118

Table 4. Performance measures for algorithms run with varied stream size

Model	Runtime± 95%CI	Memory ×10⁶± 95%CI
Stream size = approx. 250,000		
skm	2880 ± 89	1.211 ± 0.023
skmhalf	2799 ± 85	1.210 ± 0.022
sxm	3141 ± 99	1.211 ± 0.023
sxmlloyds	2989 ± 93	1.211 ± 0.024
Stream size = approx. 500,000		
skm	5818 ± 194	1.201 ± 0.022
skmhalf	5656 ± 186	1.197 ± 0.020
sxm	6317 ± 201	1.201 ± 0.022
sxmlloyds	6016 ± 195	1.196 ± 0.021
Stream size = approx. 1,000,000		
skm	11659 ± 359	1.216 ± 0.020
skmhalf	11378 ± 352	1.214 ± 0.019
sxm	12778 ± 392	1.214 ± 0.019
sxmlloyds	12136 ± 394	1.211 ± 0.018

Performance and Quality of StreamXM with Varying Stream Size.
Our experiments tested purity, sum-of-squared errors, memory usage and run-time in problems with 250,000, 500,000 and 1 million data iteams within the dataset. We explore how an increasing stream size affects the quality and performance of StreamXM against StreamKM. We ran experiments with datasets of 10 underlying clusters and 5 dimensions. In Table 3, we see sum-of-squared error scaling linearly with the number of instances across all models. Purity holds constant across all models. There is no evidence of StreamXM or StreamXM with Lloyds altering their mean k based upon the stream size.

In Table 4, we can see evidence of runtime and memory use scaling linearly across all models with the total number of instances in our data stream, with time increasing between 99.7 % and 102.4 % when doubling the number of instances for all models.

Table 5. Quality measures and clustering for algorithms run on evolving data streams

Table 6. Performance measures for algorithms run on evolving data streams

Model	Purity± 95%CI	SSQ ×10³± 95%CI	Mean k± 95%CI
	Stream shrinking from 6 clusters to 3 clusters		
skm	1.000 ± 0.000	0.408 ± 0.002	6.000 ± 0.000
sxm	0.942 ± 0.008	23.469 ± 4.549	4.564 ± 0.116
sxmlloyds	0.913 ± 0.010	29.630 ± 4.666	4.146 ± 0.128
	Stream growing from 3 clusters to 6 clusters		
skm	0.797 ± 0.008	90.654 ± 7.391	3.000 ± 0.000
sxm	0.942 ± 0.008	22.831 ± 3.931	4.560 ± 0.122
sxmlloyds	0.911 ± 0.010	31.056 ± 4.804	4.159 ± 0.134

Model	Runtime± 95%CI	Memory ×10⁶± 95%CI
	Stream shrinking from 6 clusters to 3 clusters	
skm	13504 ± 478	1.228 ± 0.018
sxm	12841 ± 442	1.227 ± 0.018
sxmlloyds	12561 ± 450	1.226 ± 0.018
	Stream growing from 3 clusters to 6 clusters	
skm	11180 ± 360	1.229 ± 0.016
sxm	12424 ± 395	1.229 ± 0.017
sxmlloyds	12155 ± 415	1.230 ± 0.016

5.2 Performance and Quality of StreamXM Run on Evolving Dataset Streams

In these experiments, we ran three algorithms: StreamXM (**sxm**), StreamXM with Lloyds (**sxmlloyds**) and StreamKM (**skm**). For StreamKM, we set k to the clustering level of the evolving dataset in the first section of data, as the change in clustering would not be known about when setting up the algorithm intially.

Examining Table 5, for the shrinking data stream, we see that StreamKM creates a very effective clustering through using $k = 6$ throughout the experiment. StreamXM delivers a very effective clustering with a mean k very close to what we should hope for (almost perfectly halfway between 3 and 6). For the growing data stream, StreamKM fails to create as effective a clustering, as three cluster centres cannot adequately capture the six clusters the data falls around in the second section of data. Both StreamXM and StreamXM with Lloyds maintain the quality of their clusterings achieved with the shrinking dataset. This situation helps highlight the value an adaptive clusterer can add when underlying concepts may change throughout the course of a stream.

From Table 6 StreamKM has a significantly slower runtime than StreamXM with Lloyds for the shrinking data stream, as it takes more time to fit six cluster centres to the data than StreamXM with Lloyds. StreamXM appears faster than StreamKM, but not significantly so in this experiment. When StreamKM has fewer clusters, it is significantly faster than both variants of StreamXM. The runtime for either variant of StreamXM is not significantly different from the other. In Table 7, we examine the change is clustering over each section of the data. As StreamXM and StreamXM with Lloyds are adaptive, we expect a significant change in clusters when the underlying true clustering changes.

5.3 Performance and Quality of StreamXM on Real-World Dataset

In this section, we examine quality and performance of running our techniques on the real-world datasets we have selected. We contrast how well they have done, and compare the overall clustering to the real data. However, we examine the fit of the clusters chosen by StreamXM variants to the data much more closely later in this section. Table 8 shows the relative quality of each clustering

Table 7. Mean clustering for each section of evolving data streams

Model	Actual clusters	Mean k	$\pm$ 95 %CI
Stream shrinking from 6 clusters to 3 clusters			
skm	6 clusters	6.000	$\pm$ 0.000
	3 clusters	6.000	$\pm$ 0.000
sxm	6 clusters	5.080	$\pm$ 0.033
	3 clusters	4.041	$\pm$ 0.033
sxmlloyds	6 clusters	4.593	$\pm$ 0.040
	3 clusters	3.746	$\pm$ 0.038
Stream growing from 3 clusters to 6 clusters			
skm	6 clusters	3.000	$\pm$ 0.000
	3 clusters	3.000	$\pm$ 0.000
sxm	3 clusters	4.040	$\pm$ 0.034
	6 clusters	5.073	$\pm$ 0.033
sxmlloyds	3 clusters	3.729	$\pm$ 0.038
	6 clusters	4.535	$\pm$ 0.042

Table 8. Quality measures and clustering: Intrusion dataset

Model	Purity	$\pm$ 95 %CI	SSQ $\times 10^3$	$\pm$ 95 %CI	Mean k	$\pm$ 95 %CI
skm	0.574	$\pm$ 0.004	1.386×10^8	$\pm 0.435 \times 10^8$	5.033	$\pm$ 0.199
skmfull	0.913	$\pm$ 0.012	212.224	$\pm$ 0.017	23.000	$\pm$ 0.000
skmhalf	0.665	$\pm$ 0.007	806.396	$\pm$ 0.010	12.000	$\pm$ 0.000
sxm	0.966	$\pm$ 0.009	114.262	$\pm$ 0.012	5.035	$\pm$ 0.152
sxmlloyds	0.568	$\pm$ 0.000	2.315×10^8	$\pm 0.479 \times 10^8$	4.858	$\pm$ 0.119

Table 9. Performance measures for algorithms: Intrusion dataset

Model	Runtime	$\pm$ 95 %CI	Memory $\times 10^6$	$\pm$ 95 %CI
skm	26442	$\pm$ 239	4.594	$\pm$ 0.010
skmfull	29814	$\pm$ 316	4.623	$\pm$ 0.012
skmhalf	27246	$\pm$ 234	4.605	$\pm$ 0.011
sxm	26975	$\pm$ 243	4.609	$\pm$ 0.011
sxmlloyds	26583	$\pm$ 232	4.608	$\pm$ 0.009

algorithm on our real-world datasets. Table 9, shows the relative performance of each clustering algorithm on our real-world datasets.

For the Intrusion dataset, StreamKM appears to have very high SSQ when using few clusters. StreamXM manages a more pure clustering with lower sum-of-squares than StreamKM with even a cluster per class available to it. The mean clustering of StreamXM variants is around 5, which would account for three dominant classes. The last two clusters could be for minor classes, or also for the major classes if they are bimodal. The overall clustering purity is impressive for StreamXM in this circumstance, especially considering a total of 23 underlying classes approximated by a mean of five cluster centres.

6 Conclusions

Our proposed algorithms, StreamXM and StreamXM with Lloyds, provide statistically supported clusterings through use of the BIC. The technique is usable on data streams through using coresets to approximate the data which they run upon. We have shown that our techniques retain similar quality of clustering, especially in terms of purity, to StreamKM, while not displaying notably different behaviour in terms of runtime nor memory use. We show that when used on data stream with appearing or disappearing concepts and real-world data, we can expect similar or better quality and performance of clustering without setting a specific level of clustering for our technique to use.

References

1. Gaber, M.M., Zaslavsky, A., Krishnaswamy, S.: Data stream mining. In: Maimon, O., Rokach, L. (eds.) Data Mining and Knowledge Discovery Handbook, pp. 759–787. Springer, US (2010)
2. Miller, Z., Dickinson, B., Deitrick, W., Hu, W., Wang, A.H.: Twitter spammer detection using data stream clustering. Inf. Sci. **260**, 64–73 (2014)
3. Hanagandi, V., Dhar, A., Buescher, K.: Density-based clustering and radial basis function modeling to generate credit card fraud scores. In: Proceedings of the IEEE/IAFE 1996 Conference on Computational Intelligence for Financial Engineering, pp. 247–251. IEEE (1996)
4. Leung, K., Leckie, C.: Unsupervised anomaly detection in network intrusion detection using clusters. In: Proceedings of the Twenty-Eighth Australasian Conference on Computer Science - Volume 38. ACSC 2005, pp. 333–342. Australian Computer Society, Inc., Darlinghurst (2005)
5. Ackermann, M.R., Märtens, M., Raupach, C., Swierkot, K., Lammersen, C., Sohler, C.: Streamkm++: a clustering algorithm for data streams. J. Exp. Algorithmics **17**, 2.4:2.1–2.4:2.30 (2012)
6. Wang, C.D., Lai, J.H., Huang, D., Zheng, W.S.: Svstream: a support vector-based algorithm for clustering data streams. IEEE Trans. Knowl. Data Eng. **25**, 1410–1424 (2013)
7. Aggarwal, C.C., Han, J., Wang, J., Yu, P.S.: A framework for clustering evolving data streams. In: Proceedings of the 29th International Conference on Very Large Data Bases-Volume 29, VLDB Endowment, pp. 81–92 (2003)
8. Cao, F., Ester, M., Qian, W., Zhou, A.: Density-based clustering over an evolving data stream with noise. In: SDM. vol. 6, SIAM, pp. 326–337 (2006)
9. Zhang, T., Ramakrishnan, R., Livny, M.: Birch: an efficient data clustering method for very large databases. In: ACM SIGMOD International Conference on Management of Data, pp. 103–114. ACM Press (1996)

10. Guha, S., Meyerson, A., Mishra, N., Motwani, R., O'Callaghan, L.: Clustering data streams: theory and practice. IEEE Trans. Knowl. Data Eng. **15**, 515–528 (2003)
11. Pelleg, D., Moore, A.: X-means: Extending k-means with efficient estimation of the number of clusters. In: Proceedings of the 17th International Conference on Machine Learning, Morgan Kaufmann, pp. 727–734 (2000)
12. Lloyd, S.: Least squares quantization in pcm. IEEE Trans. Inf. Theor. **28**, 129–137 (1982)
13. Pelleg, D., Moore, A.: Accelerating exact k-means algorithms with geometric reasoning. In: Proceedings of the fifth ACM SIGKDD International Conference on Knowledge Discovery and Data Mining, pp. 277–281. ACM (1999)
14. Manning, C.D., Raghavan, P., Schütze, H.: Introduction to Information Retrieval. Cambridge university press, Cambridge (2008)

Data Stream Mining with Limited Validation Opportunity: Towards Instrument Failure Prediction

Katie Atkinson[1], Frans Coenen[1(✉)], Phil Goddard[2], Terry Payne[1], and Luke Riley[1,2]

[1] Department of Computer Science, The University of Liverpool,
Liverpool L69 3BX, UK
coenen@liverpool.ac.uk
[2] CSols Ltd., The Heath, Runcorn, Cheshire WA7 4QX, UK

Abstract. A data stream mining mechanism for predicting instrument failure, founded on the concept of time series analysis, is presented. The objective is to build a model that can predict instrument failure so that some mitigation can be invoked so as to prevent the failure. The proposed mechanism therefore features the interesting characteristic that there is only a limited opportunity to validate the model. The mechanism is fully described and evaluated using single and multiple attribute scenarios.

Keywords: Data stream mining · Classification · Instrument failure prediction

1 Introduction

Instrument failure within scientific and analytic laboratories can lead to costly delays and compromise complex scientific workflows [14]. Many such failures can be predicted by learning a failure prediction model using some form of data stream mining. Data stream mining is concerned with the effective, real time, capture of useful information from data flows [7–9]. This necessitates the adoption of an incremental online mechanism to process the data as it becomes available. Data stream mining is often applied in domains where we wish to use data stream information for prediction purposes (for example in the case of email, the prediction of spam email versus non-spam email [3]). Thus where we wish to predict the value of some variable x in the context of a set of variables y that feature in the data stream. The principal challenges is that the data is potentially infinite and therefore only a fixed proportion can be stored.

A common application of data stream mining is the analysis of instrument (sensor) data with respect to some target objective [4,5,12]. This paper explores the idea of using data stream mining to predict the failure of the instruments (sensors) themselves. This has the novel feature that persistent validation of the data stream model is not possible, the idea is that on predicting instrument

© Springer International Publishing Switzerland 2015
S. Madria and T. Hara (Eds.): DaWaK 2015, LNCS 9263, pp. 283–295, 2015.
DOI: 10.1007/978-3-319-22729-0_22

failure appropriate maintenance is scheduled and therefore any predicted failure cannot normally be confirmed. More specifically this paper presents a probabilistic time-series analysis technique applied to data stream subsequences to predict instrument failure. Of note is the mechanism whereby significant attributes in the data stream are separated from noise attributes using a probabilistic learning approach.

2   Formalism

We assume a set of k instruments $\mathbf{Inst} = \{inst_1, inst_2, \ldots, inst_k\}$. Periodically each (operational) instrument $inst_i$ transmits a "data packet" D_i. A data packet D_i associated with an instrument $inst_i$ comprises n values $\{v_1, v_2, \ldots, v_n\}$ such that each value is correlated with a data attribute. The set of available attributes is indicated using the notation $A = \{a_1, a_2, \ldots, a_n\}$. There is a one-to-one correspondence between D_i and A $(D_i| = |A|)$. All instruments in the set $\mathbf{Inst}$ are assumed to subscribe to the same set of attributes. Periodically an instrument fails as a consequence of some fault. The conjecture is that this failure can be predicted by analysis of the attribute values.

The data packets continuously received from the collection of k instruments form a data stream. The information in the data packets is processed to produce a continuous time series. Each instrument $inst_i$ has a collection of n time series (one per data attribute), denoted by $\mathbf{T_i}$, associated with it such that $\mathbf{T_i} = \{T_{i,1}, T_{,2}, \ldots, T_{i,n}\}$. Each time series T_{ij} comprises m real-valued readings $T_{ij} = \{t_{ij_1}, t_{ij_2}, \ldots, t_{ij_m}\}$ ordered chronologically such that t_{ij_m} is the most recent. The instruments continuously generate and transmit data and thus the value for m, for each time series, increases with time. Note that at any specific time the values for m across the set of instruments $\mathbf{Inst}$ is not necessarily the same, although it will be constant with respect to the collection of time series $\mathbf{T_i}$ associated with a specific instrument $inst_i$. Given a time series $T_{ij} = \{t_{ij_1}, t_{ij_2}, \ldots\}$ for attribute j associated with instrument i the subsequence $S_{ij} = \{s_{ij_1}, s_{ij_2}, \ldots\}$ is a subsequence of T_{ij} such that $|S_{ij}| < |T_{ij}|)$. The *current subsequence* for a time series T_{ij} is the sequence $\{t_{ij_{m-p}}, \ldots, t_{ij_m}\}$ where p is some predefined subsequence length. Thus a current subsequence is the most recently received set of p values for a particular attribute j associated with a particular instrument i, we indicate this using the notation S'_{ij}.

3   Failure Prediction

From the foregoing the idea is to predict instrument failure according to the nature of the current time series subsequences associated with a particular instrument $inst_i$. We can learn the nature (shape) of subsequences that are good predictors of failure from observing the subsequences associated with instruments that have failed immediately prior to their failure. This is conceptualised as a two class classification (prediction) problem, failure versus non-failure.

There are a variety of ways of conducting time series based classification but it is generally acknowledged that the most satisfactory is the K Nearest Neighbour algorithm (KNN) [1,2] where K is the number of neighbours considered (in many cases $K = 1$ is used, hence 1NN). The main issue is the distance measure to be used [6]. The simplest measure is Euclidean distance; in which case the distance between two equal length subsequences $M = \{m_1, m_2, \ldots, m_l\}$ and $N = \{n_1, n_2, \ldots, n_l\}$ is calculated using Eq. 1. An alternative might have been to use something more sophisticated like Dynamic Time Warping (DTW) [13,15]; however, for the short time series considered with respect to the work presented in the paper, this seems unnecessary. The Euclidean based approach was therefore adopted.

$$dist(M, N) = \sum_{i=1}^{i=l} (m_i - n_i)^2 \tag{1}$$

As noted above, a particular challenge associated with the form of data stream mining considered in this paper is that we do not know which attribute is the sentinel attribute. So that we can learn which is the relevant attribute we need to allow λ instruments to fail. If we start making predictions at too early a stage in the process, instrument failure will be predicted unnecessarily causing the affected instrument to be taken off line, so that mitigating maintenance can be undertaken, when this was not required. Thus the process begins with a "learning phase" where we allow a number of instruments to fail so that we can build an effective KB for the purpose of future instrument failure prediction. This learning phase is measured in terms of the parameter λ.

Algorithm 1. Instrument Failure Prediction Engine ($main()$)

1: Input D_i
2: **if** $D_i = null$ **then**
3:     Instrument has failed $addSubsequencesToKB(S_i')$
4:     $pruneKB()$
5: **else**
6:     **for** $\forall v_{ij} \in D_i$ **do**
7:         Update time series S_{ij}' in DB by removing $S_{ij_{m-p}}'$ and adding adding v_{ij}
8:     **end for**
9:     **if** Not in learning phase **then**
10:         $prediction(inst_i)$
11:     **end if**
12: **end if**

The top level instrument failure prediction process is presented in Algorithm 1. The algorithm uses two storage structures KB (Knowledge Base) and DB (Database). The first is a set of subsequences of length p that are indicative of failure. KB is empty on startup and built up as the data stream starts to be processed. With respect to KB we used the following notations: (i) S indicates

---

Algorithm 2. Failure Prediction ($prediction(inst_i)$)

1: **for** $j = 1$ to $j = n$ **do**
2:   **for** $\forall S_j \in KB$ **do**
3:     **if** $distance(S'_{ij} \in DB, S) \leq \sigma)$ **then**
4:       Predict failure for $inst_i$
5:       Break
6:     **end if**
7:   **end for**
8: **end for**

---

the complete set of time series subsequence, each of length p, contained in KB; (ii) S_j indicates the set of subsequences belonging to attribute j ($S_j \subseteq S$), and (iii) S_{j_z} indicates a specific subsequence z associated with attribute j ($S_{j_z} \in S_j$). The second storage structure, DB, holds a set of subsequences of length p, one for each attribute j, associated with each instrument i. DB is also empty on startup and built up as the data stream starts to be processed. With respect to DB we use the following notation: (i) S'_i to indicate the set of current subsequences associated with instrument $ints_i$, and (ii) S'_{ij} to indicate the current subsequence associated with instrument $ints_i$ for attribute j. The input to Algorithm 1 is the most recent data packet D_i for instrument $inst_i$. If this is empty ($D_i = null$) then this indicates an instrument failure and KB is updated with the set of current subsequences S'_i associated with instrument $inst_i$ using the function $addSubsequenceToKB(S_{ij})$ (line 3). A KB pruning process is then applied (line 4); the significance of this will become clear later in this section. Otherwise we update the DB entry for $inst_i$ by, for each attribute j, removing the oldest value and appending the newly received value (line 7). Recall that subsequences are of length p. If we are no longer in the learning phase we then enter into the prediction phase (line 10).

The prediction process is described in Algorithm 2. During this process the set of current time series sub sequences S'_i in DB, associated with instrument $inst_i$, is compared with the corresponding subsequences contained in KB (using the Euclidean distance measured calculate using Eq. 1; however, any other similarity measure could equally well have been adopted). If the computed distance is less than or equal to a given similarity threshold σ (line 3) failure for instrument $inst_i$ is predicted (line 4) and as a result appropriate maintenance applied.

The KB update process, $addSubsequenceToKB(S'_i)$, is presented in Algorithm 3. As noted above the challenge here is that we do not know which attribute is the sentinel attribute. We assume at least one, but it may be all n attributes or some proportion between 1 and n. Each subsequence S in KB has a weighting w associated with it. This weighting is adjusted as further current time series subsequences are added to KB. Whenever a weighting associated with a subsequence falls below a given threshold γ the subsequence is removed; this is the pruning step included in Algorithm 1 (line 4). The weighting w_{j_z} for a particular KB subsequence S_{j_z} is calculated as presented in Eq. 2 where: (i) $count_{j_z}$ is a count of the number of occasions that S_{j_z} has been recorded as a result of instrument failure, and (ii) $|KB|$ is the size of KB in terms of the number of subsequences held.

$$w_{j_z} = \frac{count_{j_z}}{|KB|} \tag{2}$$

Algorithm 3. Update KB $(addSubsequencesToKB(S_i'))$

1: Input S_i'
2: **for** $j = 1$ to $j = n$ **do**
3:   **if** $S_{ij}' \in KB$ **then**
4:     $count_j = count_j + 1$
5:   **else**
6:     Add S_{ij}' to KB
7:     $count_j = 1$
8:   **end if**
9: **end for**
10: **for** $j = 1$ to $j = n$ **do**
11:   **for** $z = 1$ to $z = |S_{ij}|$ $(S_{ij} \in KB)$ **do**
12:     $weight_{j_z} = \frac{count_{j_z}}{|KB|}$ (Eq. 2)
13:   **end for**
14: **end for**

Algorithm 4. KB pruning $(prune(S_i'))$

1: **for** $j = 1$ to $j = n$ **do**
2:   **for** $z = 1$ to $j = |S_j|$ $(S_j \in KB)$ **do**
3:     **if** $weight_{j_z} < \gamma$ **then**
4:       remove S_{j_z} from KB
5:     **end if**
6:   **end for**
7: **end for**
8: **if** KB has changed **then**
9:   Recalculate weightings as per lines 10 to 14 in Algorithm 3
10: **end if**

Returning to Algorithm 3 the input is a set of current time series subsequences S_i', one subsequence per attribute in the system, associated with instrument $inst_i$. The set $S_i' = \{S_{i1}', S_{i2}', \ldots, S_{in}'\}$ is processed attribute by attribute. If S_{ij}' is already in KB its count is updated (line 4), otherwise S_{ij}' is added to KB (line 6) and its associated count set to 1 (line 7). Once the process of incorporating S_i' into KB is complete the weightings are (re)calculated for the entire contents of KB.

Algorithm 4 presents the KB pruning process whereby subsequences with associated weightings of less than γ are pruned from the KB on the grounds that they have only been infrequently associated with instrument failure and are therefore not good predictors of such failure. If such sequences were retained in *KB* this would cause erroneous instrument failure prediction causing the instrument to be taken off-line unnecessarily. Selection of the most appropriate value for γ is thus important and is discussed further in Sect. 4.

4   Evaluation

The proposed instrument failure prediction mechanism, as described in Sect. 3 above, was evaluated using a simulated environment comprising a collection of virtual instruments. The nature of this environment is presented in Sub-Sect. 4.1. The metric used to measure performance is introduced in Sub-Sect. 4.2. Evaluation was then conducted by considering two categories of scenario. The first was a simplistic scenario that considered instruments that featured only one sentinel attribute. The second considered instruments that featured a collection of attributes one of which was the sentinel attribute. The results obtained are presented and discussed in Sub-Sects. 4.3 and 4.4 respectively.

4.1   Simulation Environment

The simulation environment used for evaluation purposes assumed k instruments that featured n attributes each. The simulation operated on a loop. On each iteration a proportion of the instruments performed some sample measuring activity Whether an instrument is active or not is decided in a probability driven random manner (a probability value of $p_{sample} = 0.4$ was used). On each iteration the attribute values associated with each instrument were updated. To this end two different types of attribute were utilised: (i) activity dependent attributes and (ii) activity independent attributes. The values associated with the first were incremented/decremented on each simulation iteration whenever an instrument was active. Activity independent attributes were incremented/decremented on each iteration regardless of whether an instrument was active or not. Activity dependent attributes were designated to cause instrument failure once a particular sentinel value was reached, we refer to such attributes as *sentinel attributes*. In any given simulation the number of designated attributes that can potentially cause failure could be between 1 and n (thus between one and all); however, with respect to the evaluation presented here, scenarios featuring only one activity dependent attribute were used, this was therefore designated as the sentinel attribute. Attributes not so designated simply acted as providers of "noise".

The simulation operated using a Multi-Agent Based Simulation (MABS) environment such that each agent (instrument) communicated with a central controller agent, the Instrument Failure Prediction Engine (IFPE), which would in turn form part of a Laboratory Instrument Management System (LIMS)[1].

---

[1] A LIMS is a software system designed to manage laboratory operations. LIMS are used throughout the laboratory analysis industry.

The IFPE monitors the incoming instrument data and, using the algorithms presented in Sect. 3, predicts instrument failure. The interface between an instrument and the IFPE was enabled using a bespoke software interface called a *Dendrite*[2]. In the simulation, whenever instrument failure is detected, the affected instrument goes into a *maintenance state* for a simulation time t_{maint} after which it comes back "on stream" but with its attribute value set reset to the start values. Should the IFPE fail to detect an instrument failure in time the instrument will fail and, within the realm of the simulation, require replacement. The replacement time is given by a simulation time of $t_{replace}$ ($t_{replace} > t_{maint}$). For the evaluation reported here $t_{replace} = 20$ and $t_{maint} = 10$ was used.

4.2 Evaluation Metrics

Prediction/classification systems are typically evaluated using metrics such as precision and recall, true positive and false positive rate [11]. However in our case, because we wish to intervene prior to failure we have no information as to whether our predictions were correct or not. Instead we use an accumulated "gross profit" (GP) measure. On each iteration of the simulation, whenever an instrument conducts some sampling activity, a profit of g_{sample} is gained. Instrument maintenance incurs a cost of g_{maint} and instrument replacement a cost of $g_{replace}$ such that $g_{sample} < g_{maint} < g_{replace}$ (off course when an instrument is undergoing maintenance or is being replaced there is an additional cost because the instrument is also not earning anything).

Table 1. Comparison in terms of gross profit ($k = 20$).

p	Similarity threshold σ									
	0	1	2	3	4	5	6	7	8	9
2	**711**	706	687	657	627	593	561	528	504	480
3	691	**696**	692	683	666	650	623	601	583	557
4	650	683	**687**	677	672	663	654	639	622	605
5	575	657	**677**	675	667	660	654	645	640	627
6	470	615	659	**668**	667	661	650	644	636	633
7	351	558	634	657	**660**	658	651	644	636	632
8	248	479	595	635	650	**655**	653	646	639	632
9	167	389	540	606	635	646	**648**	645	641	634

4.3 Single Sentinel Attribute Evaluation

The aim of the single attribute evaluation was to determine the effect of different σ (similarity threshold), p (subsequence length) and k (number of instruments)

---

[2] The Dendrites software is available from CSols Ltd, http://www.csols.com.

Table 2. Comparison in terms of gross profit ($k = 40$).

p	Similarity threshold σ									
	0	1	2	3	4	5	6	7	8	9
2	**1445**	1425	1381	1319	1260	1186	1126	1064	1014	967
3	**1425**	1409	1396	1374	1338	1300	1256	1210	1167	1120
4	1381	**1400**	1395	1369	1352	1333	1312	1277	1245	1211
5	1299	1374	**1387**	1368	1343	1330	1310	1299	1285	1259
6	1156	1326	**1369**	1364	1348	1331	1306	1291	1277	1268
7	949	1250	1338	**1353**	1347	1336	1316	1291	1276	1262
8	712	1136	1292	1332	**1340**	1334	1321	1303	1284	1265
9	501	986	1223	1299	1325	**1328**	1320	1308	1291	1275

settings on the GP measure and to provide some insight into the failure prediction mechanism. This was done by running the simulation using a range of σ values ($[0, 9]$) against a range of p values ($[2, 10]$) and using $k = 20$ and $k = 40$. Recall that the minimum size of a subsequence is 2 (otherwise it will not be a sequence). For each parameter combination 1000 simulation runs were conducted, with 200 iterations for each run, and average values recorded for gross profit, number of instrument failures, number of maintained instruments and the size of KB. Throughout a single incremental, activity dependent, attribute was used with the sentinel value set to 20.

Tables 1 and 2 show the GP values obtained (best results indicated in bold font) when $k = 20$ and $k = 40$. From the tables it can be seen that there is a clear correlation between σ and p in terms of GP, for best results larger p values require a larger associate σ value. In both cases ($k = 20$ and $k = 40$) the best result, highest GP, was obtained using a combination of $\sigma = 0$ (exact matching) and $p = 2$ (GP of 711 and 1445 respectively). The lowest GP was generated using $\sigma = 9$ and $p = 9$ because prediction was at its least precise using these parameters and consequently maintenance was frequently conducted prior to this being a necessity. For reference the average GP value over 1000 repeated simulations and 200 iterations per simulation, when failure prediction and maintenance was not conducted, using $k = 20$ was 44, decreasing steadily as the simulation continued (88 when $k = 40$).

Tables 3 and 4 show comparisons of the average number of instruments that fail during the simulation runs. The number of instruments that fail decreased as the value for the σ threshold increased. This was because instruments were going into maintenance well before they would actually fail. Using $\sigma = 9$ similarity between a time series sequence in the KB and that currently associated with an instrument is measured in a fairly course manner hence more instruments go into maintenance. This is illustrated in Tables 5 and 6 which show the number of instruments maintained. From these two tables it can be seen that as the similarity threshold σ increased, the number of machines maintained increased

Table 3. Comparison in terms of number of failed machines ($k = 20$).

p	Similarity threshold σ									
	0	1	2	3	4	5	6	7	8	9
2	2	1	1	1	1	1	1	1	1	1
3	3	2	1	1	1	1	1	1	1	1
4	7	3	2	2	1	1	1	1	1	1
5	12	5	3	2	2	2	1	1	1	1
6	20	9	5	3	2	2	2	2	1	1
7	29	13	7	4	3	2	2	2	2	2
8	37	19	10	6	4	3	3	2	2	2
9	44	26	14	9	6	4	4	3	2	2

Table 4. Comparison in terms of number of failed machines ($k = 40$).

p	Similarity threshold σ									
	0	1	2	3	4	5	6	7	8	9
2	2	1	1	1	1	1	1	1	1	1
3	4	2	1	1	1	1	1	1	1	1
4	7	3	2	2	1	1	1	1	1	1
5	13	6	3	2	2	2	1	1	1	1
6	24	10	5	3	2	2	2	2	1	1
7	39	16	8	5	3	3	2	2	2	2
8	57	25	12	7	5	4	3	2	2	2
9	74	36	18	10	7	5	4	3	3	2

Table 5. Comparison in terms of number of maintained machines ($k = 20$).

p	Similarity threshold σ									
	0	1	2	3	4	5	6	7	8	9
2	60	62	63	66	68	70	73	75	76	78
3	58	61	62	64	65	66	68	70	71	73
4	55	60	62	63	64	65	66	67	68	69
5	48	57	60	62	63	64	65	66	67	67
6	40	53	58	61	63	64	65	66	67	67
7	30	47	55	59	61	63	64	65	66	67
8	21	41	51	56	59	61	63	64	65	66
9	13	33	46	53	57	60	61	63	64	65

Table 6. Comparison in terms of number of maintained machines ($k = 40$).

p	Similarity threshold σ									
	0	1	2	3	4	5	6	7	8	9
2	122	125	129	133	138	143	147	151	155	158
3	120	125	127	129	132	135	138	141	144	148
4	116	123	126	129	131	132	134	136	139	141
5	109	119	124	127	130	131	133	134	136	138
6	97	115	121	125	128	131	133	135	136	137
7	80	107	118	123	126	129	131	133	135	137
8	60	97	112	119	124	127	130	132	134	136
9	42	84	106	115	121	125	128	130	132	134

also, because the prediction becomes coarser. It can also be noted that the number of machines maintained increased as p decreased. Comparing the statistics for the number of instruments maintained given in the tables with those for the number of failed instruments, these are roughly inversely proportional. For completeness, when running the simulation without failure prediction (over 1000 simulation runs with 200 iterations per simulation) the average number of failed machines when $k = 20$ is 54, and when $k = 40$ it is 109.

Tables 7 and 8 show the recorded sizes, in terms of number of time series subsequences, of KB for $k = 20$ and $k = 40$. As anticipated the number of subsequences in KB decreases as the σ value increases. As already noted, this is because as σ is increased the prediction becomes less precise so the KB requires fewer subsequences. The number of subsequences in the KB also decreases with p; this is because as the p value is reduced the number of possible value combinations making up a time series subsequence also decreases (there are therefore fewer possible subsequences that can be included in the KB when p is small).

Table 7. Comparison in terms of KB size ($k = 20$).

p	Similarity threshold σ									
	0	1	2	3	4	5	6	7	8	9
2	2	1	1	1	1	1	1	1	1	1
3	3	2	1	1	1	1	1	1	1	1
4	6	3	2	2	1	1	1	1	1	1
5	12	5	3	2	2	2	1	1	1	1
6	20	9	4	3	2	2	2	2	1	1
7	29	13	7	4	3	2	2	2	2	2
8	37	19	10	6	4	3	3	2	2	2
9	44	26	14	9	6	4	4	3	2	2

Table 8. Comparison in terms of KB size ($k = 40$).

p	Similarity threshold σ									
	0	1	2	3	4	5	6	7	8	9
2	2	1	1	1	1	1	1	1	1	1
3	3	2	1	1	1	1	1	1	1	1
4	6	3	2	2	1	1	1	1	1	1
5	13	6	3	2	2	2	1	1	1	1
6	23	10	5	3	2	2	2	2	1	1
7	39	16	8	5	3	3	2	2	2	2
8	57	25	12	7	5	4	3	2	2	2
9	74	36	17	11	7	5	4	3	3	2

Table 9. Best parameter settings for a range of attribute set sizes, and $k = 20$. Average results obtained from 500 simulation runs per parameter permutation, 1000 iterations per simulation

# Atts. (n)	σ	p	ω	λ	# Fail. inst	# Main. inst	Final KB size	# KB Values pruned	GP
2	1	2	0.250	22	23	319	1	28	2884
3	1	2	0.225	25	26	323	1	60	2727
4	1	2	0.250	27	28	320	1	93	2701
5	1	2	0.175	31	32	325	1	137	2527
6	1	2	0.150	29	30	342	1	160	2333
7	1	2	0.125	30	31	352	1	194	2176
8	1	2	0.125	35	36	332	1	261	2308
9	1	2	0.100	36	37	352	1	300	2009
10	2	2	0.100	33	34	394	1	313	1502

4.4   Multiple Attribute Evaluation with a Single Sentinel Attribute

To evaluate the proposed instrument failure prediction mechanism in the context of multiple attributes a single sentinel attribute was used, selected at random from the set A. In the simulation the value for the remaining attributes were modelled using a sin curve with a fixed amplitude of 20 and a multiple of the attribute number as the frequency. For the "key attribute" (the attribute designated as the cause of failure) a sentinel value of 20 was again used. Experiments were conducted to investigate the effect of using a range of values for ω and λ. Recall that ω is the weighting threshold used to prune unwanted TSS values from KB; while λ is the learning window size, in other words the number of instruments that the mechanism allows to fail prior to commencing failure prediction.

Table 10. Learning window size (λ) versus Weighting threshold (ω), comparison in terms of gross profit ($k = 20$, $n = 5$, $\sigma = 1$ and $p = 2$)

ω	Learning window size (λ)							
	24	26	28	30	32	34	36	38
0.050	−14543	−14492	−13685	−13652	−13360	−13448	**−12686**	−12774
0.075	−6801	−6072	−5522	−5010	−5033	−4498	**−4460**	−4518
0.100	−1131	−1056	−529	−637	−173	−12	**22**	12
0.125	890	**1540**	1174	1377	1359	1306	1373	1245
0.150	1913	1788	2049	2097	2194	2061	**2245**	2125
0.175	2251	2291	2307	2355	**2401**	2377	2393	2285
0.200	2276	2406	2364	2292	2454	2428	**2481**	2448
0.225	−138	−140	−139	−138	−139	**−137**	−139	−138
0.250	−139	**−137**	−138	−137	−138	−138	−139	−139

Table 9 shows the mean best parameter settings for a range of numbers of attributes ($n = [1, 10]$) that serve to maximise the GP value. For each parameter combination (σ, p, ω and λ) the simulation was run 1000 times for 1000 iterations and mean values recorded. In addition to the best parameter settings, the number of failed and maintained instruments, the size of KB at the end of each run and the number of pruned KB values was also recorded as well as the final GP value. Without prediction and consequent maintenance the average number of failed machines was 281, and the average GP was -140. Note that as n was increased the number of noise attributes increased and it became harder to predict instrument failure hence the value of λ increases with n. Consequently erroneous instrument failure prediction causes larger numbers of instruments to, perhaps unnecessarily, go into maintenance. The increase in the number of instruments going into maintenance as n increases is reflected by the decreasing recorded GP values. The prediction becomes harder because the number of potential KB values increases, hence the amount of KB pruning that is undertaken also increases with n. From the table it is interesting to note that $\sigma = 1$ and $p = 2$ consistently produce the best result. It is also interesting to note that $|KB| = 1$ is consistently recorded; this is because when using $p = 2$ there are only two possible time series subsequences that can be included in KB and if $\sigma = 1$ is used one of these will then be superfluous.

Tables 10 and 11 shows the effect on GP using a range of values for ω ({0.050, 0.075, 0.100, 0.125, 0.150, 0.175, 0.200, 0.225, 0.250}) and a range of values for λ ({24, 26, 28, 30, 32, 34, 36, 38}) when $n = 5$ and $n = 10$ respectively. For the experiment $\sigma = 1$ and $p = 2$ was used because, as demonstrated above, these settings produced good results. For each parameter combination the simulation was again run 1000 times for 1000 iterations and the average GP values recorded. From the tables it can be seen that the choice of the most appropriate value for ω is important, either side of the optimum value, GP quickly starts to fall. This is particularly so with larger values of n as can be seen by comparing the two tables.

Table 11. Learning window size (λ) versus Weighting threshold (ω), comparison in terms of gross profit ($k = 20$, $n = 10$, $\sigma = 1$ and $p = 2$)

ω	Learning window size (λ)							
	24	26	28	30	32	34	36	38
0.050	−11099	−10851	−10638	−10668	−11150	−10689	**−10414**	−10499
0.075	−1838	−1735	−1058	−995	−888	−1411	**−786**	−932
0.100	1005	1191	1057	1098	1011	**1561**	1338	1314
0.125	**−137**	−139	−138	−138	−139	−139	−138	−139
0.150	−138	−139	−138	−138	**−137**	−138	−138	−139
0.175	**−138**	−139	−139	−139	−140	−138	−138	−138
0.200	−140	−140	−140	**−138**	−139	−139	−138	−139
0.225	**−138**	−139	−139	−140	−138	−140	−138	−140
0.250	−139	−138	−139	**−137**	−140	−140	−140	−139

5   Summary and Conclusions

A mechanism, founded on time series analysis, for predicting instrument failure using data stream mining has been proposed. The mechanism has been fully described. The novel feature of the mechanism is that, unlike in the case of more standard data stream mining prediction applications, there is only limited opportunity for validation. Evaluation was conducted using a simulated multi-agent based environment comprising k virtual machines communicating, via a bespoke interface called a Dendrite, with a central Instrument Failure Prediction Engine (IFPE). The mechanism uses four parameters: (i) a time series subsequence similarity threshold σ, (ii) a time series subsequence length value p, (iii) a weighting threshold ω used when pruning KB and (iv) a learning window measured in terms of the number of instruments allowed to fail λ. The presented evaluation indicated that best results are obtained when $\sigma = 1$ (almost exact matching between current time series subsequences associated with individual instruments and subsequences in KB) and $p = 2$ (the size of a subsequence). The optimum value for λ increases with n. Finding the optimum value for ω is challenging. The sensitivity associated with the ω parameter is thus a subject for future work. The intention is also to investigate the operation of scenarios where we have several sentinel attributes and alternative prediction mechanisms using (for example) dynamic classification association rule or decision tree based techniques. The advantage of these last two techniques is that they take into account the negatives as well as the positives; the mechanism that has been presented in this paper, although operating well, only predicts failure, not non-failure. Incorporating both might provide for a better predictor.

References

1. Bagnall, A., Lines, J.: An experimental evaluation of nearest neighbour time series classification. Technical report 1406.4757, Cornell University Library (2014)
2. Batista, G., Wang, X., Keogh, E.: A complexity-invarient distance measure for time series. In: Proceedings of 2011 SIAM International Conference on Data Mining, pp 699–710 (2011)
3. Grosan, C., Abraham, A.: Machine learning. In: Grosan, C., Abraham, A. (eds.) Intelligent Systems. ISRL, vol. 17, pp. 261–268. Springer, Heidelberg (2011)
4. Cohen, I., Goldszmidt, M., Kelly, T., Symons, J., Chase, J.S.: Correlating instrumentation data to system states: a building block for automated diagnosis and control. In: Proceedings of 6th Symposium on Operating Systems Design and Implementation, OSDI 2004, pp 231–244 (2004)
5. Cohen, L., Avrahami-Bakish, G., Last, M., Kandel, A., Kipersztok, O.: Real time data mining-based intrusion detection. Inf. Fusion (Spec. Issue Distrib. Sens. Netw.) 9(3), 344–354 (2008)
6. Ding, H., Trajcevski, G., Scheuermann, P., Wang, X., Keogh, E.: Querying and mining of time series daya: experimental comparison of representations and distance measures. In: Proceedings of VLDB 2008, pp 1542–1552 (2008)
7. Gaber, M.M., Zaslavsky, A., Krishnaswamy, S.: Mining data streams: a review. ACM SIGMOD Rec. 34(2), 18–26 (2005)
8. Gaber, M.M., Gama, J., Krishnaswamy, S., Gomes, J.B., Stahl, F.: Data stream mining in ubiquitous environments: state-of-the-art and current directions wiley interdisciplinary reviews. Data Min. Knowl. Disc. 4(2), 116–138 (2014)
9. Gama, J.: Knowledge Discovery from Data Streams. Chapman and Hall, California (2010)
10. Gama, J., Zliobaite, I., Bifet, A., Pechenizkiy, M., Bouchachia, A.: A survey on concept-drift adaptation. ACM Comput. Surv. 46(4), 2014 (2014)
11. Hand, D.J.: Measuring classifier performance: a coherent alternative to the area under the ROC curve. Mach. Learn. 77(1), 103–123 (2009)
12. Kargupta, H., Bhargava, R., Lou, K., Powers, M., Blair, P., Bushra, S., Dull, J.: VEDAS: a mobile and distributed data stream mining system for real-time vehicle monitoring. In: Proceedings of 2004 SIAM International Conference on Data Mining, pp 300–311 (2004)
13. Sakoe, H., Chiba, S.: Dynamic programming algorithm optimization for spoken word recognition. IEEE Trans. Acoust. Speech Signal Process. 26(1), 43–49 (1978)
14. Stein, S., Payne, T.R., Jennings, N.R.: Flexible QoS-Based Service Selection and Provisioning in Large-Scale Grids. In: UK e-Science 2008 All Hands Meeting (AHM), HPC Grids of Continental Scope (2008)
15. Vintsyuk, T.K.: Speech discrimination by dynamic programming. Kibernetika 4, 81–88 (1968)

Distributed Classification of Data Streams: An Adaptive Technique

Alfredo Cuzzocrea[1]([⊠]), Mohamed Medhat Gaber[2],
and Ary Mazharuddin Shiddiqi[3]

[1] DIA Department, University of Trieste and ICAR-CNR, Trieste, Italy
alfredo.cuzzocrea@dia.units.it
[2] School of Computing Science and Digital Media,
Robert Gordon University, Aberdeen, UK
mohamed.m.gaber@gmail.com
[3] Sepuluh Nopember Institute of Technology, Surabaya, Indonesia
amshi@its.ac.id

Abstract. Mining data streams is a critical task of actual Big Data applications. Usually, data stream mining algorithms work on resource-constrained environments, which call for novel requirements like availability of resources and adaptivity. Following this main trend, in this paper we propose a distributed data stream classification technique that has been tested on a real sensor network platform, namely, Sun SPOT. The proposed technique shows several points of research innovation, with are also confirmed by its effectiveness and efficiency assessed in our experimental campaign.

1 Introduction

Mining data streams in wireless sensor networks has many important scientific and security applications [19,20]. However, the realization of such applications is faced by two main constraints. The first is the fact that sensor nodes are battery powered. This necessitates that the running applications have a low battery footprint. Consequently, in-network data processing is the acceptable solution. The second is the resource constraints of each node in the network including memory and processing power [16].

Many applications in wireless sensor networks require event detection and classification. The use of unsupervised learning techniques has been proposed recently for this problem [24,29]. Despite the applicability of the proposed methods, they have not addressed the problem of running on resource constrained computing environments by adapting to availability of resources. The problem has been rather addressed by proposing lightweight techniques. However, this may cause the sensor node to shutdown due to the low availability of resources. Experimental results have proved that typical stream mining algorithms can cause the device to shutdown as reported in [25]. Also, the use of unsupervised learning may fail to detect events of interest due to the possibility of producing impure clusters that contain instances of two or more classes.

© Springer International Publishing Switzerland 2015
S. Madria and T. Hara (Eds.): DaWaK 2015, LNCS 9263, pp. 296–309, 2015.
DOI: 10.1007/978-3-319-22729-0_23

In this paper, we propose the use of distributed classification of data streams in wireless sensor networks for event detection and classification. The proposed technique can adapt to availability of resources and work in a distributed setting using ensemble classification. The technique is coined *RA-Class* in reference to its resource awareness capabilities. The experimental results have shown both the adaptability and high accuracy on real datasets. This paper is an extended version of the short paper [10], where we present the basic ideas that inspire our research.

The rest of the paper is organized as follows. Section 2 focuses the attention on *granularity-based approach to data stream mining*, the conceptual framework that provides the global umbrella for adaptive stream mining. The proposed technique is given in Sect. 3. Section 4 discusses the experimental results. Section 5 reviews the related work briefly. Finally, the paper is concluded in Sect. 6.

2   Mining Data Streams via Granularity-Based Approaches

Granularity-based approach to data stream mining is an adaptive resource-aware framework that can change the algorithm settings to cope with the velocity of the incoming streaming data [14,27]. It is basically a heuristics technique that periodically assesses the availability of memory, battery and processor utilisation, and in response changes some parameters of the algorithm, ensuring the continuity of the running algorithm. Accordingly, the approach has three main components:

1. *Algorithm Input Granularity (AIG)* represents the process of changing the data rates that feed the algorithm. Examples of this include sampling, load shedding, and creating data synopsis. This is a common solution in many data stream mining techniques.
2. *Algorithm Output Granularity (AOG)* is the process of changing the output size of the algorithm in order to preserve the limited memory space. In the case of data mining, we refer to this output as the number of knowledge structures. For example the number of clusters or rules. The output size could be changed also using the level of output granularity which means the less detailed the output, the higher the granularity and vice versa.
3. *Algorithm Processing Granularity (APG)* is the process of changing the algorithm parameters in order to consume less processing power. Randomisation and approximation techniques represent the potential solution strategies in this category.

The Algorithm Granularity requires continuous monitoring of the computational resources. This is done over fixed time intervals/frames that are denoted as TF. According to this periodic resource monitoring, the mining algorithm changes its parameters/settings to cope with the current consumption patterns of resources. These parameters are AIG, APG and AOG settings discussed briefly in the previous section. It has to be noted that setting the value of TF is a critical parameter for the success of the running technique. The higher the TF is, the lower the adaptation overhead will be, but at the expense of risking a high consumption of resources during the long time frame, causing the run-out of one or more of the computational resources.

The use of *Algorithm Granularity* as a general approach for mining data streams will require us to provide some formal definitions and notations. The following are definitions and notation that we will use in our discussion.

R: set of computational resources $R = \{r_1, r_2, \ldots, r_n\}$
TF: time interval for resource monitoring and adaptation.
ALT: application lifetime.
ALT': time left to last the application lifetime.
$NoF(r_i)$: number of time frames to consume the resources r_i, assuming that the
    consumption pattern of r_i will follow the same pattern of the last time frame.
$AGP(r_i)$: algorithm granularity parameter that affects the resource r_i.

According to the above, the main rule to be used to use the algorithm granularity approach is as follows:

IF $\frac{ALT'}{TF} < NoF(r_i)$
THEN SET $AGP(r_i)+$
ELSE SET $AGP(r_i)-$

Where $AGP(r_i)+$ achieves higher accuracy at the expense of higher consumption of the resource r_i, and $AGP(r_i)-$ achieves lower accuracy at the advantage of lower consumption of the resource r_i. For example, when dealing with clustering, it is computationally cheaper to allow incoming data instances in the stream to join an existing cluster with randomisation applied to which cluster the data instance would join. Ideally, the point should join a cluster that has sufficient proximity or a new cluster should be created to accommodate the new instance.

This simplified rule could take different forms according to the monitored resource and the algorithm granularity parameter applied to control the consumption of this resource. The *Algorithm Granularity* approach has been successfully applied to a number of data stream mining techniques. These techniques have been packaged in a java-based toolkit, coined *Open Mobile Miner* [15,21].

3   Classification Methods for Wireless Sensor Networks: The RA-Class Approach

RA-Class follows a similar procedure to *LWClass* proposed by Gaber et al. [16,18]. However, *RA-Class* extends *LWClass* in two different aspects:

– *RA-Class* uses all the algorithm granularity settings (input, processing and output). On the other hand, *LWClass* uses only the algorithm output granularity.

- *RA-Class* works in a distributed environment. On the other hand, *LWClass* is designed for centralized processing.

The algorithm starts by examining each incoming data record. It then determines whether the incoming record will be assigned to a specific stored entry, or will be stored as a new entry, based on the proximity of the record to the existing entries. A stored entry is basically the mean value of all the records that have been assigned to that entry. The mean value of a record is the mean value of all of its attributes. An update function provides the algorithm with information if there is a need to change the algorithm's settings. In the proposed *RA-Class*, there are three settings that can be adjusted: sampling interval, randomization factor and threshold value. A sampling interval is adjusted in response to the availability of battery charge. The randomization factor changes according to the CPU utilization. The threshold value is adjusted according to the remaining memory available. In case of low availability of resources, the algorithm granularity settings are adjusted. If the battery level drops below a preset level, the sampling interval is increased. This will reduce the energy consumed for processing incoming data streams. If CPU utilization reaches a preset critical level, the randomization factor is reduced. So when assigning an incoming record to an existing entry, the algorithm will not examine all of the currently stored entries to obtain the minimum distance. Instead, it checks randomly selected entries. Lastly, if the remaining memory decreases significantly, then the threshold value is increased to discourage the creation of new entries. This will slow down the memory consumption. The output of the *RA-Class* algorithm is a list of entries, each associated with a class label and a weight. The weight represents the number of data stream records represented by this entry. The resource-aware algorithm is shown in Algorithm 1.

In a distributed environment, there is a possibility that one of the nodes would run out of battery charge. Therefore, there must be a mechanism to handle this scenario to keep the recent list of stored entries, produced during the data stream mining process. The mechanism should enable the dying node to migrate its stored entries to one of its neighbors. The migration process is preceded by the selection of the best neighbor. This process is done by querying the current resources of the neighbors, after which the dying node calculates the best node to migrate its results to, based on the neighbor's resources. In addition, there has to be a mechanism to predict whether the current dying node's resources are sufficient to migrate all of the stored entries before the dying node dies. This is done by setting a critical limit before entering the migration stage. After obtaining the best neighbor, the dying node will start transferring its stored entries. This process is done until all of the stored entries are transferred completely. In the destination node, the migrating entries will be merged with the destination node's current entries. The merging process is done using the same mechanism as that when the *RA-Class* is processing incoming data streams. The only difference between the incoming data stream and the migrating entries is the weight. An incoming data stream will only contribute 1 to weight calculations. On the

---

Algorithm 1. RA-Class Algorithm

1: **while** data arrives **do**
2:    **for** each incoming data **do**
3:      **for** each stored entry **do**
4:        $currentDistane = newEntryValue - currentStoredEntryValue$
5:        **if** $currentDistane < threshold$ **then**
6:          **if** $currentDistane < lowestDistance$ **then**
7:           $lowestDistance = currentDistance$
8:           $lowestDistanceStoredEntryIndex = currentStoredEntryIndex$
9:          **end if**
10:        **end if**
11:        $currentStoredEntryIndex + 1$
12:      **end for**
13:      **if** $lowestDistance < threshold$ **then**
14:        **if** $newEntryLabel = lowestDistanceStoredEntryLabel$ **then**
15:          store the weighted average of the records
16:          $weightofthisentry + 1$
17:        **else**
18:          **if** $weightofthestoredentry > 1$ **then**
19:           $weightofthisentry - 1$
20:          **else**
21:           release the stored entry from memory
22:          **end if**
23:        **end if**
24:      **else**
25:        store the incoming entry as a new stored entry
26:        set the weight of the new stored entry $= 1$
27:      **end if**
28:    **end for**
29:    **if** $timer > updateTimer$ **then**
30:      do updateGranularitySettings
31:    **end if**
32: **end while**

---

other hand, a migrating entry can contribute more than 1 to weight calculations according to its weight that has been already accumulated.

In the process of deciding a class label for an incoming entry in a distributed environment, each RA-Class node needs to find the label using its own knowledge. The node labeling technique used needs to find the optimal approach to determine the closest entry. In this research, we use the K-NN algorithm with $K = 2$. The choice of assigning K the value of 2 is based on the need for fast classification of streaming inputs. The algorithm searches two stored entries that have the closest distance to the incoming unlabeled record. The result is the stored entry that has the higher weight among the two nearest neighbors. The result of this labeling process from each node is used in the labeling process of the distributed RA-Class.

We then use an ensemble approach in classifying any incoming unlabeled record. The ensemble algorithm works as an election system. Each node contributes to the election by giving a vote to each class label, while also providing an error rate. An error rate is used as a mechanism to state the *assurance level* of a vote. A lower error rate will have a higher possibility of winning the vote. In a normal condition, where all of the nodes are functioning properly, all of them contribute to the voting process. However, if one of the three nodes runs out of resources, it is not considered in the voting process as all of its results have already been migrated.

4 Experimental Assessment and Analysis

We run our experiments on the Sun SPOT sensor nodes from Sun Microsystems using SunSPOT API version 3.0. To evaluate the performance of our algorithms, we have conducted a set of experiments to assess the adaptability and the accuracy. The adaptability evaluation is used to assess the ability of the algorithm to adapt to changes in resource availability, while the accuracy evaluation is done to assess the accuracy of RA-Class. We have used two real datasets from UCI Machine Learning [2], namely, iris and vehicle. The *iris* dataset contains 150 entries, with 4 attributes, while the *vehicle* dataset contains 376 entries, with 18 attributes. The choice of the datasets was based on the low number of instances that consequently made the data stream simulator an easier task.

We have first conducted a set of experiments on a single sensor node. We have evaluated both the adaptability and the accuracy of the proposed framework and the algorithm. These experiments have been an important step to ensure the applicability of the technique before the assessment of its distributed version. To evaluate the ability of the algorithm to adapt to varying conditions of resource availability, we have used the *iris* dataset for the first set of experiments we have conducted.

Figure 1(a) shows the number of entries formed without enabling the resource-awareness module. The distance thresholds used are in the interval of [0.1, 1.0]. The figure shows that the number of entries produced declines as the threshold value increases. This is due to the fact that the increase in the threshold discourages the formation of new entries. When we have enabled the resource adaptation, the number of entries have been stabilized at around 60 entries as shown in Fig. 1(b). The memory adaptation is shown in Figs. 1(c) and (d). The figures show how the distance threshold discourages the creation of new entries that consume the memory. Figures 1(e) and (f) show how the algorithm adapts itself to different loads of CPU. We have set the criticality threshold to 40 % and randomly generated different CPU loads as shown in Fig. 1(e). Increasing the load exceeding the set threshold has resulted in the algorithm decreasing its randomization factor as shown in Fig. 1(f). However, the factor has a set lower bound of 10 % that represents the minimum acceptable value. Similarly, Figs. 1(g) and (h) demonstrate how the algorithm adapts itself by changing the sampling interval when the battery charge reaches its critical point, which has

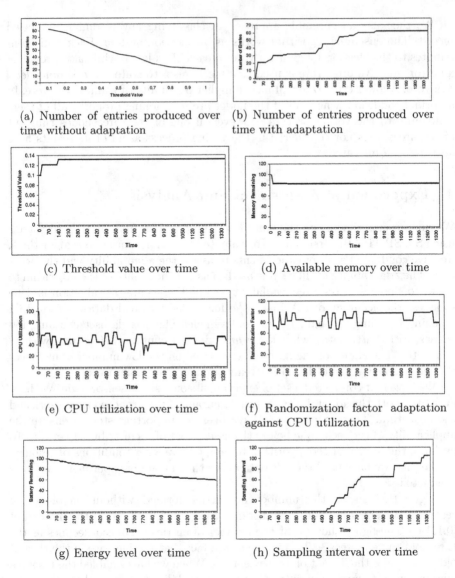

(a) Number of entries produced over time without adaptation

(b) Number of entries produced over time with adaptation

(c) Threshold value over time

(d) Available memory over time

(e) CPU utilization over time

(f) Randomization factor adaptation against CPU utilization

(g) Energy level over time

(h) Sampling interval over time

Fig. 1. Algorithm adaptability evaluation using iris dataset

been set to 80 % in this experiment. Figure 1(h) shows that the sampling interval decreases when the battery charge level reaches its critical point of 80 % as shown in Fig. 1(g).

Since memory adaptation is mostly affected by the size of the dataset, we have repeated the experiments using the larger data set, vehicle. We have set the memory critical point to 85 %. The experiments, depicted in Figs. 2(a) and (b), show clearly that as soon as the memory reached its critical point, the distance threshold has been increased to discourage the creation of new entries. This is also evident in Fig. 2(c) that shows the number of entries produced over time. The stability of the number of entries at 880 is due to releasing outliers from memory periodically.

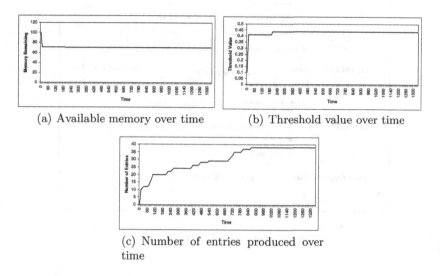

(a) Available memory over time   (b) Threshold value over time

(c) Number of entries produced over time

Fig. 2. Algorithm adaptability evaluation using vehicle dataset

The accuracy assessment on the iris dataset has been done using 10 fold cross validation and resulted in 92 % accuracy. Similarly, the vehicle dataset has produced 83 % accuracy.

In a distributed computational model, the main goal is that given a user-specified lifetime and a task such as data classification, the aim is to complete the preset lifetime and produce as accurate results as possible. The other objective is to minimize the accuracy loss in case few nodes die or stop working due to low availability of resources such as running out of battery, full of memory, and/or full of CPU utilization. Our approach is to migrate current results from a nearly-dead node to another best neighbor. There are three tasks that have to be performed before migrating the stored entries: (1) selecting the neighbor to migrate; (2) determining the time to migrate; and (3) the way to migrate. The migration scenario is shown in Fig. 3.

To examine the accuracy of the distributed RA-Class, we have divided the iris dataset into three disjoint equal in size subsets randomly. To simulate the streaming environments, we have drawn from each of these subsets a 10 times larger set than the original size. After running RA-Class on each subset, we have run the ensemble classification by voting among the results of the three classifiers using 10 fold cross validation that has resulted in an average of 91.33 % accuracy. The accuracy of each experiment is reported in Table 1.

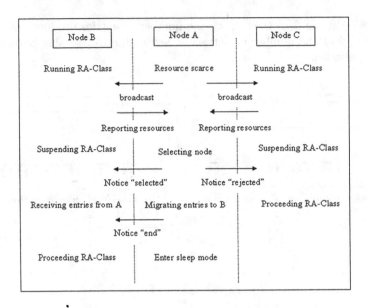

Fig. 3. Flowchart of data migration

The experiments have been repeated with result migration and merging. The reported average accuracy has been 86.67 %. The accuracy of each experiment is reported in Table 2. It is worth noting that the accuracy has not been dropped much for the ensemble choice. This provides an evidence of the applicability of our resource-adaptive framework.

The main goal of this set of experiments is to test the validity of RA-Class in a distributed environment on the real Sun SPOT devices. Similar to the simulation experiments, we use three nodes that run RA-Class, and then use the ensemble approach for the classification process. We have used the Iris dataset for this experiment. We have divided the dataset into three disjoint subsets that are equal in size. Then we have simulated 1500 data streams drawn randomly from each subset of the dataset to feed each node. We use Sun SPOT LEDs to indicate the on going process as shown in Fig. 4.

Table 1. Distributed RA-Class without migration

E1	E2	E3	E4	E5	E6	E7	E8	E9	E10
93.3 %	93.3 %	86.7 %	93.3 %	100 %	86.7 %	93.3 %	93.3 %	86.7 %	86.7 %

After performing the classification process, we have tested the accuracy of RA-Class using 15 randomly selected records from the iris dataset. This test is done on a single node functioning as a testing node. The final entries list

Table 2. Distributed RA-Class with migration

E1	E2	E3	E4	E5	E6	E7	E8	E9	E10
86.7%	80%	86.7%	93.3%	86.7%	80%	86.7%	86.7%	86.7%	93.3%

(a) Classification process

(b) Critical situation and neighbor selection

(c) Migration process

(d) Sleeping mode

Fig. 4. Distributed RA-Class on sun SPOT

of the remaining two nodes is transferred to the testing node and then the accuracy testing is performed using the ensemble algorithm. Figure 5 explains this scenario. The reason of this technique is only for efficiency purposes, so that the accuracy testing will be easier to observe. The used entries are the same ones that are used in the previous experiment. By repeating the experiments ten times, the result shows that the distributed RA-Class produced 88.0 % accuracy. The results show that in a real distributed system environment, the elaborated RA-Class remains producing a better result than a single node of RA-Class. This is due to the use of the ensemble algorithm that elevates the accuracy level.

The above experiment has not considered the migration and merging processes that clearly affect the accuracy. To measure the accuracy when the migration and merging processes take place, we have conducted the same experiment with two nodes contributing to the classification and one dying node as shown in Fig. 6 and the results have shown an average high accuracy of 84.67 % ten different runs.

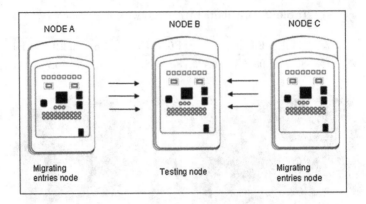

Fig. 5. Accuracy testing mechanism without migration

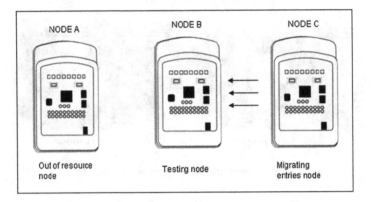

Fig. 6. Accuracy testing mechanism with migration

5   Related Work

In classical data stream mining literature there is a considerable attention towards extending popular approaches (e.g., clustering, classification, association rule mining, and so forth) as to make them working in the hard conditions defined by wireless sensor networks (e.g., [1]). *Energy efficiency* plays a central role in this context (e.g., [1]), as mining algorithms must rigorously consider energy consumptions in their core layers. Based on these main motivations, several research efforts have investigated data stream mining problems in wireless sensor networks. We briefly outline some of these approaches in the following.

Zhuang et al. [32] focus on the problem of *removing noise from data streams* as they arrive at each sensor of the target network, by exploiting both temporal and spatial information in order to reduce the stream-value-to-sink rate, hence improving the mining capabilities of given algorithms. Sheng et al. [26] and Subramaniam et al. [28] focus on the problem of *detecting outliers in wireless sensor networks*. In the first case, authors propose each sensor to represent and maintain a *histogram-based summary* of data kept in neighboring sensors, which is then propagated to the sink. This information is used by the sink for query optimization. In the second case, authors propose each sensor to represent and maintain a specialized *tree-based communication topology*, and outliers are detected by estimating the *Probability Distribution Function* (PDF) built on distributed topologies. Zhuang and Chen [31] provides a *wavelet-based method for repairing erroneous data streams* from a fixed population of sensors (e.g., due to transmission faults) by identifying them as anomalous data streams with respect to a specialised distance metrics (called *Dynamic Time Warping* – DTW) applied to data streams originated by spatially-close-to-that-population sensors. Cuzzocrea et al. investigate several approaches for supporting *data stream mining in limited-memory environments*, for both cases of dense [22] and sparse [4] data streams. In this case, the idea is to exploit popular *time-fading and landmark models* and to adapt them in the specialized case of limited-memory resources. Finally, a context of related work is represented by approaches that propose *compression paradigms* over data streams (e.g., [6,8,9]), because compression is a way for achieving better runs of typical data mining algorithms.

The resource adaptive framework has been first proposed by Gaber and Yu in [17]. Our research uses the proposed framework for adapting to variations of resource availability on a single node. The framework proposed by Gaber and Yu [17] uses three settings that are adjusted in response to the resource availability during the mining process. The input settings are termed *Algorithm Input Granularity (AIG)*. The output and processing settings are termed *Algorithm Output Granularity (AOG)*, and *Algorithm Processing Granularity (APG)* respectively. The input settings include sampling, load shedding, and creating data synopsis techniques. The output settings include number of knowledge structures created, or levels of output granularity. Changing the error rate of approximation algorithms or using randomization represent the processing granularity. The three Algorithm Granularity settings are named collectively as *Algorithm Granularity Settings (AGS)*. The framework has been applied to a data stream clustering algorithm termed as RA-Cluster. An important step towards proving the applicability of *RA-Cluster* in wireless sensor networks has been the implementation of *ERA-Cluster* by Phung et al. [23].

6   Conclusions and Future Work

The paper explored the validity of our adaptive classification technique, we termed *Resource-Aware Classification (RA-Class)*, to process data streams in wireless sensor networks. The proposed RA-Class was evaluated with regard to accuracy and resource-awareness using real datasets. The results show that RA-Class can effectively adapt to resource availabilities and improve resource consumption patterns in both single-node and distributed computing environments. The algorithm has been tested in a real testbed using the Sun SPOT sensor nodes. The results also show that the accuracy loss due to the adaptation process is limited. Future work includes applying RA-Class to a dense wireless sensor with a large number of nodes, and testing the resource-awareness framework using other data stream mining techniques. In addition to this, we plan to study further aspects of our framework, inspired by similar approaches in literature: (*i*) *fragmentation methods* (e.g., [3,7]) to gain into efficiency; (*ii*) *privacy preservation issues* (e.g., [11,12]), which are relevant for published data; (*iii*) *big data challenges* (e.g., [5,13,30]), which are really emerging at now.

References

1. Akyildiz, I.F., Su, W., Sankarasubramaniam, Y., Cayirci, E.: Wireless sensor networks: a survey. IEEE Trans. Syst. Man Cybern. Part B **38**, 393422 (2002)
2. Asuncion, A., Newman, D.J.: UCI Machine Learning Repository. Irvine, CA: University of California, School of Information and Computer Science (2007). http://www.ics.uci.edu/~mlearn/MLRepository.html
3. Bonifati, A., Cuzzocrea, A.: Efficient fragmentation of large XML documents. In: Wagner, R., Revell, N., Pernul, G. (eds.) DEXA 2007. LNCS, vol. 4653, pp. 539–550. Springer, Heidelberg (2007)
4. Cameron J.J., Cuzzocrea A., Jiang F., Leung C.K.-S.: Mining frequent itemsets from sparse data streams in limited memory environments. In: Proceedings of the 14th International Conference on Web-Age Information Management, pp. 51–578 (2013)
5. Cuzzocrea, A.: Analytics over big data: exploring the convergence of data warehousing, OLAP and data-intensive cloud infrastructures. In: Proceedings of COMPSAC 2013, pp. 481–483 (2013)
6. Cuzzocrea, A., Chakravarthy, S.: Event-based lossy compression for effective and efficient OLAP over data streams. Data Knowl. Eng. **69**(7), 678–708 (2010)
7. Cuzzocrea, A., Darmont, J., Mahboubi, H.: Fragmenting very large XML data warehouses via K-means clustering algorithm. Int. J. Bus. Intell. Data Min. **4**(3/4), 301–328 (2009)
8. Cuzzocrea, A., Furfaro, F., Mazzeo, G.M., Saccá, D.: A grid framework for approximate aggregate query answering on summarized sensor network readings. In: Meersman, R., Tari, Z., Corsaro, A. (eds.) OTM-WS 2004. LNCS, vol. 3292, pp. 144–153. Springer, Heidelberg (2004)
9. Cuzzocrea, A., Furfaro, F., Masciari, E., Sacca', D., Sirangelo, C.: Approximate query answering on sensor network data streams. In: Stefanidis, A., Nittel, S. (eds.) GeoSensor Networks, pp. 53–72. CRC Press, Boca Raton (2004)
10. Cuzzocrea, A., Gaber, M.M., Shiddiqi, A.M.: Adaptive data stream mining for wireless sensor networks. In: Proceedings of IDEAS 2014, pp. 284–287 (2014)
11. Cuzzocrea, A., Russo, V., Saccà, D.: A robust sampling-based framework for privacy preserving OLAP. In: Song, I.-Y., Eder, J., Nguyen, T.M. (eds.) DaWaK 2008. LNCS, vol. 5182, pp. 97–114. Springer, Heidelberg (2008)
12. Cuzzocrea, A., Sacc, D.: Balancing accuracy and privacy of OLAP aggregations on data cubes. In: Proceedings of DOLAP 2010, pp. 93–98 (2010)
13. Cuzzocrea, A., Sacc, D., Ullman, J.D.: Big data: a research agenda. In: Proceedings of IDEAS 2013, pp. 198–203 (2013)
14. Gaber, M.M.: Data stream mining using granularity-based approach. In: Abraham, A., Hassanien, A.E., de Leon, F., de Carvalho, A.P., Snášel, V. (eds.) Foundations of Computational, IntelligenceVolume 6. Studies in Computational Intelligence, vol. 206, pp. 47–66. Springer, Berlin (2009)
15. Gaber, M.M.: Advances in data stream mining. Wiley Interdisc. Rev.: Data Min. Knowl. Discov. **2**(1), 79–85 (2012)

16. Iordache, O.: Methods. In: Iordache, O. (ed.) Polystochastic Models for Complexity. UCS, vol. 4, pp. 17–61. Springer, Heidelberg (2010)
17. Gaber, M.M., Yu, P.S.: A holistic approach for resource-aware adaptive data stream mining. J. New Gener. Comput. **25**(1), 95–115 (2006)
18. Gaber, M.M., Zaslavsky, A., Krishnaswamy, S.: A survey of classification methods in data streams. In: Aggarwal, C.C. (ed.) Data Streams Models and Algorithms. Advances in Database Systems, pp. 39–59. Springer, Heidelberg (2007)
19. Gama, J., Gaber, M.M.: Learning from Data Streams: Processing Techniques in Sensor Networks. Springer, Berlin (2007). ISBN 1420082329, 9781420082326
20. Ganguly, A., Gama, J., Omitaomu, O., Gaber, M.M., Vatsavai, R.R.: Knowledge Discovery from Sensor Data. CRC Press, Boca Raton (2008). ISBN 1420082329, 9781420082326
21. Krishnaswamy S., Gama J., Gaber M.M.: Advances in data stream mining for mobile and ubiquitous environments. In: Proceedings of the 20th ACM International Conference on Information and Knowledge Management, pp. 2607–2608 (2011)
22. Leung, C.K.-S., Cuzzocrea, A., Jiang, F.: Discovering frequent patterns from uncertain data streams with time-fading and landmark models. In: Hameurlain, A., Küng, J., Wagner, R., Cuzzocrea, A., Dayal, U. (eds.) TLDKS VIII. LNCS, vol. 7790, pp. 174–196. Springer, Heidelberg (2013)
23. Phung N.D., Gaber M.M., Rohm, U.: Resource-aware online data mining in wireless sensor networks. In: Proceedings of the 2007 IEEE Symposium on Computational Intelligence and Data Mining, pp. 139–146 (2007)
24. Rodrigues, P.P., Gama, J., Lopes, L.: Clustering distributed sensor data streams. In: Daelemans, W., Goethals, B., Morik, K. (eds.) ECML PKDD 2008, Part II. LNCS (LNAI), vol. 5212, pp. 282–297. Springer, Heidelberg (2008)
25. Shah R., Krishnaswamy S., Gaber M.M.: Resource-aware very fast k-means for ubiquitous data stream mining. In: Proceedings of Second International Workshop on Knowledge Discovery in Data Streams, held in conjunction with the ECML/PKDD 2005, Porto, Portugal (2005)
26. Sheng, B., Li, Q., Mao, W., Jin, W.: Outlier detection in sensor networks. In: Proceedings of the 8th ACM International Symposium on Mobile and Ad Hoc Networking and Computing, pp. 219–228 (2007)
27. Stahl, F., Gaber, M.M., Bramer, M.: Scaling up data mining techniques to large datasets using parallel and distributed processing. In: Rausch, P., Sheta, A.F., Ayesh, A. (eds.) Business Intelligence and Performance Management. Advanced Information and Knowledge Processing, pp. 243–259. Springer, London (2013)
28. Subramaniam S., Palpanas T., Papadopoulos D., Kalogeraki V., Gunopulos D.: Online outlier detection in sensor data using non-parametric models. In: Proceedings of the 32nd International Conference on Very Large Databases, pp. 187–198 (2006)
29. Yin, J., Gaber, M.M.: Clustering distributed time series in sensor networks. In: Proceedings of the Eighth IEEE International Conference on Data Mining, pp. 678–687, Pisa, Italy, 15–19 December 2008
30. Yu, B., Cuzzocrea, A., Jeong, D.H., Maydebura, S.: On managing very large sensor-network data using bigtable. In: Proceedings of CCGRID 2012, pp. 918–922 (2012)
31. Zhuang, Y., Chen, L.: In-network outlier cleaning for data collection in sensor networks. In: Proceedings of the 1st International VLDB Workshop on Clean Databases, pp. 678–687 (2006)
32. Zhuang, Y., Chen, L., Wang, X., Lian, J.: A weighted average-based approach for cleaning sensor data. In: Proceedings of the 27th International Conference on Distributed Computing Systems, pp. 678–687 (2007)

New Word Detection and Tagging on Chinese Twitter Stream

Yuzhi Liang$^{(\boxtimes)}$, Pengcheng Yin, and S.M. Yiu

Department of Computer Science, The University of Hong Kong, Hong Kong, China
{yzliang,pcyin,smyiu}@cs.hku.hk

Abstract. Twitter becomes one of the critical channels for disseminating up-to-date information. The volume of tweets can be huge. It is desirable to have an automatic system to analyze tweets. The obstacle is that Twitter users usually invent new words using non-standard rules that appear in a burst within a short period of time. Existing new word detection methods are not able to identify them effectively. Even if the new words can be identified, it is difficult to understand their meanings. In this paper, we focus on Chinese Twitter. There are no natural word delimiters in a sentence, which makes the problem more difficult. To solve the problem, we derive an unsupervised new word detection framework without relying on training data. Then, we introduce automatic tagging to new word annotation which tag the new words using known words according to our proposed tagging algorithm.

Keywords: Chinese tweets · New word detection · Annotation · Tagging

1 Introduction

New social media such as Facebook or Twitter becomes one of the important channels for dissemination of information. Sometimes they can even provide more up-to-date and inclusive information than that of news articles. In China, Sina Microblog, also known as Chinese Twitter, dominates this field with more than 500 million registered users and 100 million tweets posted per day. An interesting phenomenon is that the vocabularies of Chinese tweets thesaurus have already exceeded traditional dictionary and is growing rapidly. From our observation, most of the new words are highly related to hot topics or social events. For example, the new word *"Yu'e Bao"* detected from our experimental dataset is an investment product offered through the Chinese e-commerce giant Alibaba. Its high interest rate attracted hot discussion soon after it first appeared, and without any concrete marketing strategy, *Yu'e Bao* has been adopted by 2.5 million users who have collectively deposited RMB 6.601 billion ($1.07 billion) within only half a month.

Y. Liang and P. Yin—These two authors contributed equally to this work.

© Springer International Publishing Switzerland 2015
S. Madria and T. Hara (Eds.): DaWaK 2015, LNCS 9263, pp. 310–321, 2015.
DOI: 10.1007/978-3-319-22729-0_24

Obviously, these "*Tweet-born*" new words in the Chinese setting are worthy of our attention. However, finding new words from Chinese tweet manually is unrealistic due to the huge amount of tweets posted every day. It is desirable to have an automatic system to analyze tweets. The obstacle is that Twitter users usually invent new words using non-standard rules that appear in a burst within a short period of time. Existing new word detection methods are not able to identify them effectively. Even if the new words can be identified, it is difficult to understand their meanings.

In this paper, we focus on Chinese Twitter. There are no natural word delimiters in a sentence, which makes the problem more difficult. To solve the problem, we introduce a Chinese new word detection framework for tweets. This framework uses an unsupervised statistical approach without relying on hand-tagged training data for which the availability is very limited. Then, about new word interpretation, we proposed a novel method which introducing automatic tagging to new word annotation. The tagging results represent a big step towards automatic interpretation of these new words. Such kind of tagging is not only useful in new word's interpretation, but can also help other NLP tasks such as improving machine translation performance. Our proposed approach differs from existing solutions in the following ways.

1.1 An Unsupervised Statistical Method for Detecting Out-of-Vocabulary (OOV) Words in Chinese Tweets

Unlike English and other western languages, many Asian languages such as Chinese and Japanese do not delimit words by spaces. An important step to identify new words in Chinese is to segment a sentence into potential word candidates. Existing approaches to Chinese new word detection fall roughly into two categories: supervised (or semi-supervised) method and unsupervised method.

Supervised methods transform new word detection to a tagging problem and train the classifier based on tagged training sets. Both of the two most widely used Chinese word segmentation/new word detection tool Stanford Chinese-word-segmenter (based on Conditional Random Field CRF [13]) and ICTCLAS (based on Hierarchical Hidden Markov model HHMM [11]) are using supervised method. The problem is, precision of supervised method often relies on the quality of tagged training set. Unfortunately,there is no high quality tagged dataset specifically designed for Chinese tweets so far. Meanwhile, traditional training sets cannot capture all the features of microblog crops because microblog tweets are short, informal and have multivariate lexicons. Existing solutions [9,10] for identifying new words specially designed for the Chinese Microblog word segment are still supervised machine learning methods. Thus, both suffer from the shortage of good training datasets. Unsupervised method performs Chinese word segmentation by deriving a set of context rule or calculating some statistical information from the target data.

From our study, we notice that contextual rule-based approach is not suitable for the task of detecting new words from Chinese tweets because new words emerged from Sina Microblog are rather informal and may not follow these rules while statistical method is a good solution for this problem since it can be purely data driven. Our target is to define an efficient model to detect OOV words from Chinese Twitter stream while avoiding using tagged datasets. We proposed a new word detection framework by computing the word probability of a given character sequence. This approach combines ideas from [3,7,8]. Detailed solution will be given in the later sections.

1.2 A Novel Method to Annotate New Word in Tweets by Tagging

Existing approaches for the new entity (phrase or words) interpretation include name entity recognition (NER) [4] and using the online encyclopedia as a knowledge base [2]. NER seeks to locate and classifies name entity into names of persons, organizations, locations, expressions of time, quantities, etc. [1]. However, this kind of classification cannot indicate the meaning of the new entity in detail. Another popular approach is interpreting entities by linking them to Wikipedia. This is not applicable for annotating new emerging words because most of new words will not have a corresponding/related entry in any online encyclopedias within a short period of time right after the new word's appearance.

To solve this problem, we propose a novel method which is annotating a new word by tagging it with known words. This is the first time word tagging is introduced to word annotation. Tagging is being extensively used in images (photos on facebook) [5] or articles annotation [6]. For new word tagging, the objective is to shed light on its meaning and facilitate users' better understanding. Our idea is to find out Top-K words that are most relevant to a given new word. The core issue of the problem is to derive a similarity measure between new words and their relevant words. Intuitively, words that co-occur with the new word with high frequency are more relevant with the new word. However, from our study, we found this naive definition might not be true for words in Chinese tweet. For instance, *"Mars brother"* is a nickname of *"Hua Chenyua"* (a singer) to indicate his abnormal behavior. These two terms are related but do not co-occur frequently in tweets because they can be a replacement of each other. Thus, we further quantify the similarity of two words by modeling the similarity of their corresponding contexts. The context of a word w is the surrounding text of w, roughly speaking, two words that share similar contexts are highly relevant. In this paper, we derived Context Cosine Similarity (CCS) which based on cosine similarity for the similarity measurement. The results show CCS can evaluate similarity between two words with high efficiency.

In our experiment, the approaches are evaluated with real microblog data for 7 consecutive days (2013-07-31 to 2013-08-06), which contains 3 million tweets in total. We compare our OOV detection approach with that of the most popular Chinese word segmentation tools ICTCLAS and Stanford Chinese-word-segmenter. The results show our method is competitive in OOV detection regarding to precision and recall rate. In new word tagging, we measure the accuracy of our tagging method by checking the existence of the generated tag words in corresponding Baidu Entry (Baidu Entry is an online encyclopedia like Wikipedia). The average precision is as high as 79 %.

2   New Word Detection

2.1   Definition of New Word

In order to get the new word set $S_{word}(t-t_0)$ which contains new words appears at time t but not exists at time $t_0(t_0 < t)$, we need to get the word set at time t_0 $(S_{word}(t_0))$ and the word set at time t $(S_{word}(t))$ from unsegmented tweets at time t $(S_{tweet}(t))$. For any word w extracted from $S_{tweet}(t)$, if $w \in S_{word}(t)$ and $w \notin S_{word}(t_0)$, w is regarded as a new word, otherwise w is regarded as a known word.

2.2   Word Extraction

For a set of unsegmented tweets $S_{tweet} = \{T_1, T_2, ..., T_N\}$, the first step is to extract word segments in S_{tweet}. We have discussed in the introduction that the state-of-art supervised method is not suitable for our application due to the lack of training corpus. Instead of relying on training dataset, we take a statistical approach. It is worthy to notice that Symmetrical Conditional Probability (SCP) [7] is to measure the cohesiveness of a given character sequence while Branch Entropy (BE) [8] is to measure its extent of variance. These two statistical approaches measure the possibility of s being a valid word from two perspectives such that they can complement each other in achieving accuracy. Also, we use a word statistical feature Overlap Variety [3] to further reduce the noise. For each $T \in S_{tweet}$, a probability score will be calculated for all the consecutive character sequences with length between two and four in T to measure how likely the character sequence s is a valid word based on above features.

Sequence Frequency. Sequence Frequency is an important noise filtering criteria in Chinese word segmentation. It is base on the concept that if s is a valid word, it should appear repeatedly in S_{tweet}. In this study, we only consider words with $freq(\cdot)$ larger than certain threshold T_{freq}, which is set to 15[1].

Symmetrical Conditional Probability. Symmetrical Conditional Probability (SCP) is defined to measure the cohesiveness of a given character sequence s by considering all the possible binary segmentations of s. Let n denotes the length of s, c_x denotes the xth character in s, $P(\cdot)$ denotes the possibility of the given sequence appearing in the text, which is estimated by its frequency, the SCP score of s $SCP(s)$ is as 1:

$$SCP(s) = \frac{P(s)^2}{\frac{1}{n-1}\sum_{i=1}^{n-1} P(c_1, c_i)P(c_{i+1}, c_n)} \tag{1}$$

---

[1] Here 15 is an experimental number, but this number can be evaluated by some statistical features such as mean and standardization of all the character sequences' frequency.

Branching Entropy. Symmetrical Conditional Probability (SCP) is defined to measure the cohesiveness of a given character sequence s by considering all the possible binary segmentations of s. Let n denotes the length of s, c_x denotes the xth character in s, $P(\cdot)$ denotes the possibility of the given sequence appearing in the text, which is estimated by its frequency, the SCP score of s $SCP(s)$ is as 2:

$$H(x|s) = - \sum_{x \in X} P(x|s) \log P(x|s) \tag{2}$$

We denote this $H(x|s)$ as $H_L(s)$ such that $H_R(s)$ can be defined similarly by considering the character following s. The Branch Entropy of sequence s is as 3:

$$BE(s) = \min\{H_L(s), H_R(s)\} \tag{3}$$

Word Probability Score. First of all, character sequences have extremely low BE score or SCP score will be abandoned and character sequences have extremely high BE score or SCP score can be selected as valid word directly. For the rest of words, we define an word probability score $Pr_{word}(s)$ to indicates how likely a given character sequence s is a valid word. $Pr_{word}(s)$ is calculated based on normalized BE and SCP of s as 4

$$Pr_{word}(s) = (1 + \mu)Nor(BE(s)) + Nor(SCP(s)) \tag{4}$$

We added $\mu BE(s)$ in calculating the word probability because we found BE score is more important than SCP score when defining whether character sequence s is a valid word. We set μ to 0.2 in our experiment. $Nor(BE(s))$ is the normalized BE score of s which used max-min method to perform the normalization:

$$Nor(BE(s)) = \frac{BE(s) - Min_{BE(s)}}{Max_{BE(s)} - Min_{BE(s)}} \tag{5}$$

$Nor(SCP(s))$ is the normalized SCP score of s. Experimental result shows that SCP scores of the character sequences are very uneven, so shift z score mentioned in [11] which can provide an shift and scaling zscore to normalize the majority of SCP score into $[0, 1]$ is used to perform the SCP score normalization:

$$Nor(SCP(s)) = \frac{\frac{SCP(s) - \mu}{3\sigma} + 1}{2} \tag{6}$$

Character sequence with word probability score larger than a certain threshold will be considered as a valid word. From our observation, most character sequences with $Pr_{word}(\cdot)$ larger than 0.3 are valid words. This threshold also can be evaluate with some statistical features such as mean and standard derivation of Pr_{word} for all the character sequences with length between 2 to 4.

Noise Filtering. Although we can get valid word candidate by setting a threshold on $Pr_{word}(\cdot)$, substring of a valid word exists as noise with relative high $Pr_{word}(s)$ from our observation. A dictionary will be used as knowledge base in noise filtering. The basic idea to filter this kind of noise is to consider word probability of the given character sequence and its overlapping strings [3]. For a candidate word w_0, assume its left overlapping string s_L is defined as $c_1c_2...c_kc_{k+1}...c_l, l = \{2, ..., |w_0|\}$ where $c_1...c_k$ is the k-character sequence proceeding of w_0 and $c_{k+1}...c_l$ is the first $l - k$ characters of s. Right overlapping string is defined similarly. For a noise word, it always has some overlapping strings s with $Pr_{word}(s) > Pr_{word}(w_0)$. For example, Pr_{word} (中国)$> Pr_{word}$ (国媒) (because 中国, China, is a dictionary word and Pr_{word} (中国)$=1$ while 国媒 is a wrong segment). But for a valid word, mostly $Pr_{word}(w_0)$ is larger than $Pr_{word}(s)$. For each selected candidate words w_0, let $S_{ov}(w_0)$ denotes the set of overlapping sting of w_0, w_0's overlapping score $OV(w_0)$ is then calculated as follows:

$$OV(w_0) = \frac{\sum_{s \in S_{ov}(w_0)} I(Pr_{word}(w_0) > Pr_{word}(s))}{|S_{ov}(w_0)|} \tag{7}$$

$I(\cdot)$ is the indicator function. Candidate words with $OV(\cdot)$ larger than certain threshold are rejected.

Pseudo code of new word detection process is stated in Algorithm 1.

Algorithm 1. New Word Detection

1: Let T_{freq}, T_{pr} and T_{ov} denotes the thresholds of frequency, word probability score and overlapping score respectively
2: **for all** character sequence s, $2 \leq |s| \leq 4$, s is substring of T, $T \in S_{tweet}(t)$ **do**
3:     Count occurrences of s $freq(s)$ in $S_{tweet}(t)$
4:     **if** $freq(s) \geq T_{freq}$ **then**
5:         Compute $SCP(s)$ and $BE(s)$ using formula 1 and 3
6:         Compute $Pr_{word}(s)$ based on $BE(s)$ and $SCP(s)$ using formula 4
7:         **if** $Pr_{word} > T_{Pr}$ **then**
8:             Add s to word candidate set $Cand(s)$
9:         **end if**
10:     **end if**
11: **end for**
12: **for all** $s \in Cand(s)$ **do**
13:     Compute $OV(s)$
14:     **if** $OV(s) < T_{ov}$ **then**
15:         **if** $s \in S_{tweet}(t_0)$ **then**
16:             Add s to $S_{word}(t_0)$ ($S_{word}(t_0)$ is the known word set)
17:         **else**
18:             Add s to $S_{word}(t - t_0)$ ($S_{word}(t - t_0)$ is the new word set)
19:         **end if**
20:     **end if**
21: **end for**

Here we first select all the frequent character sequences, then calculate word probability of these character sequences bases on their SCP and BE score. Finally, we filter noise by applying 7 to get the valid words.

3 New Word Tagging

Regarding to new word interpretation, our method introducing tagging by tagging new word $w_{new}(w_{new} \in S_{word}(t - t_0))$ with known word $w_{known}(w_{known} \in S_{word}(t_0))$. Words in the following two categories are potential tag words:

- Words that are highly relevant to w_{new}, i.e. its attributes, category and related Named Entities.
- Words that are semantically similar to w_{new}, i.e. synonyms.

The First category of words is important for tagging new words related to certain social events. It may include the event's related people, institutions, and microblog user's comments. Those words co-occur with w_{new} frequently. For the second category, w_{new}'s synonyms may not co-occur with w_{new} as mentioned in the previous example "*Mars brother*" and "*Hua Chenyu*". It is obvious that approaches such as picking words that co-occur with w_{new} as tagging words is rather naive. We seek to develop a similarity measure between two words that not only incorporate word co-occurrence, but can also utilize other features to deal with the above case. We found that for w_{new} and its potential tagging word w_{known}, no matter which category w_{known} belongs to, it shares similar context with w_{new}. A word w_{known}'s context is its surrounding text which may shed light on its meaning. We could simply model w_{known}'s context as the set of words co-occurring with w_{new} in $S_{tweet}(t)$.

3.1 Context Cosine Similarity

Given a new word w_{new}, the basic idea of tagging w_{new} is to find its most similar known words w_{known}. The amount of tweets is huge even just about a single topic, and the contents of tweets often cross domains. According to these characteristics, we decide to use cosine similarity to perform the similarity measurement for its efficient and domain independent. The context cosine similarity between a new word w_{new} and a known word w_{known} is computed as follow:

1. Let $D(w_1, w_2)$ denotes the pseudo document made by concatenation of all tweets containing w_1 while w_1, w_2 are excluded from the document. Compute $D(w_{new}, w_{known})$ and $D(w_{known}, w_{new})$.
2. Compute V_{new} and V_{known} where $V = \{V_1, V_2, ..., V_n\}$ is the term vector of a Document D. V_i in V is the TF-IDF weight[2] of w_i, $w_i \in S_{word}(t_0)$ and $i = \{1, 2, ..., n\}$ where n is the size of $S_{word}(t_0)$

---

[2] TF-IDF is a numerical statistic used to indicate the importance of the given word in a corpus. The score is TF × IDF, where TF is term frequency which is a normalized term count, IDF is Inverse Document Frequency which indicates the proportion of documents in the corpus containing w_i.

3. Context cosine similarity between w_{new} and w_{known} is defined as

$$Sim_{rawccs}(w_{new}, w_{known}) = \frac{V_{new} \cdot V_{known}}{|V_{new}||V_{known}|}$$

$$= \frac{\sum_{i=0}^{n} V_{newi} \cdot V_{knowni}}{\sqrt{\sum_{i=1}^{n} V_{newi}^2} \times \sqrt{\sum_{i=1}^{n} V_{knowni}^2}} \quad (8)$$

4. We get Sim_{ccs} by normalizing Sim_{rawccs} to $[0, 1]$ using max-min normalization[3] on the top 20 tag words of the new word.

Worth noting that in Step 1, we excluded w_{new} and w_{known} from $D(w_{new}, w_{known})$ and $D(w_{known}, w_{new})$ is because we assume if two words are semantically similar, their context should be similar even they co-occur with low frequency.

3.2 Choose Tag Word

When tagging a new word w_{new}, we will compute Context Cosine Similarity between w_{new} and all the known words in $S_{word}(t_0)$. For a known word w_{known}, if $Sim_{ccs}(w_{new}, w_{known})$ is larger than a threshold k, w_{known} will be selected as a tag word of w_{new}. The value of $Sim_{ccs}(w_{new}, w_{known})$ is in the range $[0, 1]$. According to Table 4, we set the threshold k to 0.5 as a balance point of the number of selected tag words and tag word precision.

4   Experiment

4.1   Dataset Setting

In this experiment, we aim at detecting newly emerged words on a daily basis. Regarding to the definition of new words, for the target day t, $S_{tweet}(t)$ is the set of tweets published on that day. Tweets published in seven consecutive days, from July 31st, 2013 to Aug 6th, 2013 are used as our input. Meanwhile, we use the tweets of May 2013 as the known word set $S_{tweet}(t_0), t_0 < t$ which serves as knowledge base.

We perform cleaning on dataset used as $S_{tweet}(t)$, where hash tags, spam tweets, tweets only contains non-Chinese characters are rejected. Table 1 shows the details of our dataset. And we store any character sequence with length between two and four in $S_{tweet}(t_0)$ to serve as the known word set $S_{word}(t_0)$ to ensure new words detected from $S_{tweet}(t)$ has never appeared in $S_{tweet}(t_0)$.

4.2   New Word Detection Result

In the new word detection experiment, we use ICTCLAS and Stanford Chinese-word-segmenter [13] to serve as our baselines. The training data used by ICT-CLAS is Peking University dataset which contains 149,922 words while training

---

[3] $Sim_{ccs} = \frac{Sim_{rawccs} - Min_{rawccs}}{Max_{rawccs} - Min_{rawccs}}$.

Table 1. List of dataset

Dataset	# of tweets	After cleaning
July 31	715,680	443,734
Aug 1	824,282	515,837
Aug 2	829,224	516,152
Aug 3	793,324	397,291
Aug 4	800,816	392,945
Aug 5	688,692	321,341
Aug 6	785,236	399,699
May	20,700,001	-

data used for CRF training is Penn Chinese Treebank which contains 423,000 words. All the words appearing in the training set will not be selected as new word. And our aim is to detect new words of certain importance as well as their relevant words, it is reasonable to focus on words with relatively high frequency. In this experiment, words appearing less than 15 times will be ignored. In addition, non-Chinese character, emotion icon, punctuation, date, word containing stop words and some common words are excluded because they are not our target.

Generally speaking, Chinese new words can be divided into several categories [12] (excluding new words with non-Chinese characters): name entity, dialect, glossary, novelty, abbreviation and transliterated words. The detected new words are classified according to these categories in our experiment. The precision of the detection result is defined as 9

$$Precision = \frac{\# \text{ of valid new words}}{\# \text{ of total new words dtected}} \tag{9}$$

The results are listed in Table 2.

Table 2. New word detection result

Category	Our method	ICTCLAS	Chinese-word-segmenter
Name entity	50	36	60
Glossary	2	1	6
Novelty	19	2	36
Abbreviation of hot topic	22	1	8
Transliterated words	0	0	5
Noise	4	5	139
Valid new words	93	40	**115**
Precision	**95.9 %**	88.9 %	45.2 %

The above results show that our method has the highest precision in detecting new words in Chinese Twitter among the three methods. Stanford Chinese-word-segmenter wins in recall. However, a large number of noise is also included in Stanford Chinese-word-segmenter's result which lowers the precision tremendously. The reason is that it uses a supervised machine learning method, for which the shortage of appropriate tagged training dataset for Chinese tweet is a fatal problem. ICTALS has an acceptable precision, but it often over segment the words which makes it fails to detect some compound words such as "*Yu'E Bao*" and "*Micro-channel Voca*".

4.3   New Word Tagging Result

As stated above, words have context cosine similarity with a new word larger than a certain threshold are selected as the new word's tags. Words such as 加油 (work hard) and 执行 (operate) are excluded in tag words manually since they are either only popular in Sina Microblog or do not have much meaning on its own. Some tagging result examples are listed as below:

- **Liu Yexing** (Name Entity. A guy from Tsinghua University become famous by attending reality show "Who's still standing" and burst their question bank)
  Tags: Zhang Xuejian(Liu Yexing's adversary), Who's still standing, Tsinghua University, Peking University, question bank, answer questions
- **Ergosterol** (Glossary. Ergosterol is a sterol found in cell membranes of fungi and protozoa, serving many of the same functions that cholesterol serves in animal cells)
  Tags: protein, amount, growth, immunity, vegetables, growing development
- **Yu'E Bao** (Novelty. Yu'E Bao is a money-market fund promoted by Alipay)
  Tags: currency, Internet, finance, fund, money management, supply-chain
- **Burst Bar event** (Abbreviation of hot topic. Pan Mengyin, a fan of Korean star G-Dragon, spread some inappropriate remarks about football stars which makes fans of the football stars get angry and attacked G-Dragon's Baidu Bar.)
  Tags: Pan Mengyin, G-Dragon, Korean star, stupid

Here we set Context Cosine Similarity threshold to 0.5 such that we can get enough tag words while achieving a relatively high precision. Table 3 shows the average number of tags and tagging precision using different similarity threshold. Among the 93 detected new words, some of them are recorded in Baidu Entry now. We randomly picked 20 recorded words from different categories to evaluate our tagging result. The precision of tagging result about a new word (w_{new}) is defined as:

$$Precision_{tag}(w_{new}) = \frac{\#\text{ of tag words hits in } w_{new}\text{ 's Baidu Entry}}{\#\text{ of } w_{new}\text{'s tag words}} \quad (10)$$

From Table 3 we can see that the number of selected tag words decreases while the tagging precision increase when Context Cosine Similarity threshold arise. That means tag word have higher context cosine similarity with the new word is more likely be the right tag of the new word.

Table 3. Word tagging result 1

Threshold	Average # of tags	Average precision
0	19.6	0.56
0.25	9.1	0.71
0.5	5.5	0.79
0.75	3	0.825

We also compared the number of tags and tagging precision of different word categories when the CCS threshold is 0.5 (See Table 4).

Table 4. Word tagging result 2

Category	# of new words	Average # of tags	Average precision
Name entity	9	6.11	0.80
Glossary	1	6.00	0.00
Novelty	4	3.00	0.96
Abbreviation of hot topic	6	6.17	0.79

An interesting phenomena is that comparing to name entity and abbreviation of hot topic, novelty have fewer number of tag words while achieves very high precision. And we failed to tag the glossary *ergosterol* precisely because a lot of tweets talking *ergosterol* are a kind of advertisement.

5   Conclusion and Future Work

In this paper, we consider the problem of detecting and interpreting new words in Chinese Twitter. We proposed an unsupervised new word detection framework which take several statistical features to derive a word probability score that can measure word-forming likelihood of a character sequence. Since this framework is a statistical approach, it could be easily applied to other languages that have similar characteristics as Chinese characters (e.g. No natural word delimiters).

Then, we used automatic tagging in new word interpretation. We derive a similarity measure between new word and its candidate tag word based on similarity of their corresponding contexts. Experiments on real datasets show the effectiveness of our approach. However,in this work, some thresholds, such as $freq(\cdot)$ and $Pr_{word}(s)$, are set by experiments and observation. In real practise, we can have a more systematic and statistical way to set some appropriate thresholds. For example, for the frequency, we can compute the mean and the standard deviation of the identified words, then set a threshold based on the mean and the standard deviation. In the future, we will try to explore an automatic way to define the parameters used in this framework and apply the language model in our research to get more accurate results.

References

1. Ritter, A., Clark, S., Etzioni, O.: Named entity recognition in tweets: an experimental study. In: Proceedings of the Conference on Empirical Methods in Natural Language Processing. Association for Computational Linguistics (2011)
2. Gattani, A., et al.: Entity extraction, linking, classification, and tagging for social media: a wikipedia-based approach. In: Proceedings of the VLDB Endowment 6.11, pp. 1126–1137 (2013)
3. Ye, Y., Qingyao, W., Li, Y., Chow, K.P., Hui, L.C.K., Yiu, S.-M.: Unknown chinese word extraction based on variety of overlapping strings. Inf. Process. Manag. **49**(2), 497–512 (2013)
4. Nadeau, D., Sekine, S.: A survey of named entity recognition and classification. Lingvisticae Investig. **30**(1), 3–26 (2007)
5. Zhou, N., et al.: A hybrid probabilistic model for unified collaborative and content-based image tagging. IEEE Trans. Pattern Anal. Mach. Intell. **33**(7), 1281–1294 (2011)
6. Kim, H.-N., et al.: Collaborative filtering based on collaborative tagging for enhancing the quality of recommendation. Electron. Commer. Res. Appl. **9**(1), 73–83 (2010)
7. Luo, S., Sun, M.: Two-character Chinese word extraction based on hybrid of internal and contextual measures. In: Proceedings of the second SIGHAN workshop on Chinese language processing, vol. 17. Association for Computational Linguistics (2003)
8. Jin, Z., Tanaka-Ishii, K.: Unsupervised segmentation of Chinese text by use of branching entropy. In: Proceedings of the COLING/ACL on Main Conference Poster Sessions. Association for Computational Linguistics (2006)
9. Wang, L., et al.: CRFs-based Chinese word segmentation for micro-blog with small-scale data. In: Proceedings of the Second CIPS-SIGHAN Joint Conference on Chinese Language (2012)
10. Zhang, K., Sun, M., Zhou, C.: Word segmentation on Chinese mirco-blog data with a linear-time incremental model. In: Proceedings of the Second CIPS-SIGHAN Joint Conference on Chinese Language Processing, Tianjin (2012)
11. Zhang, H.-P., et al.: HHMM-based Chinese lexical analyzer ICTCLAS. In: Proceedings of the Second SIGHAN Workshop on Chinese Language Processing, vol. 17. Association for Computational Linguistics (2003)
12. Gang, Z., et al.: Chinese New Words Detection in Internet. Chin. Inf. Technol. **18**(6), 1–9 (2004)
13. Tseng, H., et al.: A conditional random field word segmenter for sighan bakeoff 2005. In: Proceedings of the Fourth SIGHAN Workshop on Chinese Language Processing, vol. 171, Jeju Island (2005)

Applications of Big Data Analysis

Text Categorization for Deriving the Application Quality in Enterprises Using Ticketing Systems

Thomas Zinner[1], Florian Lemmerich[2], Susanna Schwarzmann[1], Matthias Hirth[1]([✉]), Peter Karg[3], and Andreas Hotho[1,4]

[1] Institute of Computer Science, University of Würzburg, Würzburg, Germany
matthias.hirth@informatik.uni-wuerzburg.de
[2] GESIS - Leibniz Institute for the Social Sciences, Cologne, Germany
[3] Kubus IT, Munich, Germany
[4] L3S Research Center, Hannover, Germany

Abstract. Today's enterprise services and business applications are often centralized in a small number of data centers. Employees located at branches and side offices access the computing infrastructure via the internet using thin client architectures. The task to provide a good application quality to the employers using a multitude of different applications and access networks has thus become complex. Enterprises have to be able to identify resource bottlenecks and applications with a poor performance quickly to take appropriate countermeasures and enable a good application quality for their employees. Ticketing systems within an enterprise use large databases for collecting complaints and problems of the users over a long period of time and thus are an interesting starting point to identify performance problems. However, manual categorization of tickets comes with a high workload.

In this paper, we analyze in a case study the applicability of supervised learning algorithms for the automatic identification of relevant tickets, i.e., tickets indicating problematic applications. In that regard, we evaluate different classification algorithms using 12,000 manually annotated tickets accumulated in July 2013 at the ticketing system of a nationwide operating enterprise. In addition to traditional machine learning metrics, we also analyze the performance of the different classifiers on business-relevant metrics.

1 Introduction

Today's enterprise IT infrastructure is undergoing a significant change. Employees are working with an increasing number of different applications resulting in individual configurations of their personal computers. In order to reduce maintenance of the equipment, and to simplify the installation and management of diverse applications, companies move away from the traditional personal computer architecture, where each user has her own computer running all applications for daily work. Instead, companies try to manage many devices and applications centrally to keep end user devices as simple as possible. In that direction, thin client architectures provide an universal and basic terminal for

© Springer International Publishing Switzerland 2015
S. Madria and T. Hara (Eds.): DaWaK 2015, LNCS 9263, pp. 325–336, 2015.
DOI: 10.1007/978-3-319-22729-0_25

user interaction while all applications requiring complex computations as well as data storage are carried out in the data center.

Although the overall maintenance of such an architecture is simplified compared to a purely decentralized architecture, some additional dependencies are introduced. Beside the thin client the user is interacting with, the transport network and the data center have a huge impact on the application performance and therewith on the productivity of the employees. Thus, it is of high importance for an enterprise to quickly identify slow-running applications and services, to perform a root-cause analysis, and then to improve the performance of these applications and services. A possible information source in an enterprise environment collecting complaints, problems and corresponding solutions are ticketing systems, i.e., central operational database systems used for the management of individual issues.

In this paper, we deduce the problematic applications and the application quality in an enterprise environment by using data from an internal ticketing system of a German company. Around 20.000 employees, working in $\approx$ 400 branch offices mostly located in small- to medium-sized cities, generate up to 1000 tickets per day. The proposed approach is challenging in the way, that the relevant tickets indicating performance problems have to be identified in a huge set of support tickets. A manual classification cannot be considered due to the huge amount of tickets submitted. For that reason, we present an automatic approach to identify relevant tickets based on machine learning. In that regard, we evaluate different classification algorithms using 12,000 manually annotated tickets accumulated in July 2013. From an enterprise perspective, the correct classification of a single ticket is not crucial as long as temporal trends and locally occurring performance problems are identified. Therefore, the classifier results are not only investigated by traditional machine learning metrics such as F_1-score, precision and recall, but also with respect to more business-relevant performance metrics. As an additional contribution of this paper, an anonymized version of the pre-processed dataset (as a tf-idf matrix) will be made publicly available with this paper[1].

The remainder of the paper is structured as follows: First, Sect. 2 reviews related work. Next, Sect. 3 introduces the utilized dataset and outlines the applied methodology, i.e., pre-processing procedures, classification algorithms and performance metrics. Afterwards, Sect. 4 discusses advantages and limitations of the applied approach. Finally, Sect. 5 concludes the paper with a summary and an outlook on future research directions.

2   Related Work

This section provides a brief overview of relevant related work. First, previous approaches to deducing business application quality are presented. After that, we shortly discuss text categorization with a special focus on ticket systems.

---

[1] https://github.com/lsinfo3/dataset_ticketing_system_Dawak_2015.

2.1   Identification of Application Quality via Tickets

User satisfaction with application quality – also called Quality-of-Experience (QoE) – can be defined as *the degree of delight or annoyance of the user of an application or service. It results from the fulfillment of his or her expectations with respect to the utility and / or enjoyment of the application or service in the light of the user's personality and current state* [15]. Hence, this metric is of purely subjective nature.

Typically, user satisfaction is evaluated by adjusting the application behavior while conducting a user survey. This allows to link objectively measurable parameters like the applications response time with user satisfaction. Examples for such evaluations are Staehle et al. [19] or Casas et al. [3]. Both works aim at a better understanding of the influence of varying technical parameters on the QoE. In both cases, the tests were conducted in dedicated labs with students and not in a working environment with employees. Additionally, such a methodology does not scale well in a complex IT system with a huge number of different applications.

To gain a global view of the user satisfaction in a business environment, existing user feedback can be used, e.g., from support tickets. In this direction, Mockus et al. [17] examined the satisfaction of a user group within the first three months after installation of a new software release. They evaluated service interactions like software defect reports and requests for assistance. In their study, they examined the impact of occurring problems and their frequencies. On the basis of their results, they detected predictors for customer perceived user quality. Chulani et al. [5] analyzed the relationship between development and service metrics (number of defects discovered, time to resolve a defect, severity of defects, etc.) and customer satisfaction survey metrics (overall customer satisfaction, top attributes customers look for in a product, etc.). With this knowledge, they improved the user satisfaction by handling specific reports earlier, i.e., reports that concern problems related to an attribute, of which users thought is important, but simultaneously is not satisfying.

2.2   Text Categorization of Support Tickets

Text categorization with classification algorithms is a well explored standard task for text mining and text analytics, one of its main applications being *text filtering*, cf. [20]. While machine learning techniques for text filtering have been successfully applied in various domains, e.g., spam filtering [10] and filtering of unsuitable content [4], the use of machine learning techniques in the context of support ticket systems for enterprises is only little explored. In that direction, Wei et al. applied conditional random fields to automatically extract individual units of information from ticket texts [21]. Kreyss et al. [14] applied text mining categorization technology to analyze data from the IBM Enterprise Information Portal V8.1 containing more than 15,000 product problem records. They used a proven software quality category set to categorize these problem records into different areas of interest. Medem et al. [16] used the free text of network trouble tickets, for analyzing general trends in network incidents and maintenance activities. Diao et al. used an hybrid approach of automatically learned and manually edited rules in order to improve an automated system for classifying support tickets in failure classes [7]. Recently, Altintas and Tantuk proposed an extension of an issue tracking system that aims at automatically assigning tickets to the relevant person or department [2]. By contrast, this work focuses on filtering support tickets that are of high relevance to the target application in order to derive information about the application quality in a thin client environment.

In this paper, we do not aim at contributing algorithmic advances to the field of text categorization, but apply standard techniques to our application domain, that is, the automatic classification of support tickets for enterprises. Therefore, we only outline the applied algorithms in Sect. 3.2, and refer to overview articles for more detailed explanations, see e.g., [12, 20].

3   Automatic Classification of Ticket Data

3.1   Dataset

We propose to utilize user feedback from support tickets to estimate the overall QoE of employees in a large scale business. Ticketing systems like OTRS[2] provide a central point for collecting and maintaining service and help requests in companies. Tickets cover issues ranging from simple password reset requests to notifications about severe system failures. In our case, we are interested in finding tickets indicating annoying system behavior, e.g., long application response times. Automatic approaches are required for filtering these tickets, since several hundreds to thousand tickets may be created daily.

The investigated setting origins from a German company with about 400 branch offices. For this company about 10,000 tickets are submitted consistently each month. In this work, we focus on a subset of this data consisting of around 12000 tickets, that is, all tickets submitted in July 2013. These tickets were manually categorized with a binary label. 303 tickets were labeled as positive examples, i.e., tickets reporting application quality issues and thus indicating a reduced user-perceived application quality. In the following, we use this labeled dataset as gold standard for our evaluations.

The following information is available for each ticket: Each ticket has an unique *ticket-id* that can be used for identification and a *generation date* indicating when the ticket was submitted. It further contains *details* about the reported issue. These allow the user to describe his issues in free text form. Additionally, also information about the affected *site* of the company was included.

3.2   Methodology

As common practice in text mining suggests, we utilize a *vector space approach*, cf. [18], with bag-of-words representations of the detailed ticket text. To transform the set of support tickets, we first applied standard tokenizing procedures as

---

[2] http://www.otrs.com/.

pre-processing. Additionally, stop words, i.e., very common words, were removed according to a dictionary. Furthermore, also tokens consisting of a single letter and tokens consisting of more than 25 letters were removed. In the vector representation, each document (ticket) $d \in D$ of our dataset D is represented as a feature vector $x(d) = (x_{d,1}, \ldots x_{d,m})^T$, where each element of the vector corresponds to a specific term in the document corpus. For the experiments presented in this paper, vector elements were constructed as tf–idfs (term frequency – inverted document frequency) statistics, that is, $x(x_{d,t}) = freq(t,d) \cdot \log \frac{|D|}{|d \in D : t \in d|}$. Here, $freq(t,x)$ describes the frequency of the term t in the document d, $|D|$ the number of documents in the dataset, and $|d \in D : t \in d|$ the number of documents that contain the term t. In doing so, high values of $x_{d,t}$ indicate that the term t is specifically often used in the document d. In addition, each document in our data was manually provided a binary label $y(d)$, indicating the relevance of a document (ticket) to the user performance.

To categorize tickets with an unknown label, we employ several well established techniques for text classification, cf. also [12]. In particular, we evaluate the effects of the following classification methods:

- *Naive Bayes.* This probabilistic classifier uses the well-known Bayes formula under the assumption of independence of features (terms) [9]. Even though this assumption is of course not realistic, the Naive Bayes classifier has been shown to perform relatively well for text classification in practice [8,13].
- *K-Nearest-Neighbor (KNN).* To classify a new instance, this method avoids building an explicit model, but retrieves the k most similar documents according to a distance metric instead and classifies the new document with the majority label in this neighborhood [1,6].
- *Support vector machines (SVM).* Support vector machines in their most basic form classify new instances by computing a hyperplane in the vector space that separates positive and negative instances and maximizes the distances from the hyperplane to the closest positive and negative examples [13]. Since SVMs can handle large sets of features well, they are especially suited for text classification. To enhance performance of SVMs *kernel methods* have been proposed that transform the vector space into higher dimension. In that direction, we consider in particular the anova kernel, see [11], which was chosen after some initial experiments.

3.3 Metrics

To measure the classification performance of classifiers, we considered two kinds of evaluation metrics, that is, standard metrics from machine learning, and business relevant metrics that are more directly concerned with impact in practice.

Machine Learning Metrics. Since the target label distribution in our corpus is heavily biased towards negative examples, i.e., documents that are not labeled as performance relevant, a high accuracy could be achieved by simply classifying all documents as negative. Therefore, we focus on the well known precision and recall framework. In particular, the *precision* is the fraction of correctly classified documents in the set of all documents that were classified as positive (performance relevant):

$$precision = \frac{|positive\_examples \cap examples\_classified\_as\_positive|}{|examples\_classified\_as\_positive|}.$$

By contrast, *recall* denotes the percentage of performance relevant tickets which are correctly identified as such:

$$recall = \frac{|positive\_examples \cap examples\_classified\_as\_positive|}{|positive\_examples|}.$$

Since there is usually a trade-off between these measures, the F_1 score, i.e., the harmonic mean between precision and recall, is commonly used as an overall performance metric:

$$F_1 = 2 \cdot \frac{precision \cdot recall}{precision + recall}.$$

Business-Relevant Metrics. Although text classifiers may yield a good performance with respect to scores like F_1, precision, and recall, they are not able to find all relevant tickets in general. From an enterprise view that might be fine, as long as the algorithms are capable of detecting major issues in the companies infrastructure. Of particular interest in our application context is to identify performance problems at specific branch offices (sites), and to preserve trends regarding the tickets per day. To measure classifier performances for these applications, we introduce two business-relevant metrics, namely:

- *relative amount of identified sites (rais)*: This metric depicts which share of relevant branch offices with face application performance degradations are identified by the classifier. It can be computed as:

$$rais = \frac{\#\text{correctly identified sites with performance relevant tickets}}{\#\text{all sites issuing performance relevant tickets}}.$$

  This reflects the traditional recall measure on a site level. Performance problems reported for a branch office may be due to local incidents at the branch location like disturbances of the local network or the aggregation network.
- *trend preservation*: This metric highlights if the trend of identified tickets on a per-day basis independent of the specific office is preserved. This can be formalized using the well-known Pearson's correlation coefficient. Several tickets per day, possibly from different branch offices, are an indicator for large-scale incidents, probably in a data center or the wide-area network.

4   Evaluation

The next section presents experimental results for different classifiers in our application. We first present results based on classical machine learning metrics,

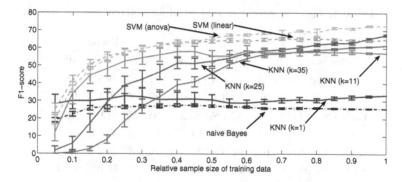

Fig. 1. Impact of differently sized random samples of the training set. In addition to the mean F_1-score (as shown by different line types and colors), confidence intervals from runs with multiple random samples are shown for different classifiers

then we discuss classifier performance with respect to the proposed business relevant metrics. All experiments were conducted in the RapidMiner[3] environment using additional scripts for pre- and postprocessing. Unless stated otherwise, we used the default parameters provided by RapidMiner. As an exception, we determined $C = 10$ as the best parameter setting for the SVMs after initial experiments. Similarly, we focused on the Euclidean distance as the distance metric for the KNN variations after initial experiments with other distance measures, e.g., the cosinus distance[4].

4.1 Performance of Machine Learning Algorithms

In this subsection, we investigate the performance of several classifiers with respect to the performance metrics *precision, recall* and F_1-*score*. For that purpose, we defined a training set consisting of the first 80 % of the data set (in chronological order) and a fixed-size test set consisting of the last 20 % of the data. The split was done with the rational, that available performance tickets are gathered and annotated to train a classifier. The trained classifier is then applied to newly generated tickets. Since the acquisition of correctly labeled training data is costly (in working hours) as it requires manual annotation of tickets, we are also specifically interested in how many labeled tickets are required by a classifier to achieve a certain performance. Therefore, we analyzed the impact of using only a randomly sampled subset of the training data. Results for the mean F_1 score as well as 80 %-confidence intervals resulting from the analysis of at least 10 different samples are shown in Fig. 1 and Table 1. It can be observed that support vector machines overall outperform Naive Bayes as well as K-nearest-neighbor (KNN) approaches. While Naive Bayes is clearly outmatched, KNN shows competitive performance if enough training data is available and if it is

---

[3] https://rapidminer.com.
[4] Due to space constraints we do not report on initial experiments performed for parameter optimization.

Table 1. F_1-score, precision, and recall for different sample sizes

	F1		Precision		Recall	
Sample size	40 %	100 %	40 %	100 %	40 %	100 %
Naive bayes	27.5 %	25.6 %	19.9 %	17.4 %	44.9 %	48.7 %
KNN (N = 1)	30.9 %	33.3 %	43.6 %	47.6 %	30.1 %	25.6 %
KNN (N = 11)	57.9 %	57.1 %	83.2 %	94.1 %	44.9 %	41.0 %
KNN (N = 25)	50.6 %	68.8 %	81.6 %	95.4 %	37.6 %	53.8 %
KNN (N = 35)	39.0 %	61.5 %	87.5 %	76.9 %	26.2 %	51.2 %
SVM (linear)	63.4 %	64.9 %	73.6 %	65.7 %	56.0 %	64.1 %
SVM (anova)	62.4 %	72.7 %	83.9 %	88.8 %	49.9 %	61.5 %

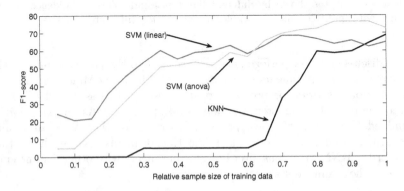

Fig. 2. Impact of differently sized subset of the training set in chronological order on the F_1-score. The investigated classifiers are highlighted with different colors

well parametrized. The best performance of KNN was achieved by $k = 25$, higher as well lower parameters for k lead to worse results. However, using less training data, SVMs clearly show the best classification results in linear as well as anova kernel variations. While these variations show similar F_1 scores for smaller sets of training data, the anova kernel version leads to better results for the full training data set. This difference seems to be specifically caused by a better precision of the classification, see Table 1. The overall superiority of support vector machines in our application is in line with previous findings for classifying text data, see [8,13]. Based on these results, we focused on the support vector machines and KNN with $k = 25$ for further experiments.

In a second series of experiments, we assume a scenario where tickets are labeled in chronological order, e.g., on a per day basis. This follows the intuition that manpower for manual annotation of tickets may be available in some time intervals, but not continuously. Varying training set sizes are again evaluated on a fixed-size test set consisting of the (chronologically) last 20 % of the data. Results for the F_1 score are shown in Fig. 2 and Table 2. It can be seen that the SVMs achieve a good F_1 score already for ≈ 40 % of the training set. For

Table 2. F_1-score, precision, and recall for different subset sizes (40 % and 80 % of the training set).

	F1		Precision		Recall	
Sample size	40 %	80 %	40 %	80 %	40 %	80 %
KNN ($N = 25$)	4.8 %	59.6 %	50.0 %	94.4 %	2.5 %	43.5 %
SVM (linear)	55.3 %	66.6 %	69.2 %	72.7 %	46.1 %	61.5 %
SVM (anova)	51.8 %	72.4 %	93.3 %	83.3 %	35.8 %	64.1 %

Table 3. Precision for all sites and precision for sites with exactly one relevant ticket within the test data set.

	Training set size: 1/3 of data		Training set size: 2/3 of data	
	$rais_{all}$	$rais_{oneticket}$	$rais_{all}$	$rais_{oneticket}$
KNN ($N = 25$)	11.1 %	2.9 %	50.0 %	31.3 %
SVM (linear)	56.8 %	25.7 %	64.2 %	43.8 %
SVM (anova)	48.1 %	22.9 %	66.1 %	43.8 %

a similar F_1 score, a training set size of 80 % is required when using the KNN approach. Details on the F_1 score, precision, and recall are provided in Table 2. Hence, we can conclude a training set that contains $\approx$ 1/3 of the whole data set is sufficient to train a SVM model that performs well in tests on the last 20 % of the data. In the following, we evaluate the impact of all classifiers on the business-relevant metrics for two training set sizes, namely $\approx$ 1/3 and $\approx$ 2/3 of the available data.

4.2 Impact of Different Algorithms on Business-Relevant Metrics

This subsection highlights the performance of the presented approach with respect to the business-relevant metrics *identified sites* and *trend preservation*. The classifiers are trained with data sets of either the first 1/3 or the first 2/3 of the data set. The relative number of identified sites reporting performance problems ($rais_{all}$) are summarized in Table 3. This table additionally shows this measure restricted to those sites that reported exactly one performance problem, noted as $rais_{oneticket}$. For all cases, a training set of 2/3 of the data results in better identification ratio. Also, SVM classifiers outperform the KNN classifier in both settings. For the smaller training set, $rais$ decreases considerably when using KNN, resulting in a poor detection of sites with application problems. Although the precision of the SVMs is reduced compared to a larger training data set, the overall detection rate remains good. However, a reliable detection of issues at specific sites cannot be guaranteed.

The results for the identification of relevant tickets on a per day basis are illustrated in Fig. 3. Figure 3a depicts the quality of the different approaches for 2/3 as testing data, respectively day 10–31. Figure 3a depicts the quality of

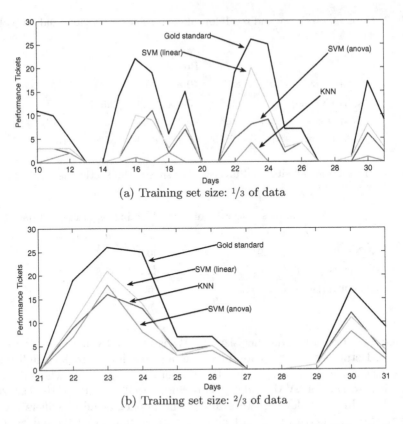

(a) Training set size: 1/3 of data

(b) Training set size: 2/3 of data

Fig. 3. Identification of relevant tickets on a per-day basis for the different classifiers (illustrated in different colors) and comparison with the gold standard

the different approaches for 1/3 as testing data, respectively day 19–31. None of the analyzed approaches is capable of identifying all relevant performance tickets per day, however, global trends are mostly preserved in most scenarios. For a training set size of 1/3, depicted in Fig. 3a, the KNN classifier performs poorly. Several days with reported performance issues are not identified. Both SVM classifiers perform better, whereas the SVM with linear kernel performs best. Further, the SVM with linear kernel is the only one capable to identify the single performance ticket occurring on day 29. In case of a training set size of 2/3, illustrated in Fig. 3b all approaches perform better, with the SVM with linear kernel performing best, and the SVM with anova kernel performing worst. The SVM with anova kernel is not capable to detect the single performance ticket on day 29, while both other approaches detect this ticket. The trend preservation is quantified using Pearson's correlation coefficient, as shown in Table 4. It can be observed, that the SVMs perform very well for our data set with respect to the business-relevant metrics. Although the KNN approach achieves a similar result quality in case of a large training data set, it clearly underestimates the

Table 4. Quantification of the trend preservation with Pearson's correlation coefficient.

	KNN (N=25)	SVM (linear)	SVM (anova)
Training set size: 1/3 of data	36.7%	90.1%	89.8%
Training set size: 2/3 of data	90.3%	96.5%	97.4%

performance problems for small training set sizes. Overall, the applied classifiers preserve the temporal trend of relevant tickets very well.

5 Conclusions

This paper featured a case study that investigated the performance of machine learning techniques for the task of identifying relevant support tickets, i.e., tickets that indicate problematic applications in a thin client architecture environment. To this end, we evaluated the performance of several classification algorithms on over 12,000 manually annotated tickets in different scenarios and differently sized training sets. Besides traditional machine learning metrics such as F_1-score, precision and recall, also business relevant metrics derived from our target application were analyzed. As a result, supervised learning methods are overall well suited to identify application problems based on ticket data. Even though classifiers cannot categorize each individual ticket accurately, major issues at specific sites and general trends can be detected. Comparing different classifiers, support vector machines outperformed other evaluated techniques in our data set. For reproducibility and further studies, the pre-processed data sets will be made publicly available in an anonymized form.

In the future, we plan to explore the combination of standard machine learning classifiers with manually created keyword lists in order to further increase the classification accuracy. In addition, the full integration of automatic classification within business processes will allow for a continuous annotation of support tickets. Finally, the binary classification of tickets could be extended into a ticket ranking system that will allow to analyze the most important tickets first.

Acknowledgement. This work is supported by the Deutsche Forschungsgemeinschaft (DFG) under Grants HO TR 257/41-1. The authors alone are responsible for the content.

References

1. Aha, D.: Lazy learning. Springer Science and Business Media, Heidelberg (1997)
2. Altintas, M., Tantug, A.C.: Machine learning based ticket classification in issue tracking systems. In: Proceedings of International Conference on Artificial Intelligence and Computer Science (AICS 2014) (2014)
3. Casas, P., Seufert, M., Egger, S., Schatz, R.: Quality of experience in remote virtual desktop services. In: Proceedings of the Workshop on Quality of Experience Centric Management (QCMAN), Ghent (2013)

4. Chandrinos, K.V., Androutsopoulos, I., Paliouras, G., Spyropoulos, C.D.: Automatic web rating: filtering obscene content on the web. In: Borbinha, J.L., Baker, T. (eds.) ECDL 2000. LNCS, vol. 1923, pp. 403–406. Springer, Heidelberg (2000)

5. Chulani, S., Santhanam, P., Moore, D., Leszkowicz, B., Davidson, G.: Deriving a software quality view from customer satisfaction and service data. In: European Conference on Metrics and Measurement (2001)

6. Cover, T., Hart, P.: Nearest neighbor pattern classification. IEEE Trans. Inf. Theory **13**(1), 21–27 (1967)

7. Diao, Y., Jamjoom, H., Loewenstern, D.: Rule-based problem classification in it service management. In: Proceedings of IEEE International Conference on Cloud Computing (CLOUD 2009), pp. 221–228. IEEE (2009)

8. Dumais, S., Platt, J., Heckerman, D., Sahami, M.: Inductive learning algorithms and representations for text categorization. In: International Conference on Information and Knowledge Management, pp. 148–155. ACM (1998)

9. Good, I.J., Hacking, I., Jeffrey, R.C., Törnebohm, H.: The estimation of probabilities: an essay on modern bayesian methods. Synthese **16**(2), 234–244 (1966)

10. Guzella, T.S., Caminhas, W.M.: A review of machine learning approaches to spam filtering. Expert Syst. Appl. **36**(7), 10206–10222 (2009)

11. Hofmann, T., Schölkopf, B., Smola, A.J.: Kernel methods in machine learning. Ann. Stat. **36**(3), 1171–1220 (2008)

12. Hotho, A., Nürnberger, A., Paaß, G.: A brief survey of text mining. Ldv Forum **20**, 19–62 (2005)

13. Joachims, T.: Text Categorization with Support Vector Machines: Learning with Many Relevant Features. Springer, Heidelberg (1998)

14. Kreyss, J., Selvaggio, S., White, M., Zakharian, Z.: Text mining for a clear picture of defect reports: a praxis report. In: Proceedings of International Conference on Data Mining, Melbourne (2003)

15. Le Callet, P., Möller, S., Perkis, A., et al.: Qualinet white paper on definitions of quality of experience. European Network on Quality of Experience in Multimedia Systems and Services (COST Action IC 1003) (2012)

16. Medem, A., Akodjenou, M.-I., Teixeira, R.: Troubleminer: mining network trouble tickets. In: Proceedings of Symposium on Integrated Network Management, Long Island (2009)

17. Mockus, A., Zhang, P., Li, P.L.: Predictors of customer perceived software quality. In: Proceedings of International Conference on Software Engineering, St. Louis (2005)

18. Salton, G., Wong, A., Yang, C.-S.: A vector space model for automatic indexing. Commun. ACM **18**(11), 613–620 (1975)

19. Schlosser, D., Staehle, B., Binzenhöfer, A., Boder, B.: Improving the qoe of citrix thin client users. In: Proceedings of International Conference on Communications, Cape Town (2010)

20. Sebastiani, F.: Machine learning in automated text categorization. ACM Comput. Surv. **34**(1), 1–47 (2002)

21. Wei, X., Sailer, A., Mahindru, R., Kar, G.: Automatic structuring of it problem ticket data for enhanced problem resolution. In: 10th IFIP/IEEE International Symposium on IM 2007 Integrated Network Management, pp. 852–855. IEEE (2007)

MultiSpot: Spotting Sentiments with Semantic Aware Multilevel Cascaded Analysis

Despoina Chatzakou[✉], Nikolaos Passalis, and Athena Vakali

Informatics Department, Aristotle University of Thessaloniki, Thessaloniki, Greece
{deppych,npassalis,avakali}@csd.auth.gr

Abstract. Given a textual resource (e.g. post, review, comment), how can we spot the expressed sentiment? What will be the core information to be used for accurately capturing sentiment given a number of textual resources? Here, we introduce an approach for extracting and aggregating information from different text-levels, namely words and sentences, in an effort to improve the capturing of documents' sentiments in relation to the state of the art approaches. Our main contributions are: (a) the proposal of two semantic aware approaches for enhancing the *cascaded phase* of a sentiment analysis process; and (b) MULTISPOT, a multilevel sentiment analysis approach which combines word and sentence level features. We present experiments on two real-world datasets containing movie reviews.

Keywords: Sentiment detection · Multilevel features

1 Introduction

Given a document, at which level can sentiment be captured? Typically, document-based sentiment analysis processes operate at a particular level, i.e. at the word or sentence level, for extracting a document's sentiment. Sentiment identification at the word level is a fine-grained approach, which emphasizes on sentimentally intense words, but its lack of awareness for words' dependencies limits its capability to accurately capture sentiments. The more coarse-grained sentence level sentiment identification permits the capturing of a document's semantic-related information since it involves ordering and words' similarity, but its higher text level processing prohibits the capturing of sentimental words' intensities. This work is motivated by the fact that sentiments' extraction out of separated sets of words or flat lined sentences leads to information loss. Here, the overall document's sentiment identification is considered as a multilevel process which aggregates both words and sentences characteristics (called features for the rest of the paper).

In machine learning, the most popular approach for sentiment analysis, the selection of appropriate features for representing a document is crucial. In sentiment identification at the word level different types of features have been introduced, which are either sentiment-based (e.g. words which express a specific sentiment), syntactic-based (e.g. part-of-speech and n-grams), or semantic-based

© Springer International Publishing Switzerland 2015
S. Madria and T. Hara (Eds.): DaWaK 2015, LNCS 9263, pp. 337–350, 2015.
DOI: 10.1007/978-3-319-22729-0_26

(e.g. semantic word vector spaces which capture the meaning of each word). Having specified the features tailored for the problem at hand, a document classification methodology is applied. When dealing with sentence-level features,the cascaded sentiment analysis is a common approach which proposes the use of two different level classifiers for extracting the overall sentiment of a document. It initially utilizes a classifier for annotating each sentence based on the expressed sentiment (sentence-level classifier) and then an additional classifier for aggregating the output of the first classifier. Up to now, most of the proposed cascaded methods ignore semantically rich features (high-level features), as for instance, the semantic closeness of two sentences in a document. As semantics deal with the meaning of a document these high-level features which are relevant to words, phrases and sentences, are promising for a more accurate capturing of a document's sentiment [4].

In this paper, the aggregation of both word and sentence level features is proposed for more accurate documents' sentiments capturing. The proposed multilevel cascaded sentiment analysis method, MULTISPOT, aims at an improved classification process due to its different text level features aggregation (i.e. word and sentence level) which are exploited in both the training and testing classification phases. The MULTISPOT method introduces an aggregation phase which considers the *Sentiment Center* (Sentcenter), the *Semantic Center* (Semcenter), where for each sentence a sentiment and semantic vector is generated, respectively, and the so called *Centerbook* (has been inspired by the computer vision field, where CodeBook learning classification approaches are characterized by improved accuracies [1]), which builds on vectors that merge sentimentally and semantically similar sentences. Such center-based approaches are chosen since the overall sentiment expressed in a document is highly impacted by all of the available sentiment and semantic information.

The sentiment and semantic vectors are produced based on the RNTN (Recursive Neural Tensor Network) [16] model which is popular for its ability in capturing more effectively the meaning of a sentence. Based on the RNTN model each sentence is defined using the representation of its words' vectors (builds upon previous word vector representations [14,15]) and a parse tree.

The contribution of this work is summarized as follows:

- **MultiSpot:** a two-level document-based classification approach which combines features extracted from different text-levels. Our motivation originates from the idea of exploiting both word and sentence level features to reach better accuracy in a cascaded sentiment analysis manner.
- **Center-Based Aggregation:** an accumulative process with sentiment and semantic relevant features' extraction, namely the *Sentcenter*, the *Semcenter* and the *Centerbook* for modelling the overall document's context. To best of the authors' knowledge, no earlier work has utilized such a *Centerbook* learning technique for texts' sentiment analysis.

2   Related Work

Document-level sentiment analysis has been realized by a large number of methods which proceed with either word or sentence level information extraction.

Word-Level Feature Extraction. The *Bag-of-Word (BoW)* model, where a document is represented as a set of its included words, has been heavily utilized. Several variations have been proposed, including the use of bigrams (e.g. [19]), the *term - frequency* (tf) and the *term - frequency inverse document - frequency* (tf-idf) (e.g. [9,10]) weighting schemes. Despite their simplicity, such approaches can reach remarkable accuracy on large documents [19], but as their models mainly focus just on the words and not on the words' positioning and ordering, semantic features are missing (e.g. cannot effectively capture the negation). To maintain the semantic similarities among words, a *semantic word space* is typically used, where each word of a document is represented as a vector [7,14,17]. The word vectors, for words with similar semantic and/or sentiment content, tend to be close (e.g. based on cosine similarity measure). However, these approaches fall behind as they fail to accurately capture the ordering of the words. Most recent efforts (such as in [6,16]) have involved words' ordering and negation capturing leading to accuracy improvements in classification.

Sentence-Level Feature Extraction. Cascaded sentiment analysis is a commonly utilized approach which exploits sentence-level information. The recognition of a document's overall sentiment involves an aggregation method, in which the mean sentiment score for a document is calculated by averaging the sentiment score of its sentences [18,21]. More advanced approaches assign different weights to sentences that may better convey information about a document's sentiment based on different aspects, such as their position, ordering, length and subjectivity. Finally, [13] involves mining initially sequential rules from the sentences' ordering, and then keeping only the top k rules as features.

MULTISPOT accumulates both word and sentence level features, while it also manages to integrate sentiment with semantic information into a tailored aggregation phase, which contributes to further enhancing the overall accuracy.

3   Multilevel Sentiment Detection

3.1   Problem Definition

Most earlier approaches suffer from the absence of the semantic context and their deficiency in capturing sentiments' evolution across a document's length. Such information loss, is due to either their unawareness of words' dependencies (ordering, positioning) or their absence of knowledge on the sentiment's formation from the words to the sentences levels. Our intuition is that by exploiting features from both levels the accuracy of a document-based sentiment analysis approach will be improved.

Next, we initially define the problem of capturing sentiments from document-level textual resources and then outline the MULTISPOT's proposed approach.

Problem 1. *Document-based sentiment analysis*

Given: *A training dataset $D = \{d_1, d_2, ..., d_n\}$, and a sentiment label for each document, $t_i \in \{pos, neg\}$;*
Learn: *A model G to predict the sentiment label for any new document $d_{test} \notin D$.*

```
Data: A set of documents D = {d₁, d₂, ..., dₙ}
Result: The expressed sentiments for each document in D.
for document dᵢ in D do
    Extract word-level features;
    Split a document d into a set of S sentences;
    for sentence sⱼ in S do
        Extract sentence-level features;
    Aggregate word and sentence level features;
    Extract docsent(d) ∈ {pos, neg};
return docsent(d) ∀ dᵢ ∈ D
```

Algorithm 1: Pseudocode for MULTISPOT

3.2 Proposed Approach

In this section we provide the pipeline of the MULTISPOT and describe its processes and characteristics.

MultiSpot Pipeline. MULTISPOT comprises of three steps: (1) *Word-level feature extraction*, which extracts word-level features from the whole document, (2) *Sentence-level feature extraction*, where having initially split the document in sentences, we derive the corresponding sentence-level features under a cascaded phase (MULTISPOT's cascaded phase), (3) *Features combination*, where features from both levels are used for training a document-based sentiment classifier.

In our case we examine two models (as features) that use word-level information, namely the unigram bag-of-words (BoW)[1] and the NB (Naive Bayes) bigrams model[2] (as being presented in [19]). **Word-level feature extraction** considers two separate phases: (i) the word to vector mapping, and (ii) the document-level vectors aggregation via a weighting scheme. In the first phase, we model a word w as a vector $v \in \mathbb{R}^k$, where each element of this vector is related to a distinct word. All the elements of v are zero except for the one that corresponds to the word w. For example, the vector $x = [0, 1, 0]$ is related to the word "*of*" in the dictionary $\{bag, of, words\}$. Then, all the vectors of the words in a document are combined into one that describes the whole document. For instance, the tf weighting scheme adds these vectors together in order to get the term frequency vector of the document. Using the previously mentioned dictionary, the term frequency vector of the phrase "bag bag words" is (2,0,1). Consequently, the word-level feature extractor f maps each word w of a document d to a vector $f(w) \in \mathbb{R}^l$ (l is the dimensionality of the output features).

Sentence level information is related to higher-level sentiment and semantic information of a sentence. This information cannot be extracted without the aid of classifiers and/or dimensionality reduction methods (e.g. deep learning). For example, a **sentence-level feature extractor** might extract features related to the sentiment and/or the objectivity of each sentence. So, the sentence level

---

[1] BoW model: represents a document as a set of its words.

[2] NB bigrams: Naive Bayes log-count ratios of bigram features.

feature extractor g maps each sentence s of a document to a vector $g(s) \in \mathbb{R}^k$ (k similar to l) using one or more sentence-level classifiers and/or dimensionality reduction methods to extract such high-level information from each sentence.

We can now define the **cascaded sentiment analysis** as a process where given a training dataset D, a sentence-level feature extractor, g, and a sentiment label for each document $t_i \in \{pos, neg\}$ learn a model, G, to predict the sentiment label for any new document d_{test}. The model G uses as input the multiset of sentence-level features of each document d, $\{g(s) | \forall s \in S_d\}$. In our case, we train a classifier to accomplish this task using the *sentiment* or the *sentiment/semantic center* of each document in D as features.

MultiSpot: Multilevel cascaded sentiment analysis exploits both word and sentence level information from a document. So, given a training dataset D, a word-level feature extractor $f(w)$, a sentence-level feature extractor $g(s)$, and a sentiment label for each document $t_i \in \{pos, neg\}$, learn a model G to predict the sentiment label for any new document $d_{test} \notin D$. If W_d is the multiset of all words in a document d and S_d is the multiset of all sentences in d, then the model G uses as input the multiset of word-level features, $\{f(w) | \forall w \in W_d\}$, and the multiset of sentence-level features, $\{g(s) | \forall s \in S_d\}$, of each document d.

Here, we use the RNTN model as a sentence-level classifier and feature extractor. It was selected since it has been proven quite powerful and it was the first one to achieve 85.4 % accuracy on binary (i.e. positive/negative) sentence-level classification. The RNTN model can classify individual sentences and produces a 5-value sentiment probability distribution vector that corresponds to the following sentiments: *1 - very negative, 2 - negative, 3 - neutral, 4 - positive, 5 - very positive*, identified by $sent_i$ (where $i = 1, ..., 5$). We define the positive sentiment of a sentence as $sent_{pos}(s) = sent_4(s) + sent_5(s)$ and the negative sentiment as $sent_{neg}(s) = sent_1(s) + sent_2(s)$ to map labels into our considered label set $\{pos, neg\}$. Moreover, during the training of the RNTN model s semantic vector is produced for each input phrase that captures its sentiment and semantic meaning. For now on we will refer to the sentiment distribution of a sentence s as $sent(s)$ and to the corresponding semantic vector as $vec(s)$.

Sentcenter: Sentiment Center Estimation. Most of the proposed cascaded sentiment analysis methods use machine learning techniques to accomplish the task of capturing a document's sentiment. So, if a training set D and the corresponding labels T are available, then we can train a classifier to proceed with this task using the sentiment center (*Sentiment center vector*) of a document or the variance of such sentences around this center (*Sentiment variance vector*). Then we can use either of such vectors as features to represent a document.

Definition 1. *Sentiment center vector (Sentcenter):*

$$sent_{center}(d) = \sum_{s \in d} sent(s) / |d|$$

where $|d|$ is the number of sentences of document d.

Definition 2. *Sentiment variance vector (Sentcenter):*

$$sent_{var}(d) = \sum_{s \in d}(sent(s) - sent_{center}(d))^2/|d|$$

which contains the squared differences from the document's center (variance).

In the experimentation section we refer to the document representation based on $sent_{center}(d)$ as *SentC* and based on $sent_{var}(d)$ as *SentC(c)*.

Semcenter: Semantic Center Estimation. Here, we propose an approach that initially represents each sentence s with a semantic vector $vec(s)$. Then, we represent a document as a vector which contains the semantic center of all sentences' vectors, with either the *Semantic center vector* or the *Semantic variance vector*. Next, we define such two possible representations.

Definition 3. *Semantic center vector (Semcenter):*

$$vec_{center}(d) = \sum_{s \in d} vec(s)/|d|$$

where $|d|$ is the number of sentences in the document d.

Definition 4. *Semantic variance center*

$$vec_{var}(d) = \sum_{s \in d}(vec(s) - vec_{center}(d))^2/|d|$$

which contains the squared differences from the document's center.

In the experimentation section we refer to the document representation based on $vec_{center}(d)$ as *SemC* and based on $vec_{var}(d)$ as *SemC(c)*.

Centerbook: Sentiment & Semantic Centered Approach. The previous approaches summarize a document's sentences by using either their sentiment or semantic centers. However, as the sentences may vary greatly in terms of their sentiment and semantic content, here we propose the *Centerbook*, an approach that describes the documents in terms of a sentence-level semantic dictionary. This semantic dictionary will contain the basic sentiment and semantic concepts that appear in the sentences of our training set. We create such dictionary by clustering the set of all sentences appearing in D, where each cluster contains similar sentences in terms of their sentiment and semantic information. Such clusters are called *centercodes* since they actually provide an encoded reference to similar sentences. The resulting cendroid of a cluster captures the corresponding sentiment and semantic information. We can represent each sentence in a document by its nearest cluster. Thus, the overall document is then modelled by the set of all centercodes.

Based on the above, the objective is to learn a Centerbook with $cNum$ centercodes. The Centerbook is presented as a matrix $P \in \mathbb{R}^{m \times cNum}$, where each column of P corresponds to a distinct centercode ($P = [c_1 \; c_2 \; c_3 \; ... \; c_{cNum}]$). Any vector $x_i \in \mathbb{R}^m$ can be reconstructed using a linear combination of these centercodes: $x_i = P sp_i + \epsilon$, where $sp_i \in \mathbb{R}^{cNum}$ is a sparse representation of x_i and ϵ is the reconstruction error (i.e. the distance between the original data and its estimate) vector.

The *Centerbook* aims to extract high level features as input to a classifier and so an encoding/decoding scheme is used. The *encoder* is responsible for mapping a sentence vector $x(s)$ to a lower dimensionality representation, while the *decoder* goes back from the high level feature representation to the original sentence vector. For going back to the original sentence there is loss of information, which is known as reconstruction error, ϵ. In Centerbook each vector of a sentence s is represented by its nearest center in the set of all the centercodes. So, the reconstruction error from a given vector is the distance between it and its representation. The encoder produces a sparse representation as each x_i is encoded by using only one center of the Centerbook. Based on the above, our objective is to compute P and each sp_i to minimize the reconstruction error:

Definition 5. *Minimization function of Centerbook:*

$$J_{Centerbook} = \sum_{i=1}^{n} ||P sp_i - x_i||_2^2$$

$$s.t. ||sp_i||_0 = 1$$

over a collection of n vectors, where $||sp||_0$ is the number of non-zero elements of sp. Each x_i is encoded using one center in our Centerbook, i.e. $||sp_i||_0 = 1$.

We use k-means clustering for extracting high-level representations from each sentence of a document. So, in our case, the minimization function is equivalent to minimizing the objective function of k-means [8]. Next, we initially present the k-mean's objective function, while then we prove that such objective function is equivalent with the *Centerbook*'s minimization of the reconstruction error.

Definition 6. *Minimization function of K-means:*

$$J_{kmeans} = \sum_{i=1}^{k} \sum_{j=1}^{|G_i|} ||c_i - x_i^{(j)}||_2^2$$

where G_i is the set of vectors of the i-th cluster, $x_i^{(j)}$ is the j-th element of G_i and c_i the centroid of the corresponding cluster.

Lemma 1. *The Centerbook's minimization function is equivalent to the minimization function of K-means.*

Proof.

$$min \ J_{Centerbook} = \sum_{i=1}^{n} ||Psp_i - x_i||_2^2 \ s.t. \ ||sp_i||_0 = 1$$

$$equiv. \ to \ min \sum_{i=1}^{n} ||c_{argmin(||c_j-x_i||_2^2)} - x_i||_2^2$$

$$= \sum_{i=1}^{cNum} \sum_{j=1}^{|G_i|} ||c_i - x_i^{(j)}||_2^2 = J_{kmean}$$

where the Centerbook contains the centers of the clusters: $P = [c_1 \ c_2 \ c_3 \ ... \ c_{cNum}]$.

The encoding function, $h(x)$, for extracting high level features from a sentence is defined as:

Definition 7. *Sentence encoding function:*

$$h_i(x) = \begin{cases} 1, & i == arg_j min(||c_j - x||_2^2) \\ 0, & otherwise \end{cases}$$

where $h_i(x)$ is the i-th element of $h(x)$ vector.

Here, we use the RNTN vectors ($vec(s)$) as input to the above method to create the Centerbook. Based on the sentence encoding, a document d is defined as a vector $code(d)$ that includes the sum of all the already encoded sentences (referred to as $C\_Book(c)$):

Definition 8. *Document representation:*

$$code(d) = \sum_{s \in S_d} h(vec(s))$$

where s is each sentence of a document d.

Finally, a binary document modelling is also defined by replacing the (positive) non-zero elements of $code(d)$ with 1 (referred to as $C\_Book(b)$):

Definition 9. *Binary document representation:*

$$code_{bin}(d) = sign(code(d))$$

where the $sign(x)$ function is applied element-wise.

The $code(d)$ or the $code_{bin}(d)$ vectors can be used as features for the classification process. Our intuition is that the first approach for modelling a document, namely $code(d)$, will result in a better classification accuracy as it provides information about the number of sentences existing in a cluster (centercode).

4   Experiments and Results

Here, we evaluate the center-based aggregation approaches for document representation and the MULTISPOT method using different proposed features in terms of their capability for accurate capturing the sentiments expressed on documents.

4.1   Datasets

We experimented on two datasets, the *Large Movie Review Dataset* (IMDB) [7] and the *Polarity Dataset v2.0* (RT-2k) [11], both of which contain movie reviews collected from the Internet Movie DataBase. These datasets were chosen due to their popularity and their suitability for comparisons in terms of the accuracy of the proposed and earlier sentiment analysis approaches. The *IMDB* dataset is composed of 25000 positive reviews, 25000 negative reviews and 50000 unlabeled reviews, while the *Polarity Dataset v2.0* is composed of 1000 positive and 1000 negative annotated reviews. Each movie review in the IMDB dataset has several sentences, while in the RT-2k dataset each review contains a single sentence.

4.2   MultiSpot Fundamental Processes

Next, we provide the specific choices for each step of the MULTISPOT approach in order to carry out the experimentation under the chosen datasets.

Word-Level Features Extraction. We use a simple BoW scheme and the NB bigrams features.

Sentence-Level Feature Extraction. We relied to the RNTN model to extract sentence-level features. The model was trained using a Sentiment Treebank [16], which contains 11,855 sentences from the dataset introduced in [12].

Clustering. We run k-mean for 15 iterations (after that our method converged). Here, due to lack of space, we present and discuss experimentation rounds with the IMDB dataset having the number of output clusters to be 200 and with the RT-2k dataset having 100 clusters. The sizes of each Centerbook were empirically selected [22]. Under our experimentation efforts we observed that around a certain number of clusters there is a little effect in the accuracy of the MULTISPOT method (see experimentation results in Sect. 4.3).

Classification. We use a linear SVM (*liblinear* library [3]) to classify each document based on the word and/or sentence level features. We conduct ten-fold cross validation to select the best SVM model. We applied a ten-fold validation as we use the SVM training protocol of [11].

4.3   MultiSpot Results and Methods Comparisons

Here, we initially evaluate the proposed center-based aggregated approaches and compare them with MULTISPOT.

Table 1. Evaluation of the proposed cascaded sentiment analysis methods.

Features	IMDB			RT-2k		
	Accuracy	Recall	F1	Accuracy	Recall	F1
SentC	84.02	80.97	83.52	83.30	86.10	83.75
SentC(c)	84.01	80.96	83.51	83.05	86.30	83.58
SemC	84.90	83.45	84.68	**85.10**	86.80	**85.35**
SemC(c)	85.27	83.53	85.01	84.90	85.50	84.99
C_Book(c)	84.76	85.05	84.80	83.15	84.10	83.34
C_Book(b)	84.54	84.64	84.50	82.75	83.60	82.90
SemC(c) + SentC	85.27	83.53	85.01	85.05	85.40	85.10
SemC(c) + C_Book(c) + SentC	**85.35**	84.30	**85.20**	84.85	87.10	85.18

Center-Based Aggregated Methods Evaluation. From Table 1, we observe that in the IMDB dataset the best accuracy is achieved through combining the proposed C_Book(c) and SentC methods with the Sentcenter approach, while in the RT-2k dataset the best accuracy is achieved with the Semcenter approach. Generally, we observe that the SemC and SemC(c) approaches which use semantic features, always provide better classification results than the semantic-free, Sentcenter (i.e. SentC), approach.

MultiSpot Evaluation. To combine word level with sentence level features initially we evaluate MULTISPOT (see Table 2) with the BoW model (as word level features). For the IMDB dataset term frequency weighting scheme was utilized, while for the RT-2k the binary weighting scheme.

By comparing Tables 1 and 2 we observe that combining word and sentence level features always increases significantly the classification accuracy. More specifically, the combination of all the proposed sentence level features with the BoW word level features succeeds an 1.71 % improvement for the IMDB dataset and 1.9 % for the RT-2k dataset.

Table 2. Evaluation of the MULTISPOT method using BoW features.

Features	IMDB			RT-2k		
	Accuracy	Recall	F1	Accuracy	Recall	F1
BoW	87.77	88.01	87.80	87.15	88.40	87.31
BoW + SentC	88.99	89.01	88.99	87.45	88.70	87.60
BoW + SemC(c)	89.36	89.18	89.34	88.20	89.60	88.36
BoW + C_Book(c)	89.29	89.26	89.29	88.85	90.80	89.06
BoW + SentC + SemC(c)	89.38	89.22	89.36	88.25	89.70	88.42
BoW + SentC + SemC(c) + C_Book(c)	**89.48**	89.19	**89.45**	**89.05**	91.50	**89.31**

Table 3. Evaluation of the MULTISPOT method using NB-bigram features.

Features	IMDB			RT-2k		
	Accuracy	Recall	F1	Accuracy	Recall	F1
NB bi	91.43	92.13	91.49	89.45	90.80	89.59
NB bi+ SentC	91.72	91.77	91.73	90.00	91.30	90.13
NB bi+ SemC(c)	91.76	91.49	91.73	90.90	91.90	90.99
NB bi+ C_Book(c)	91.72	91.90	91.73	**91.30**	93.50	**91.49**
NB bi+ SentC + SemC(c)	**91.78**	91.53	**91.76**	90.85	91.90	90.95
NB bi+ SentC + SemC(c) + C_Book(c)	91.60	91.33	91.58	90.65	93.10	90.87

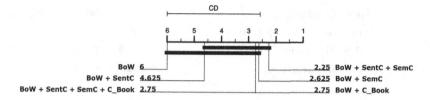

Fig. 1. Nemenyi post hoc test

We have also experimented with the choice of the NB bigrams for examining how the "strength" of the word-level features affects the accuracy gain in a multilevel setup (see Table 3). We've used the same number of folds and parameters in the SVM classifier in the RT-2k dataset as in [19], to ensure the comparability of the results. We observe again that through combining word and sentence level features the overall classification accuracy is increased by 0.35 % in the IMDB case and by 1.85 % in the RT-2k.

The next experiments support our claim that the multilevel cascaded sentiment analysis approach can improve the accuracy of any document-based sentiment classifier which uses word-level information. Initially, we use *Friedman's*[3] test to confirm that the accuracy improvement succeeded with the MULTISPOT is statistically significant. We use such test to detect statistical differences among the results of the proposed methods and the word-level methods. The tests were performed using both datasets. The Friedman test showed that the effect of MULTISPOT for sentiment classification is statistically significant at a significance level of 5 %. Thus, we can reject the null hypothesis: *Multilevel cascaded sentiment analysis does not increase the accuracy of the baseline classifier (BoW).*

Based on this rejection, the *Nemenyi post hoc*[4] test was used to compare all classifiers with each other and spot the methods that are statistically better. Figure 1 shows that the combination of BoW with the Sentcenter (SentC) and

---

[3] It is used to test differences between different samples.

[4] Explores the groups of data that differ after a statistical test of multiple comparisons, e.g. the Friedman test.

Table 4. Comparison of MULTISPOT with the state-of-the-art approaches.

Method	IMDB	RT-2k
MultiSpot method		
BoW + SentC(c) + SemC(c)	89.38	88.25
BoW + SentC(c) + SemC(c) + C_Book(c)	89.48	89.05
NB bi + C_Book(c)	91.72	**91.30**
NB bi + SentC(c) + SemC(c)	**91.78**	90.85
State-of-the-art approaches		
Full + Unlabeled + BoW [7]	88.89	88.90
BoW SVM [11]	-	87.15
tfΔidf [9]	-	88.10
Appr. Taxonomy [20]	-	**90.20**
Word Repr. RBM + BoW [2]	89.23	-
NB SVM bigrams [19]	91.22	89.45
Paragraph Vector [6]	**92.58**	-

Semcenter (SemC) features is statistically better than a simple BoW classifier. Thus, we can safely conclude that the accumulation of word and sentence level features improves the accuracy of a cascaded sentiment analysis approach.

Table 4 provides a comparison of the MULTISPOT with the state-of-the-art approaches (we report only the best results of each approach). Our approach appears to outperform the state-of-the-art in the RT-2k dataset by 1.1 %. In the IMDB dataset we achieved 91.78 % accuracy, which even though it is 0.8 % less than the most successful approach, it outperforms all the other methods. Here, due to space limits we did not have the opportunity to combine our method with the paragraph vector method. However, our intuition is that such a combination would lead to better classification accuracy. We also observe that MULTISPOT's accuracy is in all cases improved when the NB binary features are used which demonstrates the importance of a careful selection of word-level features.

Centerbook Analysis. We examined how the number of clusters and the number of the available training data affect the quality of the Centerbook results. Due to lack of space, here we present the results obtained using the IMDB dataset. Figure 2a shows the effect of the number of clusters on the classification accuracy, with the best accuracy achieved at around 100 and 200 clusters. Figure 2b shows how the number of the available training data affect the quality of the Centerbook. We observe that the accuracy of the Centerbook method is not significantly affected (slowly increases as more training data is available) by the number of the training data. So, even with a small amount of training data the Centerbook approach will succeed to effectively capture a document's sentiment.

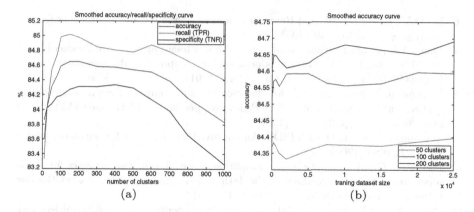

Fig. 2. (2a) Effect of the number of clusters on Centerbook's quality (2b) Effect of different training dataset sizes on Centerbook's quality

5 Conclusions and Future Work

In this work, we elaborate the problem of efficiently capturing the sentiment expressed on textual resources. The contributions of this work are the following:

- we present MULTISPOT, which exploits both word and sentence level features for acquiring a better sense of a document's sentimental content;
- two novel semantic related approaches (i.e. *Semcenter* and *Centerbook*) which further assist in better spotting a document's overall sentiment.

The conducted experiments have shown that the enhancement of a cascaded sentiment analysis approach with semantic aware features significantly increases its accuracy. Also, the initial hypothesis that the combination of word level with sentence level features provides better capturing of the sentiment expressed in a text is confirmed, as we observe that in all cases such a combination outperforms document-based word level approaches.

References

1. Coates, A., Ng, A.Y.: Learning feature representations with K-means. In: Montavon, G., Orr, G.B., Müller, K.-R. (eds.) Neural Networks: Tricks of the Trade, 2nd edn. LNCS, vol. 7700, pp. 561–580. Springer, Heidelberg (2012)
2. Dahl, G.E., et al.: Training restricted boltzmann machines on word observations (2012). arXiv preprint arXiv:1202.5695
3. Fan, R.-E., et al.: LIBLINEAR: a library for large linear classification. Mach. Learn. Res. **9**, 1871–1874 (2008)
4. Godbole, N., et al.: Largescale sentiment analysis for news and blogs. In: Proceedings of the Conference on Weblogs and Social Media (ICWSM). Citeseer (2007)
5. Lazebnik, S., et al.: Beyond bags of features: spatial pyramid matching for recognizing natural scene categories. In: Computer Vision and Pattern Recognition, vol. 2, pp. 2169–2178. IEEE (2006)

6. Le, Q.V., et al.: Distributed Representations of Sentences and Documents (2014). arXiv preprint arXiv:1405.4053

7. Maas, A.L., et al.: Learning word vectors for sentiment analysis. In: Proceedings of the 49th Annual Meeting of the Association for Computational Linguistics: Human Language Technologies, Vol. 1, pp. 142–150 (2011)

8. MacQueen, J.: Some methods for classification and analysis of multivariate observations. In: Proceedings of 5-th Berkeley Symposium on Mathematical Statistics and Probability, pp. 281–297 (1967)

9. Martineau, J., et al.: Delta TFIDF: An improved feature space for sentiment analysis. In: ICWSM (2009)

10. Paltoglou, G., et al.: A study of information retrieval weighting schemes for sentiment analysis. In: Proceedings of the 48th Annual Meeting of the Association for Computational Linguistics, pp. 1386–1395 (2010)

11. Pang, B., et al.: A sentimental education: Sentiment analysis using subjectivity summarization based on minimum cuts. In: Proceedings of the 42nd annual meeting on Association for Computational Linguistics, p. 271 (2004)

12. Pang, B., et al.: Seeing stars: Exploiting class relationships for sentiment categorization with respect to rating scales. In: Proceedings of the 43rd Annual Meeting on Association for Computational Linguistics, pp. 115–124 (2005)

13. Shiyang, W., et al.: Emotion classification in microblog texts using class sequential rules. In: 28th AAAI Conference on Artificial Intelligence (2014)

14. Socher, R., et al.: Semantic compositionality through recursive matrix-vector spaces. In: Proceedings of the 2012 Joint Conference on Empirical Methods in Natural Language Processing and Computational Natural Language Learning, pp. 1201–1211 (2012)

15. Socher, R., et al.: Semi-supervised recursive autoencoders for predicting sentiment distributions. In: Proceedings of the Conference on EMNLP, pp. 151–161 (2011)

16. Socher, R., et al.: Recursive deep models for semantic compositionality over a sentiment treebank. In: Proceedings of the Conference on EMNLP, pp. 1631–1642. Citeseer (2013)

17. Turney, P.D., et al.: From frequency to meaning: Vector space models of semantics. Artif. Intell. Res. **37**(1), 141–188 (2010)

18. Wang, H., et al.: Sentiment classification of online reviews: using sentence-based language model. JETAI **26**(1), 13–31 (2014)

19. Wang, S., et al.: Baselines and bigrams: Simple, good sentiment and topic classification. In: Proceedings of the 50th Annual Meeting of the Association for Computational Linguistics: Short Papers, Vol. 2, pp. 90–94 (2012)

20. Whitelaw, C., et al.: Using appraisal groups for sentiment analysis. In: Proceedings of ACM Conference on Information and knowledge management, pp. 625–631. ACM (2005)

21. Zhang, C., et al.: Sentiment analysis of chinese documents: from sentence to document level. JASIST **60**(12), 2474–2487 (2009)

22. Zhang, W., et al.: Learning non-redundant codebooks for classifying complex objects. In: Proceedings of the 25th Conference on Machine learning, pp. 1241–1248. ICML (2009)

Online Urban Mobility Detection
Based on Velocity Features

Fernando Terroso-Saenz[✉], Mercedes Valdes-Vela,
and Antonio F. Skarmeta-Gomez

Department of Information and Communication Engineering,
Computer Science Faculty, University of Murcia, Murcia, Spain
{fterroso,mdvaldes,skarmeta}@um.es

Abstract. The study of the mobility models that arise from the city dynamics has become instrumental to provide new urban services. In this context, many proposals applied an off-line learning on historical data. However, at the dawn of the Big Data era, there is an increasing need for systems and architectures able to process data in a timely manner. The present work introduces a novel approach for online mobility model detection along with a new concept for trajectory abstraction based on velocity features. Finally, the proposal is evaluated with a real-world dataset.

1 Introduction

The detection of the underlying models of a city dynamics plays an indispensable role in order to come up with more reliable and innovative urban services. Such mobility models allow to provide better location-based services [1] and detect actual or potential problems in a city's transportation infrastructure.

In that sense, the endless improvement of location sensors included in common handheld devices has incredibly facilitated the collection of a massive amount of high-resolution mobility datasets which, in turn, has enabled the development of many solutions for mobility pattern detection based on the crowdsensing paradigm [2].

A common approach for that goal consists of applying different machine learning techniques to the aforementioned datasets in order to extract meaningful mobility patterns or statistical models [3]. Since most of these solutions need the whole dataset in advance, they can be regarded as off-line methods. Nontheless, on the verge of the Big Data era, which has recently emerged as one of the most important new technological topics [4], there exist a pressing need for novel systems capable of detecting new and relevant knowledge in nearly real time.

Another important drawback of existing solutions is that they frequently rely on space-partitioning [5] and road-map approaches [6] in order to make up the abstraction of the mobility models. Such approaches do not fully leverage the fact that people tend to move freely in certain situations and, hence, their movements are not constrained by pre-defined paths or road networks.

© Springer International Publishing Switzerland 2015
S. Madria and T. Hara (Eds.): DaWaK 2015, LNCS 9263, pp. 351–362, 2015.
DOI: 10.1007/978-3-319-22729-0_27

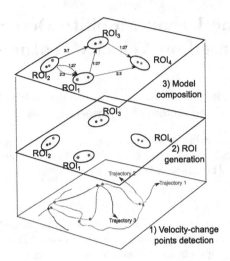

Fig. 1. Overview of the trajectory abstraction and model generation in the 2D space.

Bearing in mind the two aforementioned problems, the present work introduces an innovative approach for the detection of the mobility model of a city by means of crowdsensing procedures. The underlying idea for the model extraction is that a person's trajectory can be uniquely identified by means of its velocity features, that is, the points where its speed or direction remarkably changes. As Fig. 1 depicts, on the basis of these points, it is possible to automatically extract Regions of Interest (ROIs) of a city where the regular trajectories of its inhabitants usually change their direction, such as crossroads, roundabouts or crossings along with its frequent origins and destination like residential areas or business parks.

Next, as Fig. 1 shows, the mobility model of a city is abstracted as a multigraph comprising a set of different sequences of these areas covered by the its inhabitants. This new approach is suitable to capture the movement of people who freely moves through large areas or with poor road-map coverage solving the limitations of previous space-partitioning and map-based solutions.

Finally, in order to generate such a model in real time, an architecture comprising a thin client embedded in a personal mobile device and a central server has been defined. Such an architecture intends to collect large amounts of location data from city dwellers and timely process it by minimizing the communication load between the mobile devices and the back-end server.

The remainder of the paper is structured as follows. An overview about mobility mining is put forward in Sect. 2. Section 3 is devoted to describing in detail the logic structure of the proposed system. Then, Sect. 4 discusses the main results of the experiments. Finally, the main conclusions and the future work are summed up in Sect. 5.

2   Related Work

Since the two key contributions of the proposal focus on the abstraction of the mobility model and the real-time generation of such a model, an overview about the current state of the art of both disciplines is put forward.

2.1   Model Abstraction

When the mobility model is generated on the basis of the spatio-temporal trajectories (regarded as sequential data of timestamped locations) covered by a target group of people, it is necessary to adapt such trajectories to a more suitable representation for pattern matching. In this context, existing works can be classified into three different alternatives.

Firstly, representation based on gridding spatial partion consists of dividing the area of interest into squared cells of the same size, so each trajectory is represented as a sequence of cells according to the sequence of its locations [5,7–9]. A major drawback of this type of representation is that defining a proper grid granularity for the whole area of interest is not trivial.

A second trend makes use of cartography to represent a trajectory as the sequence of streets (or segments) a person has covered [6,10–12]. Although this approach seems suitable for vehicular trajectories, people sometimes move through areas which are poorly covered by street-maps (or not covered at all, like open sea) which makes rather difficult the mandatory map-matching process.

Finally, our work is enclosed in a third trend that represents trajectories as the sequences of Regions of Interest (ROIs). In this context, most works define ROIs as frequently-visited stops [13–15]. Nonetheless, our solution enlarges this stop-based concept by also considering regions where other trajectory's velocity features, like bearing, vary. Besides, existing methods generally make use of primitive grilled spatial partitions to generate the ROIs so they might inherit the above-mentioned problems of grilled-based abstractions. On the contrary, our proposal does not require any previous spatial partition or external street-map database to undertake the trajectory abstraction stage, thus it does not suffer from the aforementioned drawbacks of previous approaches.

2.2   Probabilistic Model Generation

Once trajectories have been converted into sequences of representative cells/ segments/regions, approaches for the next step have mainly followed two different trends, trajectory pattern mining and probabilistic model generation.

On the one hand, trajectory pattern mining has been widely studied, and works within this discipline can be classified into three lines of research, namely frequent item mining, trajectory clustering and graph-based trajectory mining [3].

On the other hand, during the last years a host of studies have put forward novel approaches to generate probabilistic mobility models without explicitly creating trajectory patterns [5,6]

For this discipline, the present work studies the incremental generation of the probabilistic model in real time instead of carrying out an off-line training. In this context, a similar approach was already studied in [9]. Nonetheless, such mechanism relies on a travel time database previously generated. On the contrary, our solution does not require any apriori information generating the probabilistic model from the scratch.

Regarding the particular model, our work provides a novel multigraph-based structure to represent such model. Thus, unlike previous tree-based approaches [13], each ROI is represented by a unique vertice, whereas the multi-edges contain frequency features extracted from the historical trajectories. As we will see later, such a representation allows to easily integrate new trajectories to the model.

3   Online Mobility Model Generation

This section is devoted to explain in detail the key features of the proposal.

3.1   Architecture Overview

One of the goals of the present work is to compose the mobility model of a city by means of a crowdsensing approach. This way, in order to profit from the sensing capabilities that personal mobile devices now provide, the system architecture, depicted in Fig. 2, has been split into two different parts, (1) a client side that runs on the mobile device of each user whose main goal is to collect and pre-process his/her location trace and (2) a centralized server in charge of the tasks which are too expensive to be executed on mobile platforms.

Both parts collaborate with each other in order to carry out the two processing steps of the system, (1) the abstraction of the raw trajectories from the users before (2) use them to compose the mobility model of their city. These two phases are put forward in the following sections.

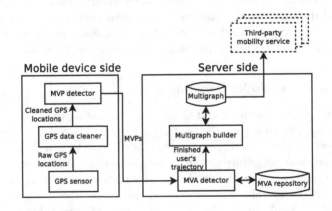

Fig. 2. Client-server structure.

3.2   Trajectory Abstraction

Trajectory Abstraction Format. The proposed ROI-based abstraction of a trajectory makes use of its velocity features. These velocity-based ROIs have been named *meaningful velocity areas* (MVAs). Depending on the particular feature taken into account, it is possible to distinguish between *meaningful turn areas* (MTAs) for the direction feature and *meaningful stop areas* (MSAs) for speed.

As a result, the abstraction of a trajectory T^u of a user u takes the form, $T^u_{abs} = \{MSA_o \rightarrow MTA_j \rightarrow ... \rightarrow MTA_{j+n} \rightarrow MSA_d\}$ where

- MSA_o represents the origin of T^u.
- $MTA_j, ..., MTA_{j+n}$ ($n \geq 0$) are the intermediate MTAs covered by T^u.
- MSA_d stands for the final destination of T^u.

Trajectory Abstraction Process. The online abstraction of a trajectory involves two steps.

1. *Meaningful Velocity Episodes Extraction.* Firstly, its trace of spatiotemporal points is analysed to detect *Meaningful Velocity Points* (MVPs) representing remarkable episodes of direction changes or low speed.
2. *MVA Detection.* For each MVP, a second step detects if it is included in any MVA. If that is the case, the MVA is appended as a new element of the trajectory abstraction.

This two steps are carried out by the mobile device and the server side in a collaborative way as it is explained next.

MVP Extraction. This abstraction step is undertaken by the *GPS data cleaner* and *MVP detector* modules in the mobile side (see Fig. 2).

GPS Data Cleaner. The present version of system only relies on GPS as the positioning feed to extract the trajectories' points. However, under certain conditions, a GPS sensor may return erroneous or irrelevant locations [16]. Hence, it is necessary to perform a filtering of the GPS measurements.

This cleaning is done by means of a distance-based filter. This way, if the distance between two consecutive GPS measurements is within a range $[d_{min}, d_{max}]$ then it means the new measurement is neither irrelevant nor an outlier. Otherwise, the measurement is filtered out.

MVP Detector. The key goal of this module is to detect the MVPs that arises in the ongoing trajectory.

For the detection of meaningful turns, we have followed navigational approach as turns in a trajectory are reflected as variations of its bearing. For example, changing the direction from north to east implies a trajectory bearing decrease from 360° to 90°.

Consequently, the turn-based MVPs of a trajectory are detected as meaningful increments of decrements of its underlying bearing. In a nutshell, a meaningful bearing variation $\mathcal{B}_{var}$ can be perceived as a sequence of consecutive *filtered GPS measurements* $\mathcal{B}_{var} = \{fgps_i \rightarrow fgps_{i+1} \rightarrow ... \rightarrow fgps_{i+n} \rightarrow fgps_{i+n+1}\}$ ($n \geq 0$) accomplishing two conditions.

1. the bearing steadily increases (decreases) for each pair of consecutive measurements $fgps_j, fgps_{j+1} \in \mathcal{B}_{var}$
2. $|fgps_i.bearing - fgps_{i+n+1}.bearing| \geq \triangle_{dir}$ where $\triangle_{dir}$ is a predefined threshold.

If both conditions are fulfilled then a new MVP reporting a meaningful turn is created.

Concerning the stop episodes, this module assumes that if a person remains stationary during a certain period of time then the *GPS measurements* generated by the GPS sensor will be discarded by the distance filter's lower boundary. As a result, no new *filtered GPS measurements* will be made up during that period. Hence, if it is detected a long time period $\mathcal{P}_{stop}$ with no new *filtered GPS measurements* a new MVP reporting a stop of the ongoing trajectory is generated.

Finally, the MVPs are eventually sent to the server for further processing.

MVA Detection. This step is undertaken by the *MVA detector* module (see Fig. 2) which incrementally generates the MVAs of the target city at the same time it receives the MVPs of the users.

The rationale for the MVA detection is that, a MVA can be regarded as a particular region where many MVPs from different citizens' trajectories occur quite close to each other.

As a result, a density-clustering approach has been followed where each cluster with a high density of MVPs is considered a MVA. In short, density-based clustering is based on the concept of *local neighbourhood* $\mathcal{N}$ of a point p, that is, the number of points that are within a certain straight distance *Eps* to p,

$$\mathcal{N}(p) = \{q \in \mathcal{S} \mid dist(p, q) \leq Eps\}$$

where $\mathcal{S}$ is the set of available points. If $|\mathcal{N}(p)|$ is over a certain threshold *MinPoints*, then $\mathcal{N}(p)$ is considered as a cluster. Furthermore, $\mathcal{N}(p)$ is *density-joinable* to $\mathcal{N}(q)$ (p $\neq$q) if $\mathcal{N}(p) \cap \mathcal{N}(q) \neq \emptyset$

The density clustering algorithm applied for MVA detection is a slightly modified version of the online *landmark discovery algorithm* (LDM) described in [17].

This algorithm has been adapted so that, for this work, takes as input a MVP and returns the MVA such point belongs to. In particular, the *MVA detector* component executes two isolated instances of the algorithm, one for the turn-MVPs and another for the stop-MVPs. Whilst the former generates MTAs, the latter composes the MSAs.

Finally, when a MVP of a trajectory $\mathcal{T}^u$ is mapped to a MVA, this area is appended as a new element of its abstraction $\mathcal{T}^u_{abs}$. In case of a MSA, the system infers that the user u has reached his/her destination and, thus, $\mathcal{T}^u$ has been completed. This way, the abstraction of this just-finished trajectory is delivered to the *multigraph builder* module in order to update the probabilistic model (see Fig. 2).

3.3 Model Composition

The probabilistic model of a city's dynamics is encoded as a directed multigraph $\mathcal{G}_{stat}$ comprising the statistical information from the historical trajectories. This way, each vertice of $\mathcal{G}_{stat}$ represents a unique MVA and the directed edges contain frequency features of the trajectories.

As we can see from Fig. 3, the multigraph approach allows to connect the same pair of MVAs with different directed edges where each one is labelled with the frequency of a particular type of trajectory. This way, a trajectory is encoded in $\mathcal{G}_{stat}$ as a sequence of *exclusive* edges connecting the MVA vertices in the same order than its abstraction.

$\mathcal{G}_{stat}$ is updated by the *multigraph builder* on the basis of the abstractions of all the trajectories completed by the users $\mathcal{T}^u_{abs}$. This is done by means of a two-step procedure.

– Firstly, the module checks if $\mathcal{T}^u_{abs}$ is already *fully included* in $\mathcal{G}_{stat}$. A trajectory is considered as *fully included* in $\mathcal{G}_{stat}$ if and only if all its MVAs are already vertices of $\mathcal{G}_{stat}$ and there exists a trajectory identifier whose associated edges connect these MVAs in the same order than the abstraction. If that is the

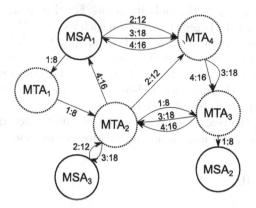

Fig. 3. Example of the multigraph structure comprising 4 different trajectories, $\mathcal{T}^1_{abs} = \{MSA_1 \rightarrow MTA_1 \rightarrow MTA_2 \rightarrow MTA_3 \rightarrow MSA_2\}$, $\mathcal{T}^2_{abs} = \{MSA_3 \rightarrow MTA_2 \rightarrow MTA_4 \rightarrow MSA_1\}$, $\mathcal{T}^3_{abs} = \{MSA_1 \rightarrow MTA_4 \rightarrow MTA_3 \rightarrow MTA_2 \rightarrow MSA_3\}$, $\mathcal{T}^4_{abs} = \{MSA_1 \rightarrow MTA_4 \rightarrow MTA_3 \rightarrow MTA_2 \rightarrow MSA_1\}$. Each edge is labelled with the tuple {trajectory-id:frequency}.

case, the identifier of such trajectory is extracted. Otherwise, a new identifier is generated.

- Secondly, the frequency attribute of each edge associated to that identifier is incremented. In case of a new identifier, a new set of edges connecting the MVAs are created. During this step, if the incoming abstraction comprises a new MVA, a new vertice representing this new area is also generated.

This multigraph approach allows that a completed trajectory is reflected in the model by only inserting each new MVA vertice once. On the contrary, in a tree-based approach a new trajectory is inserted as one or multiple branches each one comprising several nodes [13]. Furthermore, this lightweight update method makes possible to incrementally integrate *on the fly* new trajectories. This particular features are quite useful so as to smoothly adapt to changes of mobility routines in the city of interest. This is not possible in previous off-line processes where the probabilistic model is hard-coded beforehand.

3.4   General Workflow

To sum up, the general flow of information of this architecture is as follows.

To begin with, the *MVP detector*, running in the different mobile devices carried by the users, endlessly processes the timestamped locations returned by the device's GPS. The main goal of this module is to detect if the person carrying the device has been stopped for a certain period of time or has performed a meaningful turn. In any of both cases, the component generates a MVP indicating the location of the meaningful velocity episode and sends it to the server side.

This server side processes the MVPs emitted by all the mobile devices in order to eventually generate the mobility model. For that goal, the *MVA detector* firstly performs a density-based clustering on each received MVP so as to discover the MVA comprising it. In case the MVP is mapped to a MVA, it is appended to the trajectory's abstraction of the user who has emitted such MVP. This way, this side stores the abstraction of the ongoing trajectories of all the users. When any of such trajectories has finished, its whole MVA abstraction is used to update the multigraph-based model.

On the whole, the real time processing of all the trajectories allow to keep up to date the model and include any new trajectory as soon as it is completed.

4   Experiments

In order to state a comprehensive view of our proposal, we evaluated the system with a subset of *GeoLife dataset* [18] whose trace is shown in Fig. 4. This subset falls into a large area of Beijing city (China) and its details are listed in Table 1.

The evaluation was conducted on a PC running a Ubuntu 12.04 with 2 GiB of memory, Intel(R) Core 2 at 2.66 GHz and *Java Runtime Environment* 7.0. Table 2 summarizes the default configuration of the system.

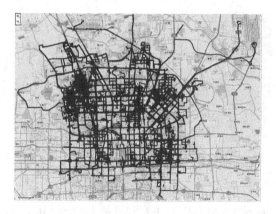

Fig. 4. Digital trace of the dataset under study.

Table 1. Dataset details.

Total			Per route	
Locations	Routes	Time period	Locations	Time length
2950411	7481	29-4-2006→6-4-2012	394	∼ 27 m

Table 2. System default configuration.

Parameter	Module	Value
d_{min}, d_{max}	GPS data cleaner	100 m, 800 m
Δ_{dir}	MVP detector	45°
$\mathcal{P}_{stop}$	MVP detector	5'
$Eps, minPoints$	MVA detector	300 m, 15

In this preliminary evaluation, the main objective was to study the effect of the MVA *size* in the system. In this context, the *size* of a MVA is mainly defined by the *Eps* and *minPoints* parameters of the LDA. The former indicates the radius of the area whereas the latter defines its actual density of points. Figure 5 depicts the number of MVAs generated by the system along with the average distance between them for different $Eps \times minPoints$ configurations, 100×5, 300×15, 600×30, 1000×50, 2000×100, 3000×150. Unsurprisingly, there exists a inverse correlation between the selected size and the number and average distance of the MVAs generated by the system.

From the results depicted in Fig. 5, we can conclude that the MVA size is instrumental in order to define the *granularity* of the model as it has a direct impact on its number of areas.

Moreover, since the real time capabilities is a foremost feature of the proposal, the impact of the MVA size on the time latency of the system was also studied.

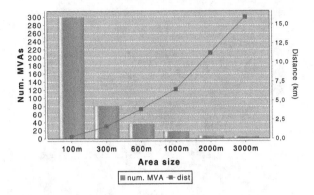

Fig. 5. Total number of MVAs (MVAs + MSAs) and average distance among them for different sizes.

Table 3. Time latency for different $Eps \times minPoints$ configurations.

100×5	300×15	600×30	1000×50	2000×100	3000×150
1.48 ms	1.44 ms	1.16 ms	1.02 ms	0.55 ms	0.49 ms

Hence, Table 3 shows the system's time latency to detect if a MVP belongs to any MVA for several $Eps \times minPoints$ configurations.

Regarding the results of the aforementioned table, the inverse correlation between the MVA size and the system's time latency is because larger MVAs involves less number of areas (as it has been previously stated for Fig. 5). This number is a key factor in the LDA performance as the more MVAs the more time required to map a MVP to a MVA. However, it is important to remark

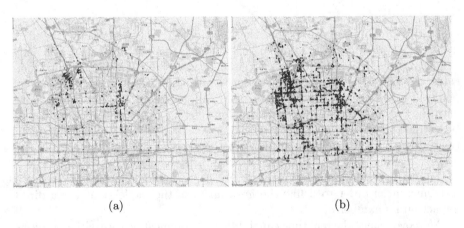

(a)   (b)

Fig. 6. MVAs generated by the system. (a) Centroids of the MSAs. (b) Centroids of the MTAs.

that the latency increases in a linear factor with respect to the MVA size. This is a paramount requirement in order to ensure the scalability of the proposal.

Finally, we also studied the *consistency* of the MVAs so as to know whether the spatial regions they covered actually contained potential stops of turns of a trajectory. In this context, Fig. 6 depicts the MVA centroids generated by the 300×15 configuration. As we can see, the MSA centroids (Fig. 6a) are organized in two major areas, one at the nort-west and the other as a long line at the center of the city. The former comprises different colleges and student accommodations, whereas the latter includes long avenues in the city center with several public facilities. In both cases, this regions are consistent with the fact that they represent potential origins and destinations of trajectories. Regarding MTA centroids (Fig. 6b), they are more spread covering many intersections of the road network. This is consistent with the MTA meaning.

5   Conclusions and Future Work

In this day and age, mobility mining has attracted great attention from the research community. Although it is possible to find several methodologies in this domain, they are limited in terms of application and adaptation to free-movement and dynamic scenarios.

The present work introduces an innovative solution for domains where the user movement is not constrained by a road network. Moreover, its online model generation can adapt to users' mobility shifts making it an interesting solution for those domains where the regeneration of an off-line model is not feasible.

In the end, further work will focus on developing crowdsensing services on top of this model to provide final users with novel solutions for their common mobility problems. As a matter of fact, a route predictor based on the probabilistic model will be developed so as to provide city dwellers with better travel suggestions.

Acknowledgment. The research leading to these results has received funding from the Fundación Séneca Programme for Helping Excellent Research Groups TIN2014-52099-R.

References

1. Krner, C., May, M., Wrobel, S.: Spatiotemporal modeling and analysis-introduction and overview. KI - Knstliche Intelligenz **26**(3), 215–221 (2012)
2. Guo, B., Yu, Z., Zhou, X., Zhang, D.: From participatory sensing to mobile crowd sensing. In: 2014 IEEE International Conference on Pervasive Computing and Communications Workshops (PERCOM Workshops), pp. 593–598. IEEE (2014)
3. Lin, M., Hsu, W.-J.: Mining GPS data for mobility patterns: a survey. Pervasive Mob. Comput. **12**, 1–16 (2014)
4. Marz, N., Warren, J.: Big Data: Principles and Best Practices of Scalable Realtime Data Systems. Manning Publications Co., Shelter Island (2015)

5. Zhou, J., Tung, A.K., Wu, W., Ng, W.S.: A "semi-lazy" approach to probabilistic path prediction in dynamic environments. In: Proceedings of the 19th ACM SIGKDD International Conference on Knowledge Discovery and Data Mining, KDD 2013, pp. 748–756. ACM, New York, NY, USA (2013)
6. Krumm, J., Gruen, R., Delling, D.: From destination prediction to route prediction. J. Location Based Serv. **7**(2), 98–120 (2013)
7. Xue, A., Zhang, R., Zheng, Y., Xie, X., Huang, J., Xu, Z.: Destination prediction by sub-trajectory synthesis and privacy protection against such prediction. In: 2013 IEEE 29th International Conference on Data Engineering (ICDE), pp. 254–265, April 2013
8. He, W., Li, D., Zhang, T., An, L., Guo, M., Chen, G.: Mining regular routes from GPS data for ridesharing recommendations. In: Proceedings of the ACM SIGKDD International Workshop on Urban Computing, UrbComp 2012, pp. 79–86. ACM, New York, NY, USA (2012)
9. Krumm, J.: Real time destination prediction based on efficient routes. Technical report, SAE Technical Paper (2006)
10. Krumm, J.: Where will they turn: predicting turn proportions at intersections. Pers. Ubiquit. Comput. **14**(7), 591–599 (2010)
11. Ziebart, B.D., Maas, A.L., Dey, A.K., Bagnell, J.A.: Navigate like a cabbie: probabilistic reasoning from observed context-aware behavior. In: Proceedings of the 10th International Conference on Ubiquitous Computing, UbiComp 2008, pp. 322–331. ACM, New York, NY, USA (2008)
12. Simmons, R., Browning, B., Zhang, Y., Sadekar, V.: Learning to predict driver route and destination intent. In: Intelligent Transportation SystemsConference, ITSC 2006, pp. 127–132. IEEE, September 2006
13. Chen, L., Lv, M., Ye, Q., Chen, G., Woodward, J.: A personal route prediction system based on trajectory data mining. Inf. Sci. **181**(7), 1264–1284 (2011)
14. Giannotti, F., Nanni, M., Pinelli, F., Pedreschi, D.: Trajectory pattern mining. In: Proceedings of the 13th ACM SIGKDD International Conference on Knowledge Discovery and Data Mining, KDD 07, pp. 330–339. ACM, New York, NY, USA (2007)
15. Jeung, H., Shen, H.T., Zhou, X.: Mining trajectory patterns using hidden Markov models. In: Song, I.-Y., Eder, J., Nguyen, T.M. (eds.) DaWaK 2007. LNCS, vol. 4654, pp. 470–480. Springer, Heidelberg (2007)
16. Zhang, J., Goodchild, M.F.: Uncertainty in Geographical Information. CRC Press, Boca Raton (2002)
17. Terroso-Saenz, F., Valds-Vela, M., Campuzano, F., Botia, J.A., Skarmeta-Gmez, A.F.: A complex event processing approach to perceive the vehicular context. Inf. Fusion **21**, 187–209 (2015)
18. Zheng, Y., Xie, X., Ma, W.-Y.: Geolife: a collaborative social networking service among user, location and trajectory. IEEE Data Eng. Bull. **33**(2), 32–39 (2010)

Big Data

Partition and Conquer: Map/Reduce Way of Substructure Discovery

Soumyava Das[(⊠)] and Sharma Chakravarthy

ITLAB and CSE Department, University of Texas at Arlington, Arlington, Texas
soumyava.das@mavs.uta.edu, sharma@cse.uta.edu

Abstract. Transactional data mining (decision trees, association rules etc.) has been used to discover non trivial patterns in unstructured data. For applications that have an inherent structure (such as social networks, phone networks etc.) graph mining is useful as mapping such data into an unstructured representation will lead to loss of relationships. Graph mining finds use in a plethora of applications: analysis of fraud detection in transaction networks, finding friendships and other characteristics are to name a few. Finding interesting and frequent substructures is central to graph mining in all of these applications. Until now, graph mining has been addressed using main memory, disk-based as well as database-oriented approaches to deal with progressively larger sizes of applications.

This paper presents two algorithms using the Map/Reduce paradigm for mining interesting and repetitive patterns from a partitioned input graph. A general form of graphs, including directed edges and cycles are handled by our approach. Our primary goal is to address scalability, solve difficult and computationally expensive problems like duplicate elimination, canonical labeling and isomorphism detection in the Map/Reduce framework, without loss of information. Our analysis and experiments show that graphs with hundreds of millions of edges can be handled with acceptable speedup by the algorithm and the approach presented in this paper.

1 Introduction

Substructure discovery is the process of discovering substructure(s) as a connected subgraph in a graph (or a forest) that best characterizes a concept based on some criterion. Due to increasing size of graphs, there has been some effort on partitioned approach to substructure discovery as in [15,22], but they do not discuss expansion of substructures across partitions. In order to analyze large graphs and discover interesting substructures within a meaningful response time, the input data needs to be processed using a paradigm that can leverage partitioning and distribution of data on a large scale. Map/Reduce provides a powerful parallel and scalable programming framework to process large data sets.

Briefly, in order to detect interesting concepts in a graph (or a forest), an iterative algorithm is used that: generates all substructures of increasing sizes (starting from substructure of size one that has one edge), counts the number

© Springer International Publishing Switzerland 2015
S. Madria and T. Hara (Eds.): DaWaK 2015, LNCS 9263, pp. 365–378, 2015.
DOI: 10.1007/978-3-319-22729-0_28

of identical (or similar) substructures and applies a metric to rank the substructures. This process is repeated until a given substructure size is reached or there are no more substructures to generate. Although main memory based data mining algorithms exist, they typically face two problems with respect to scalability: (i) storing the entire graph (or its adjacency matrix) in main memory may not be possible and (ii) the computational space requirements of the algorithm may exceed available main memory. Hence it is important to partition the graph such that the individual partitions do not encounter these two problems. Mining using these partitions is tricky and needs effective partition management strategies. Moreover each parallel approach requires synchronization of computations across partitions and proper key/value pairs need to be formulated in the Map/Reduce paradigm to accomplish proper synchronization and lossless computation.

The contributions of this paper are:

- Two approaches to partition management for graph mining and their analysis
- Corresponding Map/Reduce algorithms
- Formulation of key/value pairs for isomorphism detection and duplicate removal
- Extensive experimental analysis of algorithms using diverse data sets to validate intuitive conjectures for scalability, I/O usage and response time behavior.

Roadmap: The rest of the paper is organized as follows: Sect. 2 describes the related work. Section 3 presents the problem definition and a brief overview on graph mining. Sections 4 and 5 detail our algorithmic approach and problem solving methodologies. Section 6 discusses experiments and their analysis while Sect. 7 concludes the paper.

2   Related Work

Due to vast literature on graph mining and analysis, we limit our related work discussion to substructure discovery. For more detailed surveys see [11,14]. Main memory approaches [3,8,13,23] required loading the entire graph in main memory and have some inherent limitations: handling large patterns, huge candidate set generation and multiple scans of the database. Disk-based graph mining techniques [4,21] were developed but suffered from random access problems. Database-oriented approaches [6,20] required sorting columns (in row based DBMSs) for important graph operations making it expensive. Recent advancements [2,7,18] have been shown to be effective in a cloud architecture. Parallel paradigms like Giraph [1] works for vertex centric algorithms while Naiad [19] works for dynamic graphs. But here we are interested in static graphs or snapshots of dynamic graphs, and also use edge based methodologies thereby choosing Map/Reduce as our distributed paradigm. Our goal in this paper is to explore alternative approaches to substructure generation, isomorphism detection and counting using the Map/Reduce paradigm. Our perspective on the proposed problem is novel and is also quite different from the work in the literature. A major chunk of subgraph mining have also looked at graph mining using commodity servers, with increasing memory sizes and heuristics (like timed-out explorations [9]) to fit the graph in a single machine [9,10]. Other approaches [12] focus on a database of small graphs instead of a large single graph. However our approach is distributed processing of a large graph across a commodity cluster with limited availability of main memory, making our problem different from the ones studied in literature.

3   Preliminaries and Problem Definition

In this paper, we focus on directed labeled graphs where node and edge labels are not assumed to be unique. We introduce few definitions used in the rest of the paper.

Graph Isomorphism: *Formally, graphs $G_1 = (V_1, E_1)$ and $G_2 = (V_2, E_2)$ are* **isomorphic** *if (i) $|V_1| = |V_2|$ and $|E_1| = |E_2|$, (ii) there is a bijection (one to one correspondence) f from V_1 to V_2, (iii) there is a bijection g from E_1 to E_2 that maps each edge (u,v) in E_1 to (f(u), f(v)) in E_2. We use isomorphism to detect identical (or exact) substructure patterns.*

Minimum Description Length (MDL): Minimum description length (MDL) is an information theoretic metric that has been shown to be domain independent and highlights the importance of substructure on how well it can compress the entire graph. Both frequency of the subgraph and its structure have a bearing on compression. MDL is detailed in [13,20].

Problem Definition: *Given a directed labeled graph as input, the general problem is to find the substructure(s) that reduces the graph best using the MDL principle. The subgraph size can be optionally used to find the best substructure up to that size.* The goal is to develop algorithms that are amenable to partitioning. For the current work, we have chosen the Map/Reduce paradigm. This entails generating substructures iteratively, counting them and using the MDL principle as a heuristic to limit the number of substructures used for the next iteration. The role of isomorphism in the algorithm is twofold: prune duplicates (that are generated while expanding a substructure) and group non-duplicate but similar substructures and count them. The algorithm needs to achieve two objectives: (i) scalable to handle graphs of any size and connectivity and (ii) has acceptable (close to linear) speedup with increasing number of machines.

4   Our Approach

We shall first discuss the representation of input graph followed by strategies to manage partitions for parallel processing.

4.1   Input Graph Representation

Our input graph is represented as a sequence of unordered edges (or 1-edge substructure) including its direction. Each edge is completely represented by a 5 element tuple $<$ *edge label, source vertex id, source vertex label, destination vertex id, destination vertex label* $>$. Figure 1 shows our input graph and its representation. Our method can be easily extended for undirected graphs using two directed edges instead of an undirected edge. However such a representation doubles the number of edges in the graph. To keep the number of edges same, and use less space, a sixth element can be introduced into the current edge representation indicating the nature of the edge (directed or undirected). Our representation can also be extended for multiple edges by adding an edge identifier for each edge. These extensions are used appropriately during the substructure expansion phase.

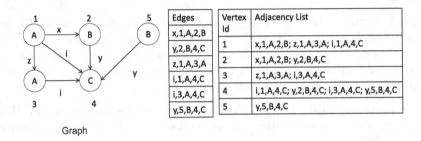

Edges		Vertex Id	Adjacency List
x,1,A,2,B			
y,2,B,4,C		1	x,1,A,2,B; z,1,A,3,A; i,1,A,4,C
z,1,A,3,A		2	x,1,A,2,B; y,2,B,4,C
i,1,A,4,C		3	z,1,A,3,A; i,3,A,4,C
i,3,A,4,C		4	i,1,A,4,C; y,2,B,4,C; i,3,A,4,C; y,5,B,4,C
y,5,B,4,C		5	y,5,B,4,C

Fig. 1. An example graph, graph representation and adjacency lists

4.2   Partition Management

Any graph algorithm requires an initial set of vertices/edges and an adjacency list of the graph to progressively generate bigger substructures. Hence the adjacency list needs to be available where the graph is processed. Since we are using a partition-based approach, each partition of the graph needs its adjacency list for processing. We assume that each adjacency list partition can be held in memory. Below we present two partition management techniques that we propose in this paper. Note that our focus is not on partitioning (which is a research problem in itself) but on effective management of partitions for developing scalable parallel mining algorithms.

Arbitrary Partitioning: In one pass of the graph, the edges of a graph are grouped into required number of non overlapping partitions. Any partitioning scheme can be used for this purpose. Each such partition is called a *substructure partition*. For each substructure partition, its adjacency partition needs to be generated as well. The *adjacency partition* contains the adjacency list of all nodes present in that substructure partition. Each substructure partition is expanded in parallel and uses the corresponding adjacency partition. Alternatives to using adjacency partition have been shown to be inefficient in the literature [17]. Although substructure partitions are disjoint, adjacency partitions are not. The quality of initial partitioning is governed by the number of vertices replicated across adjacency partitions. The quality of initial partitioning can be improved by using partitioning approaches such as METIS [16].

Need for Partition Update: Expansion of a 1-edge substructure to a 2-edge substructure (and beyond), adds new edges (and hence new vertices) to the substructure partition. The new vertices, if not present in the current adjacency partition, needs to be added along with their adjacency list in the current partition after each iteration. Since we do not know a mapping between vertex and an adjacency partition, update of an adjacency partition may need us to load all the other (p-1) partitions (in the worst case and (p-1)/2 in the average case) from the distributed file system. With graphs of bigger sizes such an update phase has a huge bearing on the I/O.

One way to eliminate the above overhead is to avoid updating adjacency partitions. We still have to expand substructures on all nodes. An alternative is to keep adjacency partitions fixed and expand a substructure in multiple substructure partitions using a fixed adjacency partition. This will avoid the I/O intensive adjacency partition update after each iteration but will require grouping substructures differently and sending them to appropriate partitions for expansion.

Range Partitioning: We generate substructure partitions and ranged-based adjacency partitions (non overlapping) using vertex id in one pass. The adjacency partitions need not correspond to substructure partitions. Now, as the adjacency partition for each node id is unique, in the first iteration 1-edge substructures are grouped based on the adjacency partition needed to expand it. In this scheme a substructure may need adjacency lists from multiple partitions. The same substructure is now routed to multiple substructure partitions and they will be expanded on different nodes using the adjacency partition during each iteration. As a result, the adjacency partitions do not change across iterations, and the *same* adjacency partition is loaded for expansion, in each iteration, thereby avoiding the costly update phase. However overhead is incurred for routing the same substructure to multiple partitions. Routing substructures to multiple partitions will affect the shuffle. Mappers shuffle data to all reducers anyway and hence increase in shuffle is likely to be significantly less than the update cost of each adjacency partition. Hence we believe that the range partitioning system should perform better than its counterpart. Below we develop algorithms for the above two approaches and compare their performance against intuitive analysis.

5   Map/Reduce Based Substructure Discovery in Graphs

Finding the best substructures that compress the graph entails: generating substructures of increasing sizes (starting from an edge or 1-edge substructure) and counting isomorphic (identical) substructures. As each substructure is expanded independently, we are interested in grouping isomorphic substructures. However we need to deal with duplicates that are generated as a byproduct of expansion.

Duplicates must be eliminated to avoid wrong count. Below we define two notions of canonical instance and canonical substructure to distinguish duplicates and isomorphism in this paper.

Canonical Instance: In a parallel substructure discovery setting, isomorphic structures across different processors need to be grouped for counting and hence need to be ordered in a unique way. We employ a lexicographic ordering on edge label to arrange edges in a substructure in a unique way. If there are multiple edges with the same edge label in a substructure, they are ordered on the source vertex label. If source vertex label is also same, they are further ordered on the destination vertex label. If edge label, and vertex labels are also identical, then source and destination vertex ids can be used for ordering. Hence, a substructure can be uniquely represented using the above lexicographic order of 1-edge components. We call this a *canonical k-edge instance*. Intuitively, two duplicate k-edge substructures must have the same ordering of vertex ids and thereby the same canonical k-edge instance.

Canonical Substructure: However, it is easy to see that this canonical instance is not the same for isomorphic substructures and hence cannot be used for substructure identification and matching. The problem is the presence of vertex ids in the canonical instance whereas isomorphic match is done without using vertex id. A *canonical k-edge substructure*, derived from canonical k-edge instance is composed of k components where *each* component is a 5 element tuple as $< edge\ label, source\ vertex\ label, destination\ vertex\ label, from\ information, to\ information >$[1]. The *to information* and *from information* are the relative positions of the unique vertex ids in the order of their appearance in the canonical instance. Intuitively, both isomorphic substructures and duplicates must have the same relative ordering among the vertices present in the substructure and hence the same canonical substructure.

5.1   M/R Algorithm Using Arbitrary Partitions (dynamicAL-SD)

For this algorithm, a mapper expands a substructure by one edge in all possible ways in each iteration while a reducer counts isomorphic substructures and updates adjacency partitions for use in the next iteration.

Substructure Expansion by Mapper: We draw our inspiration from the Subdue way of unconstrained graph expansion. Each processor (mapper), uses a substructure partition and the corresponding adjacency partition for expanding a canonical instance by adding an edge. The substructure partition is read as mapper input (one canonical instance at a time) while the corresponding adjacency partition is kept in memory. Each expanded instance maintains the lexicographic order. We derive canonical substructure from the expanded canonical instance to group isomorphic instances across multiple mappers. A *combiner*

---

[1] The canonical k-edge substructure can not distinguish pathological substructures where all node and edge labels in a single substructure are identical and hence cannot distinguish bigger substructures which have a partial substructure with identical labels.

is used to remove duplicates in a mapper. Removing duplicates in a combiner has the same cost as removal in a reducer, but improves the shuffle cost by not emitting duplicates in the same partition to the reducers. Reducers still remove duplicates across partitions.

Algorithm 1 details our subgraph expansion routine in the mapper. For the first iteration, the input key is the line number and the input value is the 1-edge instance. For the subsequent iterations, the value is the k-edge canonical instance. Line 3 loads the adjacency partition in memory. Lines 8 to 17 expands the canonical instance and derives the canonical substructure from it. Creating canonical substructure from instance requires a hash table to identify relative positions of unique vertices (line 8). The hash table is local and removed after expanding a canonical instance (line 19).

Algorithm 1. *Mapper of dynamicAL-SD Algorithm*

1: **class Mapper**
2: **function** SETUP
3:     Map = load corresponding adjacency partition
4: **end function**
5: **function** MAP(*key, value*)
6:     vMap = hashtable to hold unique vertex info
7:     **for** each vertex id v in value **do**
8:         vMap.put(v,null)
9:     **end for**
10:    **for** each vertex id v in vMap **do**
11:        aL = get adjacency list of v from Map
12:        **for** each 1-edge substructure e in aL **do**
13:            newValue = if e is not in value expand value by adding e
14:            if ($u = e.getNewVertex$) not in vMap
15:                vMap.put(u,null)
16:            update vMap with positions from newValue
17:            newKey = generate canonical substructure
18:            emit(newKey,newValue) to combiner
19:            delete u from vMap
20:        **end for**
21:    **end for**
22:    delete vMap
23: **end function**

Duplicate Elimination and Frequency Counting by Reducer: The reducer receives the duplicates and isomorphic substructures across mappers grouped on the canonical substructure. Once the duplicates are removed, the reducer needs to count the isomorphic substructures for MDL. However, after duplicate removal, the number of elements with the reducer does not reflect the proper count of discriminative isomorphic canonical instances due to overlap (instances may overlap on vertices/edges). Finding the maximum independent subset (MIS) from the set of isomorphic instances is a NP complete problem.

We therefore use a single node-based support measure [5] to estimate the count of isomorphic instances[2]. Once the number of non-overlapped instances is estimated, MDL computation is done for each canonical substructure (and not for each canonical instance).

To limit future expansions only on high quality substructures, each reducer stores the canonical substructures with the top-k MDL values. Intuitively, this prunes the unimportant candidates and uses the best canonical substructures (and all their instances) for future expansion. The reducer uses a notion of beam (of size k) to store the instances with the top k MDL values. The user has the flexibility to override the system-specified beam size. We also handle ties in top-k MDL values by storing all the instances. Remember that, expansion may introduce vertices not present in partition. Hence, after the best instances in a reducer have been evaluated, the reducer updates the adjacency partition by adding the adjacency list of the new vertices.

We depict our reducer in Algorithm 2. Line 5 allocates a beam of size k as a hashmap to hold MDL values and all instances with that MDL value. Lines 8 to 14 remove duplicates and find instances with top k MDL values. The reducer also keeps an additional data structure for keeping the new vertices needed by adjacency partition for future expansion. The update continues from lines 18 to 24 by loading all other adjacency partitions, one at a time, till all new vertices and their adjacency partitions are found. All adjacency partition updates are happening in parallel. After the update, line 27 emits all the instances in the beam for next iteration. Mappers will calculate the canonical substructure from this expanded instance in the next mapper, the reducer output key is null while the value is each canonical instance in the hashmap.

5.2   M/R Algorithm Using Range-Based Partitions (staticAL-SD)

We now introduce another algorithm (staticAL-SD) which follows the range-based partition management strategy as discussed in Sect. 4.2. A closer look at this partitioning management strategy reveals two grouping criteria among substructures: instances across different processors must be routed to use the adjacency partition for expansion and the expanded instances then need to be grouped (based on their canonical substructure) for MDL computation. See that these two processes are sequential: we can only count after expansion. Hence each iteration of staticAL-SD requires two chained Map/Reduce jobs. The first Map/Reduce job aids in expansion of a substructure while the second Map/Reduce job removes duplicates and finds the top-k substructures with the best MDL value.

Expansion in First Map/Reduce Job: Given a single canonical instance, a mapper uses the range information (from range-based partition management) to route a canonical instance to use appropriate adjacency partitions. With fixed

---

[2] This measure returns a superset of substructures using MIS and we can use MIS with additional cost (if needed) over it.

Algorithm 2. *Reducer of dynamicAL-SD Algorithm*

1: **class Reducer**
2: **function** SETUP()
3:     cId = corresponding adj partition id
4:     load adjMap //load adj partition cId
5:     beamMap = null //map to store best substructures
6: **end function**
7: **function** REDUCER(*key*, *values*)
8:     create Set *isoSet* to store isomorphs
9:     **for** each canonical k-edge instance ks in values **do**
10:         add ks to isoSet //remove duplicates
11:     **end for**
12:     c = count(substructures in isoSet)
13:     mdl = MDL(c,#vertices and #edges in key)
14:     update beamMap with mdl
15: **end function**
16: **function** CLEANUP()
17:     newV = vertices in beamMap not in adjMap
18:     **for** each partition p in hdfs **do**
19:         if p != cId load partition p
20:         **for** each vertex id vId in newV **do**
21:             if p has vId
22:             add adjacency list of vId in adjMap
23:             remove vId from newV
24:             if newV is empty write adjMap back in HDFS
25:         **end for**
26:     **end for**
27:     emit (null, each instance in beamMap)
28: **end function**

---

adjacency partitions, the range information also remains the same across iterations. All canonical instances needing the *same* adjacency partition are therefore hashed to the *same* reducer for expansion. The reducer with *that* adjacency partition in memory expands all canonical instances only on vertex ids present in *that* partition.

Algorithm 3 depicts the first Map/Reduce job for expansion. Each mapper loads the range information only once (line 3). Note that an instance can have multiple vertices in the same adjacency partition(s). The function in line 8 stores unique partition ids for vertices in an instance ensuring that an instance is routed only once to appropriate partitions for expansion. The mapper emits each unique partition id along with the canonical instance. Expansion in reducer logic is same as the mapper in dynamicAL-SD.

Second Map/Reduce Job for Duplicate Removal and Frequency Counting: The expanded canonical substructure from the first Map/Reduce job should be grouped for isomorphism checking. In Map/Reduce environment, the reducer output cannot be an input to another reducer forcing us to use

Algorithm 3. First Map/Reduce job for substructure expansion in staticAL-SD

```
 1: class Mapper
 2: function SETUP()
 3:     R = load range information of adj partitions
 4: end function
 5: function MAP(key, value)
 6:     PIds = Set for unique partition ids for an instance
 7:     for vertex–id v in value do
 8:         PIds.addPartitionId(v,R)
 9:     end for
10:     for partition–id p in PIds do
11:         emit(p, value)
12:     end for
13: end function
14: class Reducer
15: function REDUCE(key, values)
16:     Load partition key from HDFS
17:     for each k-edge canonical instance ks in values do
18:         expand ks to (k+1)-edge canonical instance
19:         emit new key and new value similar to mapper in dynamicAL-SD
20:     end for
21: end function
```

an intermediate mapper. Moreover we still need to generate the canonical substructure from the expanded canonical instance. Instead of generating canonical substructures in the reducer for expansion, we do it in this intermediate mapper. Irrespective of where the canonical substructures are generated they require the same computation time. If canonical substructures are generated in the reducer of counting phase, they need to be sent from reducer to intermediate mapper and then again to the reducers of counting phase, incurring more I/O. Again, we use a combiner with the mapper to remove duplicates in a partition. The reducer now groups instances by the canonical substructure, removes duplicates across partitions and counts isomorphic substructures for evaluating MDL. It is worth mentioning that newer paradigms like Spark [24] with caching technique will save considerable disk I/O for loading partitions in both the approaches.

Analysis of Approaches: In dynamicAL-SD, updates happen in parallel, making the system perform as fast as the partition taking the longest time to update. Generally the worst update time involves loading the partition to be updated and all other partitions, making it equivalent to loading the entire graph. As the update cost is proportional to graph size, improvement between staticAL-SD and dynamicAL-SD should improve with graph size. Connectivity of graphs also influences performance of our algorithms. The adjacency partition size is likely to grow as substructures are expanded and new nodes are added to the substructure. This is likely to be the case for dense graphs where each node is connected to more nodes on the average. The heavy interconnection between partitions in a densely connected graph will also lead to an increase in shuffle cost. But it does not increase with each iteration unlike adjacency partition size. Another factor affecting performance is duplicates. With iterations, duplicates increase as the same k-edge substructure can be generated from k smaller substructures (each missing an edge). Duplicates, if generated across partitions, are shuffled to reducers increasing shuffle cost. As the substructure size increases with each iteration, the size of key/value pairs increase, which coupled with the frequency of substructures influence the shuffle cost. As an anti-monotone metric controls the frequency, the shuffle cost does not affect the time as much as updates, making staticAL-SD a better choice. Even with increasing processors, dynamicAL-SD ends up with a constant update phase, prompting staticAL-SD to outperform it in speedup as well. As is typical of partitioned approaches, the overhead of managing partitions may dominate if the partition sizes are small.

6 Experimental Analysis

In this section we experimentally evaluate staticAL-SD and dynamicAL-SD. All experiments are conducted using Java on a Hadoop cluster with 1 front-end server and 17 worker nodes each having a 3.2 GHz Intel Xeon CPU, 4 GB of RAM and 1.5 TB of local disk. The server has same specification but with 3 TB of local disk. Each node was running Hadoop version 1.0.3 on ROCKS Cluster 6.3 operating system and connected by gigabit Ethernet to a commodity switch.

Datasets: We experiment on several real world and synthetic datasets to establish the effectiveness and scalability of our approach. Table 1 shows the datasets that we use to evaluate our approach. For both the Orkut and LiveJournal datasets, we have randomly assigned a category to the user nodes from a set of distinct 100 categories following a Gaussian distribution. Subgen, a synthetic graph generator is also used which allows embedding small graphs with user defined frequency in a single graph giving better control in generating graphs of different sizes and characteristics.

Correctness and Efficiency: The correctness of both our algorithms are verified by running them on small synthetic graphs generated by Subgen. Subgen was used to generate small graphs with predefined embedded substructures from size 20V/40E to 10KV/20KE and Subdue [13], staticAL-SD and dynamicAL-SD when run on them discovered the same number of substructures. To compare individual performance of dynamicAL-SD and staticAL-SD algorithms we run 5

Table 1. Datasets for experiments

Name	Nodes	Edges	Use
LiveJournal (snap.stanford.edu/data/com-LiveJournal.html)	3.99M	34.68M	scalability
Orkut (snap.stanford.edu/data/com-Orkut.html)	3.07M	117.18M	scalability
Subgen (ailab.wsu.edu/subdue)	800K	1600K	correctness

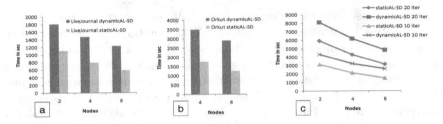

Fig. 2. Comparisons of running time of graphs using staticAL-SD and dynamicAL-SD

iterations of both algorithms with a beam of size 4 on the LiveJournal and Orkut graphs. The average of 5 runs is taken to avoid cold start issues. Our results in Fig. 2a indicate that in LiveJournal, staticAL-SD outperforms dynamicAL-SD by 38.8 %, 46.1 % and 52.1 % with 2, 4 and 8 reducers respectively while the trend remained same in Orkut with 49.8 % and 56.8 % improvement for 4 and 8 reducers in Fig. 2b. Orkut overwhelms main memory with 2 processors and is run on 4 and 8 reducers only. The increase in improvement is attributed to decrease in computations on increasing number of processors, and a constant partition update time in each reducer. The Orkut graph needed more time for update and demonstrates the effectiveness of staticAL-SD with increasing graph size. To study the effect of shuffle with increasing iterations, we compared the performances of both the algorithms for 10 and 20 iterations on the LiveJournal dataset as shown in Fig. 2c. Though there is some increase in shuffle time for bigger substructures, but update still dominates the running time making staticAL-SD approximately 33 % faster than its counterpart.

Scalability: Our results in Fig. 2 indicate that staticAL-SD has a speedup of 39.4 % from 2 to 4 reducers in LiveJournal and 34.4 % from 4 to 8 reducers as compared to dynamicAL-SD's 22.9 % and 19.3 % respectively. StaticAL-SD on Orkut also shows the same trend with 39.7 % speedup as compared to DynamicAL-SD's 20.2 % from 4 to 8 reducers. DynamicAL-

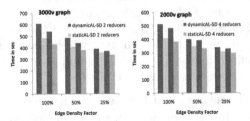

Fig. 3. Running times of algorithms on graphs with varying density

SD has a lower speedup due to its constant update phase at each reducer. Note that with doubling number of reducers, the time for mining is not exactly halved. This is due to two reasons. First, with doubling the number of reducers, each reducer emits the instances with local top-k MDL values to find the instances with global top-k best MDL values. So the amount of data handled by each reducer is not necessarily halved. Second, as the reducer output records increase, the I/O cost increases. Typically in a Map/Reduce system the I/O is defined as the sum of map input records, map output records, reduce input records and reduce output records. Increasing number of reducers coupled with the notion of beam in each reducer increase the reducer output in each iteration thereby not generating a linear speedup.

Varying Edge Density: We want to analyze how our algorithms perform for graphs of varying densities (where number of edges vary and the number of vertices are fixed). For a graph with n nodes, we generate an *edge density factor (edf)* considering the completely connected graph with n(n-1) edges to be 100 % dense and the linear graph with n-1 edges to be least dense graph. We generate graphs with 2000 and 3000 vertices with an edf of 25 %, 50 % and 100 % to analyze how our method works for graphs with varying densities. For each of these graphs we run our algorithms for 4 iterations. Figure 3 shows the results with varying density and number of reducers. The X-axis shows the edf for all the graphs while the Y-axis represents the time in seconds. For completely connected graphs, even with high shuffle cost, staticAL-SD is still 20 % faster than dynamicAL-SD. With graphs of decreasing density and hence decreasing size, the update cost reduces proportionately making staticAL-SD to be only 9 % better. As expected, small sized partitions lead to poor speedup (average of 6 % in 2000v and 25 % edf) than bigger graphs as seen already in Fig. 2. The experiments confirm our earlier analysis of staticAL-SD and dynamicAL-SD. Our experiments cover important aspects of graph mining: scalability for graph size, performance based on connectivity and size of substructures. In all categories, experimental evaluation concurs with our earlier analysis.

7   Conclusion

In this paper, we have shown how substructure discovery algorithms can be made to work efficiently using the Map/Reduce paradigm. Since partitioning is the key to scalability of graph mining, we proposed an approach where the shuffle can be effectively used for keeping the adjacency partition fixed. As future work, extending these algorithms for mining inexact patterns is underway. Constrained expansions to prevent duplicate generation in a processor, better load distribution and managing partitions in better ways are being investigated.

References

1. http://giraph.apache.org/
2. Foto, N., Afrati, D.F., Jeffrey, D.U.: Enumerating subgraph instances using map-reduce. Technical report, Stanford University, December 2011
3. Agrawal, R., Srikant, R.: Fast Algorithms for Mining Association Rules. In: Very Large Data Bases, pp. 487–499 (1994)
4. Alexaki, S., Christophides, V., Karvounarakis, G., Plexousakis, D.: On storing voluminous RDF descriptions: the case of web portal catalogs. In: International Workshop on the Web and Databases, pp. 43–48 (2001)
5. Bringmann, B., Nijssen, S.: What is frequent in a single graph? In: Washio, T., Suzuki, E., Ting, K.M., Inokuchi, A. (eds.) PAKDD 2008. LNCS (LNAI), vol. 5012, pp. 858–863. Springer, Heidelberg (2008)

6. Chakravarthy, S., Pradhan, S.: DB-FSG: an SQL-based approach for frequent subgraph mining. In: Bhowmick, S.S., Küng, J., Wagner, R. (eds.) DEXA 2008. LNCS, vol. 5181, pp. 684–692. Springer, Heidelberg (2008)
7. Das, S., Chakravarthy, S.: Challenges and approaches for large graph analysis using map/reduce paradigm. In: Bhatnagar, V., Srinivasa, S. (eds.) BDA 2013. LNCS, vol. 8302, pp. 116–132. Springer, Heidelberg (2013)
8. Deshpande, M., Kuramochi, M., Karypis, G.: Frequent sub-structure-based approaches for classifying chemical compounds. In: IEEE International Conference on Data Mining, pp. 35–42 (2003)
9. Elseidy, M., Abdelhamid, E., Skiadopoulos, S., Kalnis, P.: GRAMI: frequent subgraph and pattern mining in a single large graph. PVLDB 7(7), 517–528 (2014)
10. Fiedler, M., Borgelt, C.: Subgraph support in a single large graph. In: ICDM Workshops, pp. 399–404 (2007)
11. Han, J., Cheng, H., Xin, D., Yan, X.: Frequent pattern mining: current status and future directions. Data Min. Knowl. Discov. 15(1), 55–86 (2007)
12. Hill, S., Srichandan, B., Sunderraman, R.: An iterative mapreduce approach to frequent subgraph mining in biological datasets. In: BCB, pp. 661–666 (2012)
13. Holder, L.B., Cook, D.J., Djoko, S.: Substucture Discovery in the SUBDUE System. In: Knowledge Discovery and Data Mining, pp. 169–180 (1994)
14. Jiang, C., Coenen, F., Zito, M.: A survey of frequent subgraph mining algorithms. Knowl. Eng. Rev. 28(1), 75–105 (2013)
15. Jiang, W., Vaidya, J., Balaporia, Z., Clifton, C., Banich, B.: Knowledge discovery from transportation network data. In: ICDE 2005, pp. 1061–1072, April 2005
16. Karypis, G., Kumar, V.: Multilevel k-way partitioning scheme for irregular graphs. J. Parallel Distrib. Comput. 48, 96–129 (1998). Elsevier
17. Lin, J., Schatz, M.: Design patterns for efficient graph algorithms in MapReduce. In: 16th ACM SIGKDD Conference, pp. 78–85 (2010)
18. Liu, Y., Jiang, X., Chen, H., Ma, J., Zhang, X.: MapReduce-based pattern finding algorithm applied in motif detection for prescription compatibility network. In: Advanced Parallel Programming Technologies, pp. 341–355 (2009)
19. Murray, D.G., McSherry, F., Isaacs, R., Isard, M., Barham, P., Abadi, M.: Naiad: a timely dataflow system. In: ACM SIGOPS 24th Symposium on Operating Systems Principles, SOSP 2013, pp. 439–455 (2013)
20. Padmanabhan, S., Chakravarthy, S.: HDB-subdue: a scalable approach to graph mining. In: Pedersen, T.B., Mohania, M.K., Tjoa, A.M. (eds.) DaWaK 2009. LNCS, vol. 5691, pp. 325–338. Springer, Heidelberg (2009)
21. Pei, J., Han, J., Mortazavi-Asl, B., Pinto, H., Chen, Q., Dayal, U., Hsu, M.-C.: PrefixSpan,: mining sequential patterns efficiently by prefix-projected pattern growth. In: ICDE, pp. 215–224 (2001)
22. Rathi, R., Cook, D.J., Holder, L.B.: A serial partitioning approach to scaling graphbased knowledge discovery. In: Russell, I., Markov, Z. (eds.) FLAIRS Conference, pp. 188–193. AAAI Press, Menlo Park (2005)
23. Yan, X., Han, J.: gSpan: graph-based substructure pattern mining. In: IEEE International Conference on Data Mining, pp. 721–724 (2002)
24. Zaharia, M., Chowdhury, M., Franklin, M., Shenker, S., Stoica, I.: Spark: cluster computing with working sets. In: HotCloud (2010)

Implementation of Multidimensional Databases with Document-Oriented NoSQL

M. Chevalier, M. El Malki[✉], A. Kopliku, O. Teste, and R. Tournier

IRIT 5505, Université de Toulouse, 118 Route de Narbonne,
31062 Toulouse, France
{Max.Chevalier,Mohammed.ElMalki,Arlind.Kopliku,Olivier.Teste,
Ronan.Tournier}@irit.fr

Abstract. NoSQL (Not Only SQL) systems are becoming popular due to known advantages such as horizontal scalability and elasticity. In this paper, we study the implementation of data warehouses with document-oriented NoSQL systems. We propose mapping rules that transform the multidimensional data model to logical document-oriented models. We consider three different logical translations and we use them to instantiate multidimensional data warehouses. We focus on data loading, model-to-model conversion and cuboid computation.

1 Introduction

NoSQL solutions have proven some clear advantages with respect to relational database management systems (RDBMS) [14]. Nowadays, the research attention has moved towards the use of these systems for storing "big" data and analyzing it. This work joins our previous work on the use of NoSQL solutions for data warehousing [3] and it joins substantial ongoing works [6, 9, 15]. In this paper, we focus on one class of NoSQL stores, namely document-oriented systems [7].

Document-oriented systems are one of the most famous families of NoSQL systems. Data is stored in collections, which contain documents. Each document is composed of key-value pairs. The value can be composed of nested sub-documents. Document-oriented stores enable more flexibility in schema design: they allow the storage of complex structured data and heterogeneous data in one collection. Although, document-oriented databases are declared to be "schema less" (no schema needed), most uses convey to some data model.

When it comes to data warehouses, previous work has shown that it can be instantiated with different logical models [10]. We recall that data warehousing relies mostly

---

M. El Malki—This study is supported by the ANRT funding under CIFRE-Capgemini partnership.

S. Madria and T. Hara (Eds.): DaWaK 2015, LNCS 9263, pp. 379–390, 2015.
DOI: 10.1007/978-3-319-22729-0_29

on the multidimensional data model. The latter is a conceptual model,[1] and we need to map it in document-oriented logical models. Mapping the multidimensional model to relational databases is quite straightforward, but until now there is no work (except of our previous [3]) that considers the direct mapping from the multidimensional conceptual model to NoSQL logical models (Fig. 1). NoSQL models support more complex data structures than relational model i.e. we do not only have to describe data and the relations using atomic attributes. They have a flexible data structure (e.g. nested elements). In this context, more than one logical model is candidate for mapping the multidimensional model. As well, the evolving needs might demand for switching from one model to another. This is the scope of our work: NoSQL logical models and their use for multidimensional data warehousing.

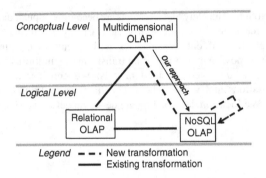

Fig. 1. Translations of a conceptual multidimensional model into logical models.

In this paper, we focus on multidimensional data models for data warehousing. We compare three translations of the conceptual model at logical document-oriented model level. We provide formalism that enables us to define the mapping from the conceptual model to the logical model. Then, we show how we can instantiate data warehouses in document-oriented systems. Our studies include the load of data, the conversions model-to-model and the computation of pre-aggregate OLAP cubes.

Our motivations are multiple. The implementation of OLAP systems with NoSQL systems is a new alternative [6, 14]. It is justified by the advantages such as more scalability. The increasing scientific research in this direction demands for formalization, common-agreement models and evaluation of different NoSQL systems.

The paper is organized as follows. The following section studies the state of the art. In Sect. 3, we formalize the multidimensional data model and OLAP cuboids. Then, we focus on formalisms and definitions of document-oriented models. In Sect. 4, we show experiments.

---

[1] The conceptual level consists in describing the data in a generic way regardless the information technologies whereas the logical level consists in using a specific technique for implementing the conceptual level.

2   State of the Art

Considerable research has focused on the translation of data warehousing concepts to relational R-OLAP logical level [2, 5]. Multidimensional databases are mostly implemented using RDBMS technologies. Mapping rules are used to convert structures of the conceptual level (facts, dimensions and hierarchies) into a logical model based on relations. Moreover, many researchers [1] have focused on the implementation of optimization methods based on pre-computed aggregates (also called materialized views, or OLAP cuboids). However, R-OLAP implementations suffer from scaling-up to very large data volumes (i.e. "Big Data"). Research is currently under way for new solutions such as using NoSQL systems [14]. Our approach aims at revisiting these processes for automatically implementing multidimensional conceptual models directly into NoSQL models.

Other studies investigate the process of transforming relational databases into a NoSQL logical model (bottom part of Fig. 1). In [12], an algorithm is introduced for mapping a relational schema to a NoSQL schema in MongoDB [7], a document-oriented NoSQL database. However, either these approaches not consider the conceptual model of data warehouses because they are limited to the logical level, i.e. transforming a relational model into a documents-oriented model. In [11] the author proposes an approach to optimize schema in NoSQL.

There is currently no approach that automatically and directly transforms a multidimensional conceptual model into a NoSQL logical model. It is possible to transform multidimensional conceptual models into a logical relational model, and then to transform this relational model into a logical NoSQL model. However, this transformation using the relational model as a pivot model has not been formalized as both transformations were studied independently of each other. The work presented here is a continuation of our previous work where we study and formalize the implementation of data warehouses with NoSQL systems [3]. Our previous work considers two NoSQL models (one column-oriented and one document oriented). This article focuses only on document-oriented systems; we analyze three data models (with respect to 1); we consider all cross-model mappings; we improve the formalization and we provide new experiments.

3   Multidimensional Conceptual Model and Olap Cube

3.1   Conceptual Multidimensional Model

To ensure robust translation rules we first define the multidimensional model used at the conceptual level [8, 13].

A **multidimensional schema**, namely E, is defined by (F^E, D^E, $Star^E$)

- $F^E = \{F_1,..., F_n\}$ is a finite set of facts,
- $D^E = \{D_1,..., D_m\}$ is a finite set of dimensions,

- $Star^E: F^E \rightarrow 2^{DE}$ is a function that associates facts of F^E to sets of dimensions along which it can be analyzed (2^{D^E} is the *power set* of D^E).

A **dimension**, denoted $D_i \in D^E$ (abusively noted as D), is defined by (N^D, A^D, H^D)

- N^D is the name of the dimension,
- AD = a1D,…,auDuidD,AllD $A^D = \{a_1^D, \dots, a_u^D\} \cup \{id^D, all^D\}$ is a set of dimension attributes,
- $^{HD} = \{$H1D,…,HvD$\}$ $H^D = \{H_1^D, \dots, H_v^D\}$ is a set of hierarchies.

A **hierarchy**, denoted $H_i \in H^D$, is defined by ($N^{Hi}, Param^{Hi}, Weak^{Hi}$)

- N^{Hi} is the name of the hierarchy,
- $Param^{Hi} = <$id^D,p_1^{Hi},…,p_{vi}^{Hi},AllD$>$ $Param^{Hi} =< id^D, p_{v_1}^{H_i}, \dots, p_{v_i}^{H_i}, All^D >$ is an ordered set of $v_i + 2$ attributes which are called parameters of the relevant graduation scale of the hierarchy, $\forall k \in [1..v_i]$, p_k^{Hi} $p_k^{Hi} \in A^D$,
- $Weak^{Hi}:Param^{Hi} \rightarrow 2$AD - ParamHi $2^{A^D - param^{Hi}}$ is a function associating with each parameter possibly one or more weak attributes.
- A **fact**, denoted $F \in F^E$, is defined by (N^F, M^F)
- N^F is the name of the fact,
- ^{M}F = $\{$f1(m1F),…,fv(mvF)$\}$ $M^F = \{f_1(m_1^F), \dots, f_v(m_v^F)\}$ is a set of measures, each associated with an aggregation function f_i.

3.2   The OLAP Cuboid

The **pre-aggregate view** or **OLAP cuboid** corresponds to a subset of aggregated measures on a subset of analysis dimensions. OLAP cuboids are often pre-computed to turn frequent analysis of data more efficient. Typically, we pre-compute aggregate functions on given interest measures grouping on some analysis dimensions. The OLAP cube $O = (F^O, D^O)$ derived from E is formally composed of

- $F^O = (N^{Fo}, M^{Fo})$ a fact derived from $F \in F^E$ with $N^{Fo} = N^F$ a subset of measures.
- $D^O = Star(F^O) \subseteq D^E$ a subset of dimensions.

If we generate OLAP cuboids on all combination of dimension attributes, we have an **OLAP cube lattice**.

Illustration: Let's consider an excerpt of the star schema benchmark [12]. It consists in a monitoring of a sales system. Orders are placed by customers and the lines of the orders are analyzed. A line consists in a part (a product) bought from a supplier and sold to a customer at a specific date. The conceptual schema of this case study is presented in Fig. 2.

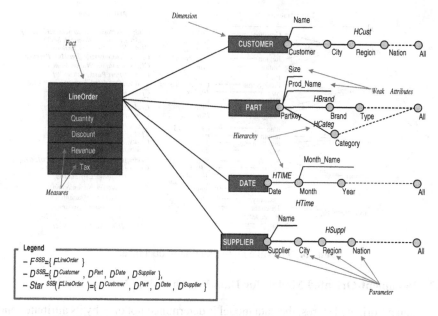

Fig. 2. Graphical notations of the multidimensional conceptual model

From this schema, called E^{SSB}, we can define cuboids, for instance:

- $(F^{LineOrder}, \{D^{Customer}, D^{Date}, D^{Supplier}\})$,
- $(F^{LineOrder}, \{D^{Customer}, D^{Date}\})$.

4 Document-Oriented Modeling of Multidimensional Data Warehouses

4.1 Formalism for Document-Oriented Data Models

In the document-oriented model, data is stored in collections of documents. The structure of documents is defined by attributes. We distinguish **simple attributes** whose values are atomic from **compound attributes** whose values are documents called **nested documents** or **sub-documents**. The **structure** of document is described as a set of paths from the document tree. The following example illustrates the above formalism. Consider the document tree in Fig. 3 describing a document d_i of the collection C with identifier $id:v_i$.

A document $C[id:v_i]$ is defined by:

- C: the collection where the document belongs. We use the notation $C[id:v_i]$ to indicate a document of a collection C with identifier $id:v_i$.
- P: all paths in the tree of the document. A path p is described as $p = C[id:v_i]\{k_1:k_2...k_n:a\}$ where $k_1, k_2, ... k_n$ are keys within the same path ending at the leaf node with the value of a simple attribute.

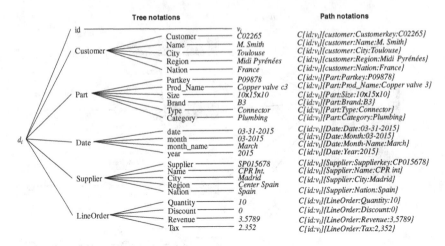

Fig. 3. Tree-like representation of documents

4.2 Document-Oriented Models for Data Warehousing

In document-oriented stores, the data model is determined not only by its attributes and values, but also by the path to the data. In relational database models, the mapping from conceptual to logical is more straightforward. In document-oriented stores, there are multiple candidate models that differ on the collections and structure. No model has been proven better than the others and no mapping rules are widely accepted. In this section, we present three approaches of logical document-oriented modeling. These models correspond to a broad classification of possible approaches for modeling multi-dimensional cuboid (one fact and its star). In the first approach, we store all cuboid data in one collection in a flat format (without sub-document). In the second case, we make use of nesting (embedded sub-documents) within one collection (richer expressivity). In the third case, we consider distributing the cuboid data in more than one collection. These approaches are described below.

MLD0: For a given fact, all related dimensions attributes and all measures are combined in one document at depth 0 (no sub-documents). We call this the flat model noted MLD0.

MLD1: For a given fact, all dimension attributes are nested under the respective attribute name and all measures are nested in a subdocument with key "measures". This model is inspired from [3]. Note that there are different ways to nest data, this is just one of them.

MLD2: For a given fact and its dimensions, we store data in dedicated collections one per dimension and one for the fact. Each collection is kept simple: no sub-documents. The fact documents will have references to the dimension documents. We call this model MLD2 (or shattered). This model has known advantages such as less memory usage and data integrity, but it can slow down interrogation.

4.3   Mapping from the Conceptual Model

The formalism that we have defined earlier allows us to define a mapping from the conceptual multidimensional model to each of the logical models defined above. Let $O = (F^O, D^O)$ be a cuboid for a multidimensional model for the fact F^O (M^O is a set of measures) with dimensions in D. N^F and N^D stand for fact and dimension names.

The above mappings are detailed below. Let C be a generic collection, C^D a collection for the dimension D and C^F a collection for a fact F^O. The Table 1 shows how we can map any measure m of F^O and any dimension of D^O into any of the models: MLD0, MLD1, MLD2.

Table 1. Mapping rules from the conceptual model to the logical models

Conceptual	Logical		
	MLD0	**MLD1**	**MLD2**
$\forall D \in D^O, \forall a \in A^D$	$a \rightarrow C[id]\{a\}$	$a \rightarrow C[id]\{N^D:a\}$	$a \rightarrow C^D[id]\{a\}$ $id^D \rightarrow C^F[id']\{a\}$ (*)
$\forall m \in M^F$	$m \rightarrow C[id]\{m\}$	$m \rightarrow C[id]\{N^F:m\}$	$m \rightarrow C^F[id']m$

(*) The two identifiers $C^F[id']$ and $C^D[id]$ are different from each other because the document collections are not same C^D et C^F.

Conceptual Model to MLD0: To instantiate this model from the conceptual model, these rules are applied:

- Each cuboid O (F^O and its dimensions D^O) is translated in a collection C.
- Each measure $m \in M^F$ is translated into a simple attribute (i.e. $C[id]\{m\}$)
- For all dimension $D \in D^O$, each attribute $a \in A^D$ of the dimension D is converted into a simple attribute of C (i.e. $C[id]\{a\}$)

Conceptual Model to MLD1: To instantiate this model from the conceptual model, these rules are applied:

- Each cuboid O (F^O and its dimensions D^O) is translated in a collection C.
- The attributes of the fact F^O will be nested in a dedicated nested document $C[id]\{N^F\}$. Each measure $m \in M^F$ is translated into a simple attribute $C[id]\{N^F:m\}$.
- For any dimension $D \in D^O$, its attributes will be nested in a dedicated nested document $C[id]\{N^D\}$. Every attribute $a \in A^D$ of the dimension D will be mapped into a simple attribute $C[id]\{N^D:a\}$.

Conceptual Model to MLD2: To instantiate this model from the conceptual model, these rules are applied:

- Each cuboid O (F^O and its dimensions D^O), the fact F^O is translated in a collection C^F and each dimension $D \in D^O$ into a collection C^D.
- Each measure $m \in M^F$ is translated within CF as a simple attribute (i.e. $C^F[id']\{m\}$)
- For all dimension $D \in D^O$, each attribute $a \in A^D$ of the dimension D is mapped into C^D as a simple attribute (i.e. $C^D[id]\{a\}$), and if $a = id^D$ the document C^F is completed by a simple attribute $C^F[id']\{a\}$ (the value reference of the linked dimension).

5 Experiments

Our experimental goal is to validate the instantiation of data warehouses with the three approaches mentioned earlier. Then, we consider converting data from one model to the other. In the end, we generate OLAP cuboids and we compare the effort needed by model. We rely on the **SSB +** benchmark that is popular for generating data for decision support systems. As data store, we rely on **MongoDB** one of the most popular document-oriented system.

5.1 Protocol

Data: We generate data using the SSB + [4] benchmark. The benchmark models a simple product retail reality. It contains one fact table "LineOrder" and 4 dimensions "Customer", "Supplier", "Part" and "Date". This corresponds to a star-schema. The dimensions are hierarchic e.g. "Date" has the hierarchy of attributes [d_date, d_month, d_year]. We have extended it to generate raw data specific to our models in JSON file format. This is convenient for our experimental purposes. JSON is the best file format for Mongo data loading. We use different scale factors namely sf = 1, sf = 10, sf = 25 and sf = 100 in our experiments. The scale factor sf = 1 generates approximately 10^7 lines for the LineOrder fact, for sf = 10 we have approximately 10^8 lines and so on. In the **MLD2**model we will have ($sf \times 10^7$) lines for LineOrder and quite less for the dimensions.

Data loading: Data is loaded into MongoDB using native instructions. These are supposed to load data faster when loading from files. The current version of MongoDB would not load data with our logical model from CSV file, thus we had to use JSON files.

Lattice computation: To compute the pre-aggregate lattice, we use the aggregation pipeline suggested as the most performing alternative by Mongo itself. Four levels of pre-aggregates are computed on top of the benchmark generated data. Precisely, at each level we aggregate data respectively on: the combination of 4 dimensions all combinations of 3 dimensions, all combinations of 2 dimensions, all combinations of 1 dimension, 0 dimensions (all data). At each aggregation level, we apply aggregation functions: *max, min, sum* and *count* on all dimensions.

Hardware. The experiments are done on a cluster composed of 3 PCs (4 core-i5, 8 GB RAM, 2 TB disks, 1 Gb/s network), each being a worker node and one node acts also as dispatcher.

5.2 Results

In Table 2, we summarize data loading times by model and scale factor. We can observe at scale factor SF1, we have 10^7 lines on each line order collections for a 4.2 GB disk memory usage for MLD2 (15 GB for MLD0 and MLD1). At scale factors SF10 and SF100 we have respectively 10^8 lines and 10^9 lines and 42 GB (150 GB MLD0 and MLD1) and 420 GB (1.5 TB MLD0 and MLD1) for of disk memory usage. We observe that memory usage is lower in the MLD2 model. This is explained by the absence of redundancy in the dimensions. The collections "Customers", "Supplier", "Part" and "Date" have respectively 50000 records, 3333 records, 3333333 records and 2556 records.

Table 2. Loading times by model into MongoDB

	MLD0	MLD1	MLD2
SF = 1 10^7 lines	1306 s/15 GB	1235 s/15 GB	1261 s/4.2 GB
SF = 10 10^8 lines	16680 s/150 GB	16080 s/150 GB	4320 s/42 GB
SF = 25 25.10^7 lines	46704 s/375 GB	44220 s/375 GB	10980 s/105 GB

In Fig. 4, we show the time needed to convert data of one model to data of another model with SF1. When we convert data from MLD0 to MLD1 and vice versa conversion times are comparable. To transform data from MLD0 to MLD1 we just introduce a depth of 1 in the document. On the other sense (MLD1 to MLD0), we reduce the depth by one. The conversion is more complicated when we consider MLD0 and MLD2. To convert MLD0 data into MLD2 we need to split data in multiple tables: we have to apply 5 projections on original data and select only distinct keys for dimensions. Although, we produce less data (in memory usage), we need more processing time than when we convert data to MLD1. Converting from MLD2 to MLD0 is the slowest process by far. This is due to the fact that most NoSQL systems (including MongoDB) do not support joins (natively). We had to test different optimization techniques hand-coded. The loading times fall between 5 h to 125 h for SF1. It might be possible to optimize this conversion further, but the results are illustrative of the jointure issues in MongoDB.

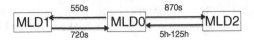

Fig. 4. Inter-model conversion times

In Fig. 5, we sumarize experimental observations concerning the computation of the OLAP cuboids at different levels of the OLAP lattice for SF1 using data from the model MLD0. We report the time needed to compute the cuboid and the number of records it contains. We compute the cuboids from one of the "upper"-in hierarchy cuboids with less records, which makes computation faster.

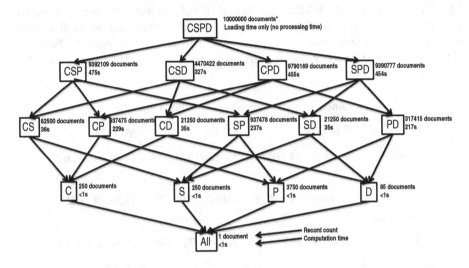

Fig. 5. Computation time and records by OLAP cuboid

We observe as expected that the number of records decreases from one level to the lower level. The same is true for computation time. We need between 300 and 500 s to compute the cuboids at the first level (3 dimensions). We need between 30 s and 250 s at the second layer (2 dimensions). We need less than one second to compute the cuboids at the third and fourth level (1 and 0 dimensions).

OLAP computation using the model MLD1 provides similar results. The performance is significantly lower with the MLD2 model due to joins. These differences involve only the layer 1 (depth one) of the OLAP lattice, cause the other layers can be computed from the latter. We do not report this results for space constraints.

Observations: We observe that we need comparable times to load data in one model with the conversion times (except of MLD2 to MLD0). We also observe reasonable times for computing OLAP cuboids. These observations are important. At one hand, we show that we can instantiate data warehouses in document-oriented data systems. On the other, we can think of pivot models or materialized views that can be computed in parallel with a chosen data model.

6   Conclusion

In this paper, we have studied the instantiation of data warehouses with document-oriented systems. We propose three approaches at the document-oriented logical model. Using a simple formalism, we describe the mapping from the multidimensional conceptual data model to the logical level.

Our experimental work illustrates the instantiation of data warehouses with each of the three approaches. Each model has its weaknesses and strengths. The shattered model (MLD2) uses less disk memory, but it is quite inefficient when it comes to answering queries with joins. The simple models MLD0 and MLD1 do not show significant performance differences. Passing from one model to another is shown to be easy and comparable in time to "data loading from scratch". One conversion is significantly non-performing; it corresponds to the mapping from multiple collections (MLD2) to one collection. Interesting results are also met in the computation of the OLAP lattice with document-oriented models. The computation times are reasonable enough.

For future work, we will consider logical models in column-oriented models and graph-oriented models. After exploring data warehouse instantiation across different NoSQL systems, we need to generalize across logical model. We need a simple formalism to express model differences and we need to compare models within each paradigm and across paradigms (document versus column).

References

1. Bosworth, A., Gray, J., Layman, A., Pirahesh, H.: Data cube: A relational aggregation operator generalizing group-by, cross-tab, and sub-totals. Technical report MSRTR-95-22, Microsoft Research, February 1995
2. Chaudhuri, S., Dayal, U.: An overview of data warehousing and OLAP technology. SIGMOD Rec. **26**, 65–74 (1997)
3. Chevalier, M., malki, M.E., Kopliku, A., Teste, O., Tournier, R.: Implementing multidimensional data warehouses into NoSQL. In: 17th International Conference on Entreprise Information Systems, April 2015
4. Chevalier, M., El Malki, M., Kupliku, A., Teste, O., Tournier, R.: *Benchmark for OLAP on NoSQL technologies, comparing NoSQL multidimensional data warehousing solutions.* In: 9th International Conference on Research Challenges in Information Science (RCIS). IEEE (2015)
5. Colliat, G.: Olap, relational, and multidimensional database systems. SIGMOD Rec. **25**(3), 64–69 (1996)
6. Cuzzocrea, A., Song, I.Y., Davis, K.C.: Analytics over large-scale multidimensional data: The big data revolution!. In: 14th International Workshop on Data Warehousing and OLAP DOLAP 2011, pp. 101–104. ACM (2011)
7. Dede, E., Govindaraju, M., Gunter, D., Canon, R.S., Ramakrishnan, L.: Performance evaluation of a MongoDB and hadoop platform for scientific data analysis. In: 4th ACM Workshop on Scientific Cloud Computing Science Cloud 2013, pp.13–20. ACM (2013)
8. Dehdouh, K., Boussaid, O., Bentayeb, F.: Columnar NoSQL star schema benchmark. In: Ait Ameur, Y., Bellatreche, L., Papadopoulos, G.A. (eds.) MEDI 2014. LNCS, vol. 8748, pp. 281–288. Springer, Heidelberg (2014)

9. Golfarelli, M., Maio, D., Rizzi, S.: The dimensional fact model: A conceptual model for data warehouses. Int. J. Coop. Inf. Syst. **7**, 215–247 (1998)

10. Kimball, R., Ross, M.: The Data Warehouse Toolkit: The Complete Guide to Dimensional Modeling, 2nd edn. Wiley, New York (2002)

11. Mior, M.J.: Automated schema design for NoSQL databases. In: SigMOD (2014)

12. O'Neil, P., O'Neil, E., Chen, X., Revilak, S.: The star schema benchmark and augmented fact table indexing. In: Nambiar, R., Poess, M. (eds.) TPCTC 2009. LNCS, vol. 5895, pp. 237–252. Springer, Heidelberg (2009)

13. Ravat, F., Teste, O., Tournier, R., Zuruh, G.: Algebraic and graphic languages for OLAP manipulations. IJDWM **4**(1), 17–46 (2008)

14. Stonebraker, M.: New opportunities for new SQL. Commun. ACM **55**(11), 10–11 (2012). http://doi.acm.org/10.1145/2366316.2366319

15. Zhao, H., Ye, X.: A practice of TPC-DS multidimensional implementation on NoSQL database systems. In: Nambiar, R., Poess, M. (eds.) TPCTC 2013. LNCS, vol. 8391, pp. 93–108. Springer, Heidelberg (2014)

A Graph-Based Concept Discovery Method for n-Ary Relations

Nazmiye Ceren Abay[1], Alev Mutlu[2]([✉]), and Pinar Karagoz[1]

[1] Department of Computer Engineering,
Middle East Technical University, Ankara, Turkey
{ceren.abay,karagoz}@ceng.metu.edu.tr
[2] Department of Computer Engineering, Kocaeli University, Kocaeli, Turkey
alev.mutlu@kocaeli.edu.tr

Abstract. Concept discovery is a multi-relational data mining task for inducing definitions of a specific relation in terms of other relations in the data set. Such learning tasks usually have to deal with large search spaces and hence have efficiency and scalability issues. In this paper, we present a hybrid approach that combines association rule mining methods and graph-based approaches to cope with these issues. The proposed method inputs the data in relational format, converts it into a graph representation, and traverses the graph to find the concept descriptors. Graph traversal and pruning are guided based on association rule mining techniques. The proposed method distinguishes from the state-of-the art methods as it can work on n-ary relations, it uses path finding queries to extract concepts and can handle numeric values. Experimental results show that the method is superior to the state-of-the art methods in terms of accuracy and the coverage of the induced concept descriptors and the running time.

Keywords: Concept discovery · Graph · Path · Support · Confidence

1 Introduction

Multi-relational data mining (MRDM) is concerned with discovering hidden patterns in multiple tables from a relational database [1]. One of the most commonly addressed tasks in MRDM is concept discovery where the problem is inducing the logical definitions of a specific relation, called *target relation*, in terms of other relations, called *background knowledge*. Concept discovery can be considered a predictive learning task where the input is a set of positive and negative instances that belong to the target relation, background knowledge, maximum length of the concept descriptors and the quality metrics the concept descriptors should satisfy; and output is a set of logical definitions of the target relation. Target relation instances are generally ground facts, background knowledge is represented either intensionally or extensionally and concept descriptors are usually in the form of Horn clauses. Concept discovery has been applied in various

© Springer International Publishing Switzerland 2015
S. Madria and T. Hara (Eds.): DaWaK 2015, LNCS 9263, pp. 391–402, 2015.
DOI: 10.1007/978-3-319-22729-0_30

domains such as biochemistry [2], bioinformatics [3], environmental sciences [4] and promising results have been reported.

Logic-based, more specifically Inductive Logic Programming (ILP)-based [5], and graph-based approaches are two competitors in concept discovery. ILP-based approaches benefit from the powerful data representation framework of first order logic and the easily interpretable concept descriptors. On the other hand such approaches have to deal with large search spaces, local maxima and local minima issues [6]. Graph-based approaches also provide powerful mechanisms to represent relational data and have heavily studied algorithms that can be used to find concepts in a graph. Graph-based approaches in concept discovery usually have to deal with computationally expensive algorithms like subgraph isomorphism [7] and path profiling [8], i.e. counting similar paths.

In this work, we present a graph-based method for concept discovery. It is a path finding-based method where frequently appearing paths are considered as concept descriptors. The proposed method inputs a relational database that contains the target instances and the background knowledge, converts it into a graph database, extracts all paths that originate from target instances and output those paths that appear frequently as concept descriptors.

The proposed method distinguishes from state-of-the-art methods as it can work on n-ary relations and can handle numeric attributes. The experimental results show that the proposed method can induce the same set of concept descriptors in shorter running time compared to an ILP-based concept discovery system and is compatible with state-of-the-art graph-based concept discovery systems.

This rest of the paper is organized as follows. In Sect. 2, we formally introduce the concept discovery process and introduce some graph-based concept discovery systems. In Sect. 3, we present the proposed method. In Sect. 4, we present the experimental results. Section 5 concludes the paper with an overview and future work.

2   Background and Related Work

Given a set of target examples $\mathcal{E}$ and background knowledge $\mathcal{B}$, concept discovery is defined as finding a hypothesis $\mathcal{H}$, expressed in some concept descriptor language $\mathcal{L}$, such that $\mathcal{H}$ is complete and consistent with respect to $\mathcal{B}$ and $\mathcal{E}$. $\mathcal{H}$ is called complete with respect to $\mathcal{B}$ and $\mathcal{E}$ if it explains all of the positive target instances. $\mathcal{H}$ is called consistent with respect to $\mathcal{B}$ and $\mathcal{E}$ if it explains none of the negative target instances. Due to the noisy nature of the input data, completeness and consistency properties are relaxed, respectively, as to explain as many positive examples as possible and as few negative examples as possible.

ILP and graph-based approaches are two competitors in concept discovery. ILP-based concept discovery systems represent the data within first order logic framework and concept descriptors are derived either by applying a substitution to the clause and adding a literal to the body of the clause (case with the bottom-up approach) or by applying an inverse substitution to the clause and removing a literal from the body of the clause (case with the top-down approach) [9]. Graph-based concept discovery systems rely on the assumption that concepts should appear as frequent subgraphs (case with the substructure-based approaches) or as frequent path that connect certain nodes (case with the path finding-based approaches). In addition to the structure that graph-based concept discovery systems look for, they also differ in the way they represent the data.

Relational Paths-based Learning (RPBL) [10] is path finding-based concept discovery. It performs breadth first search on a graph called SIS to extract paths. RPBL works for binary relations only, nodes in SIS represent binary facts and edges connect nodes that have a common argument value. RPBL outputs paths with the highest F1-measure as concept descriptors.

Mode Directed Path Finding [11] extends Relational path-finding algorithm for saturated bottom clauses. It builds a hypergraph from a saturated bottom clause where nodes represent facts and edges connect nodes with respect to mode declarations. To find concept descriptors, it enumerates all hyperpaths and chooses those with the highest F1 score as concept descriptors.

The Hybrid Graph-based method (Hybrid) [12] combines properties of substructure-based and path-finding-based methods. Different than other concept discovery studies, it does not look for paths on a already built graph but instead it generates paths while constructing the graph. The proposed approach is limited with binary relations, nodes hold relation constants and edges connects nodes whose values form a fact.

Subdue [13] is a general purpose substructure-based tool in graph-based concept discovery problem. It uses constrained beam search to find repetitive substructures and it evaluates them based on the Minimum Description Length principle. Subdue system discovers the best substructure which is repetitive in data.

SubdueCL is a version of Subdue method and it contains core functions of Subdue [14]. SubdueCL differs from Subdue in terms of its learning and evaluation mechanism. It evaluates quality of substructure based on uncovered positive examples and covered negative examples. SubdueCL inputs both positive and negative examples in graph.

DIFFER [15] is proposed to overcome limitations of graph mining methods in learning from structured data. DIFFER inputs structured data and transforms it into graph based canonical representation by method called finger prints. Finger prints have ability to combine isolated substructures and detecting similar substructures on graph. DIFFER outputs a text file on .arff format that contains set of features for given structured data.

The method we propose incorporates methods of association rule mining to prune the search space by eliminating infrequent and weak concept descriptors. Similar pruning mechanism have been embodied in [12,16].

The proposed method differs from [10,12] in the sense that it can work on n-ary relations. Different than [11], the proposed method does not need mode declaration which require expertise to define. The proposed method considers all of the target instances at once, hence it avoids target instance ordering problem which is the case with [10, 17].

The proposed method is implemented in Java and Neo4j [1] is employed as the graph database engine. Paths from the graph are extracted using functionality provided by Neo4j.

3   The Proposed Approach

In this section, we firstly introduce our data representation method, then present the proposed method. To illustrate these steps we employ a subset of the *mutagenesis* data set as a running example. The data set is provided in Table 1, where the *molecule* is the concept to be learned.

Table 1. The *mutagenesis* data set

Target Concept Instances	Background Knowledge
molecule(d12, true)	lumo(d12, -1.627)
molecule(d19, false)	lumo(d19, -1.478)
	max_charge(d12, 0.82)
	max_charge(d19, 0.832)
	inda(d12, 0)
	inda(d19, 0)
	atm(d12, d12_1, c, 22, -0.111)
	atm(d19, d19_6, c, 22, -0.097)

3.1   Graph Representation for n-Ary Relations

In the proposed method, we construct a directed, labeled, simple graph to represent data. While constructing the graph we follow the semantics of the relations and construct the graph accordingly. Our observations suggest that (1) binary relations usually assert a property of an entity, and (2) n-ary relations assert some properties that hold between two distinct entities. For example background knowledge *lumo(d12, -1.627)* states the energy level of the lowest unoccupied molecular orbital of drug d12 is -1.627, while background knowledge *atom(d12, d12_1, c, 22, -0.111)* states that in drug d12, atom number 1 is a carbon atom of QUANTA type 22 with a partial charge of -0.111.

To represent the knowledge that falls into first class, we create two nodes to represent the entity and its attributes and link these two nodes with an edge labeled after the relation name. To represent the knowledge that falls into the second class, we create two nodes to represent the entities, and link them with an edge labeled after the relation that holds between these two entities. To embed

---

[1] http://neo4j.com.

the information that holds between two entities, we enhance the edges with bundled properties and store the information as bundled properties of edges. Use of bundled properties are supported by several tools [18,19]. In Fig. 1 we illustrate graph representations of background knowledge instances *lumo(d12, -1.627)* and *atm(d12, d12_1, c, 22, -0.111)*.

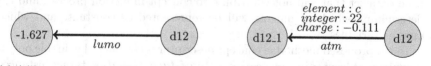

(a) (1)Representation of binary relations   (b) (2)Representation of n-ary relations

Fig. 1. Graph representation of binary and n-ary relations

To represent the entire data set as a graph

(1) We create a distinct node called *root*
(2) We create as many distinct nodes as the number of the classes target instances belong to, and connect them to the *root* node via special edge labeled *is*
(3) We connect the graph representation of each target instance to the class node the target instance belongs to vi the special edge labeled as *is*.

Figure 2 depicts the graph representation of the data set given in Table 1.

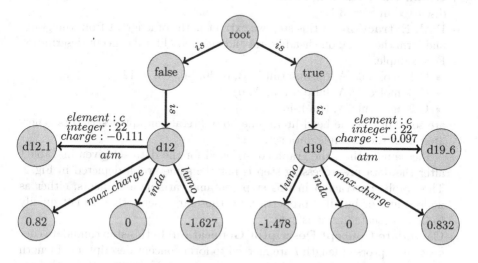

Fig. 2. Graph for sample *mutagenesis* data set

3.2 Path-Based Concept Discovery Under N-Ary Relations Graph

The proposed method inputs a set of target instances, a set of background facts, minimum support, minimum confidence and maximum rule length parameters. Figure 3 shows steps of the proposed graph-based concept discovery process.

(1) **Data Preprocessing:** This step is concerned with (1) removing relations from data set that will not contribute to concept induction process, and (2) determining which arguments will be represented as constants and which ones as variables.

   In the proposed method, a concept descriptor is considered valuable only if it explains at least $min\_sup \times cardinality\_of\_target\_relation$ target instances. It is possible that some relations in a data set contain information for less than $min\_sup \times cardinality\_of\_target\_relation$ target instances. In this step we remove such relations from the data set.

   While inducing the concept descriptors, an argument of a relation is represented either as a constant or as a variable. In this step we determine which arguments will appear as constants and which as variables. An alphanumeric argument of a relation is represented as a constant value only if appears at least $min\_sup \times cardinality\_of\_relation$ number of times in the relation. Due to the nature of numeric arguments, it is a very rare occasion that the same numeric value appears more than $min\_sup \times cardinality\_of\_relation$ times in a relation. For this reason to represent numeric attributes, we sort them in ascending order, split them into portions of size $\frac{1}{min\_sup}$, and pick bordering values as slot representatives.

(2) **Graph Construction:** In this step the data is transformed into a graph as discussed in Sect. 3.1.

(3) **Path Extraction:** In this step we extract paths of length 1 from the graph and form the initial one head literal and one body literal concept descriptors. For example:
   - C_1: molecule(A, false):-atm(A, B, o, 40, [-0.383, -0.128])
   - C_2: molecule(A, false):-inda(A, 0)
   - C_3: molecule(A, true):-inda(A, 0)

   are some of the one head literal one body literal concept descriptors generated from the example data set.

   Please note that, the graph constructed for the data set given in Table 1 after the data preprocessing step is not the same graph depicted in Fig. 2. The graph constructed in this step contains variable arguments, either as node values or bundled properties, for those arguments that do not qualify to be represented as constants.

(4) **Candidate Concept Descriptor Generation:** In this step candidate concept descriptors of length l are merged to form concept descriptors of length $l+1$ based on the Apriori principle. Two concept descriptors of length l are merged if they have the same head literal and differ exactly by only one body literal. For example C_1 can be merged with C_2 but not with C_3.

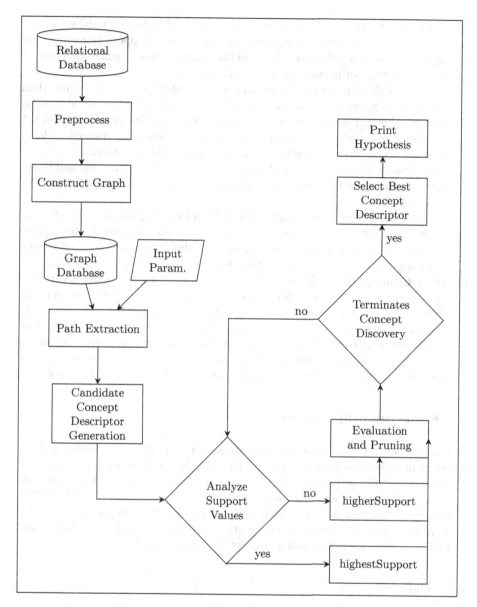

Fig. 3. Flowchart of the proposed method

(5) **Candidate Concept Descriptor Evaluation:** Concept descriptors generated at Step 4 are pruned based on their support and confidence values. Support of a concept descriptor is defined as a proportion of frequency of given concept descriptor over total number of concept descriptors. Confidence of the concept descriptor is defined as a proportion of target instances

covered by concept descriptor over the number of instances induced by the body of concept descriptor. In the proposed method, support value of a concept descriptor is calculated by counting paths, while confidence value of concept descriptor is calculated via SQL queries.

Concept descriptors that have support and confidence values higher than minimum support and minimum confidence, respectively, are added to solution set. Concept descriptors with support values less than minimum support and confidence value higher then minimum confidence are merged to form concept descriptors of length $l+1$ as described in Step 4. Concept descriptors of with both support and confidence values less than thresholds are pruned. The algorithm loops through steps 4 and 5 until maximum concept descriptor is reached.

Two different concept descriptors of length l may represent different argument values with the same variable argument. If two such concept descriptors are merged to form a candidate concept descriptor of length $l+1$, such variables need to be renamed. This step also handles such renaming operations.

(6) **Best Concept Descriptor Selection:** After all solution clauses are generated, the best one is selected based on the f-score [20]. If the best concept descriptor leaves more than $min\_sup \times cardinality\_of\_target\_relation$ target instances the induction process restarts with the preprocessing step. The algorithm terminates when either the number of unexplained target instances fall below $min\_sup \times cardinality\_of\_target\_relation$ or no more concept descriptors can be generated to model the unexplained target instances.

4   Experimental Analysis

In this section, we firstly describe the data sets used in the experiments and the experimental settings. Next, we present the experimental results and discuss the performance of the proposed approach in comparison to CRIS and two state-of-the-art graph based concept discovery systems, namely SubdueCL and DIFFER. We compare the proposed approach with CRIS as both systems follow similar concept descriptor generation and evaluation mechanisms, and to SucdueCL and DIFFER as they are graph-based concept discovery systems.

4.1   Data Sets

The proposed method is tested on three different data sets, namely *Eastbound* [21], *Mutagenesis* [22] and *PTE-1* [23]. The experiments are conducted on a Intel Core 2.53GHz processor computer with 4 GB memory.

Eastbound is about the problem of learning whether the train is eastbound or westbound based on train properties. *Mutagenesis* is a biochemical data set where the problem is to classify drugs as mutagenic or non-mutagenic based on their molecular properties. *PTE-1* is also a biochemical data set where the problem is to learn if a compound is carcinogenic or not. The properties of the data sets are given in Table 2 along with minimum support, minimum confidence and max depth parameters used in the experiments. Minimum support, minimum confidence and maximum rule length parameters are determined based on the reference studies.

Table 2. Data set properties and the experimental settings

Data Set	# Relations	# Facts	Min. Sup	Min. Conf	Max. Length
Eastbound	13	196	0.1	0.6	3
Mutagenesis	13	16544	0.1	0.7	3
PTE-1	28	23850	0.1	0.7	3

The proposed method constructs graphs that represent the given data sets. Table 3 shows graph properties in terms of number of nodes and number of relations. In graphs, we store fewer number of facts in comparison to Table 2 because we eliminate unqualified tuples. Graphs are simplified version of given data sets.

Table 3. Graph properties

Data Set	# Nodes	# Relations
Eastbound	197	196
Mutagenesis	11388	22401
PTE-1	15509	22234

4.2 Experimental Results

In this section, we firstly present the amount of time spent in the preprocessing step, i.e. transforming the relation data into graph database and finding feasible argument values, and time spent for inducing the concept descriptors. We also compare the running time of the proposed method with that of CRIS. The experimental results on running time presented in Table 4 show that the preprocessing time is negligible compared to running time of the concept rule induction. When compared to the running time of CRIS[2], experimental results show that the proposed method runs magnitudes faster than CRIS.

In order to analyze the quality of the concept descriptors induced by the proposed method, we compare the coverage and the accuracy of the concept descriptors to those of CRIS, SubdueCL [14] and DIFFER [15]. Coverage is defined as the fraction of the number of target instances of test data set explained by the induced concept descriptors over the all target instances. Accuracy is

---

[2] Although the running time of the proposed method and that of CRIS are not directly comparable, as the experiments are conducted on different computers, we provide them to provide intuition. In addition, these results are obtained from a recent study hence the configurations of computers are expected to be similar.

Table 4. Running time (Time format hh:mm:ss)

Data Set	Proposed Method		CRIS
	Preprocessing	Rule Induction	
Eastbound	00:00:01	00:00:12	00:11:36
Mutagenesis	00:00:02	00:02:45	03:42:00
PTE-1	00:00:03	01:18:41	05:17:00

defined as the ratio of sum of negative and positive instances that are correctly explained by the induced concept descriptors over the sum of true positive, true negative, false negative and true negative instances. In order to find the false positive and false negative instances, the test data sets are extended with their duals under CWA.

Coverage results are reported in Table 5. As the experimental results show the proposed method is superior to the other concept discovery systems in terms of coverage.

Table 5. Comparison of coverage results

Data Set	Proposed Method	CRIS
Eastbound	%100	%100
Mutagenesis	%79	%53
PTE-1	%87	NA

Lastly, in Table 6, we report the predictive accuracy of the proposed method and compare it with those of CRIS, SubdueCL, and DIFFER. For the *Eastbound* data set, the proposed method achieves better accuracy than CRIS and DIFFER. For the *Mutagenesis* data set, the proposed method demonstrates considerably better accuracy than DIFFER, while achieves the same predictive accuracy with CRIS. For the $PTE-1$ data set, the proposed method has a better predictive accuracy than SubdueCL and DIFFER, while it performs slightly worse than CRIS.

Table 6. Comparison of accuracy results

Data Set	Proposed Method	CRIS	SubdueCL	DIFFER
Eastbound	%100	%70	NA	%80
Mutagenesis	%84.6	%85	NA	%80.61
PTE-1	%85.2	%88	%61.54	%65.25

As seen in the experimental results, the proposed method is superior to other systems when compared in terms of running time. It is also compatible with the state-of-the-art concept discovery systems in terms of coverage and accuracy.

5   Conclusion

In this paper, we propose a new graph-based approach for concept discovery. The method is hybrid of both path finding approaches and techniques of association rule mining. In addition, it differs from the current state-of-the-art knowledge discovery systems, by handling constants in preprocessing step. Also, it eliminates weak and infrequent concept descriptors at early steps provide efficiency on running time and accuracy. We have tested our approaches on several benchmark problems and experiments show that the proposed method has better performance in terms of accuracy, coverage and running time.

IAs the future work, the proposed method can be improved in terms of calculation of confidence values. Instead of calculating confidence on relational database, we may try to calculate confidence values by querying on Neo4j.

References

1. Džeroski, S., De Raedt, L.: Multi-relational data mining: the current frontiers. SIGKDD Explor. Newsl. **5**(1), 100–101 (2003)
2. Doncescu, A., Waissman, J., Richard, G., Roux, G.: Characterization of biochemical signals by inductive logic programming. Knowl.-Based Syst. **15**(1–2), 129–137 (2002)
3. Turcotte, M., Muggleton, S., Sternberg, M.J.E.: Generating protein three-dimensional fold signatures using inductive logic programming. Comput. and Chem. **26**(1), 57–64 (2002)
4. Dzeroski, S., Jacobs, N., Molina, M., Moure, C., Muggleton, S., Laer, W.V.: Detecting traffic problems with ILP. In: Page, D.L. (ed.) ILP 1998. LNCS, vol. 1446, pp. 281–290. Springer, Heidelberg (1998)
5. Muggleton, S., De Raedt, L.: Inductive logic programming: theory and methods. J. Logic Program. **19**, 629–679 (1994)
6. Duboc, A.L., Paes, A., Zaverucha, G.: Using the bottom clause and mode declarations in FOL theory revision from examples. Mach. Learn. **76**(1), 73–107 (2009)
7. Cook, D.J., Holder, L.B.: Substructure discovery using minimum description length and background knowledge. J. Artif. Intell. Res. (JAIR) **1**, 231–255 (1994)
8. Ball, T., Larus, J.R.: Efficient path profiling. In: Melvin, S.W., Beaty, S. (eds.) Proceedings of the 29th Annual IEEE/ACM International Symposium on Microarchitecture, MICRO 29, Paris, France, 2–4 December 1996, pp. 46–57. ACM/IEEE Computer Society (1996)
9. Lavrac, N., Dzeroski, S.: Inductive Logic Programming: Techniques and Applications, vol. 10001. Routledge, New York (1993)
10. Gao, Z., Zhang, Z., Huang, Z.: Learning relations by path finding and simultaneous covering. In: 2009 WRI World Congress on Computer Science and Information Engineering, vol. 5, pp. 539–543. IEEE (2009)
11. Ong, I.M., de Castro Dutra, I., Page, D.L., Santos Costa, V.: Mode directed path finding. In: Gama, J., Camacho, R., Brazdil, P.B., Jorge, A.M., Torgo, L. (eds.) ECML 2005. LNCS (LNAI), vol. 3720, pp. 673–681. Springer, Heidelberg (2005)
12. Mutlu, A., Karagoz, P.: A hybrid graph-based method for concept rule discovery. In: Bellatreche, L., Mohania, M.K. (eds.) DaWaK 2013. LNCS, vol. 8057, pp. 327–338. Springer, Heidelberg (2013)

13. Holder, L.B., Cook, D.J., Djoko, S., et al.: Substucture discovery in the subdue system. In: KDD workshop, pp.169–180 (1994)
14. Gonzalez, J., Holder, L., Cook, D.J.: Application of graph-based concept learning to the predictive toxicology domain. In: Proceedings of the Predictive Toxicology Challenge Workshop (2001)
15. Karunaratne, T., Böstrom, H.: Differ: a propositionalization approach for learning from structured data. Mutagenesis **80**(88.86), 76–92 (2006)
16. Kavurucu, Y., Senkul, P., Toroslu, I.H.: Concept discovery on relational databases: new techniques for search space pruning and rule quality improvement. Knowl.-Based Syst. **23**(8), 743–756 (2010)
17. Richards, B.L., Mooney, R.J.: Learning relations by pathfinding. In: AAAI, pp. 50–55 (1992)
18. Robinson, I., Webber, J., Eifrem, E.: Graph Databases. O'Reilly Media Inc., Sebastopol (2013)
19. Schling, B.: The Boost C++ Libraries. XML Press (2011)
20. Goutte, C., Gaussier, É.: A probabilistic interpretation of precision, recall and F-score, with implication for evaluation. In: Losada, D.E., Fernández-Luna, J.M. (eds.) ECIR 2005. LNCS, vol. 3408, pp. 345–359. Springer, Heidelberg (2005)
21. Larson, J., Michalski, R.S.: Inductive inference of vl decision rules. ACM SIGART Bull. **63**, 38–44 (1977)
22. Srinivasan, A., Muggleton, S.H., Sternberg, M.J., King, R.D.: Theories for mutagenicity: a study in first-order and feature-based induction. Artif. Intell. **85**(1), 277–299 (1996)
23. Srinivasan, A., King, R.D., Muggleton, S.H., Sternberg, M.J.: The predictive toxicology evaluation challenge. In: IJCAI Citeseer, vol. 1, pp. 4–9 (1997)

Exact Detection of Information Leakage
in Database Access Control

Farid Alborzi[1]([✉]), Rada Chirkova[1], and Ting Yu[2]

[1] Computer Science Department, North Carolina State University,
Raleigh, NC, USA
falborz@ncsu.edu, chirkova@csc.ncsu.edu
[2] Qatar Computing Research Institute, Doha, Qatar
tyu@qf.org.qa

Abstract. Elaborate security policies often require organizations to restrict user data access in a fine-grained manner, instead of traditional table- or column-level access control. Not surprisingly, managing fine-grained access control in software is rather challenging. In particular, if access is not configured carefully, information leakage may happen: Users may infer sensitive information through the data explicitly accessible to them in centralized systems or in the cloud.

In this paper we formalize this *information-leakage problem,* by modeling sensitive information as answers to "secret queries," and by modeling access-control rules as views. We focus on the scenario where sensitive information can be deterministically derived by adversaries. We review a natural data-exchange based inference model for detecting information leakage, and show its capabilities and limitation. We then introduce and formally study a new inference model, view-verified data exchange, that overcomes the limitation for the query language under consideration.

Keywords: Privacy and security in cloud intelligence · Data exchange

1 Introduction

In many data-intensive applications, access to sensitive information is regulated by *access-control policies* [7], which determine, for each user or class of users, the stored information that they are permitted to access. For instance, an employee working in the communications department of a company may be granted access to view his salary in the employee database, but not any other employee's salaries. At the same time, an employee of the human-resources department may be permitted to view the salaries of all the employees in the communications department. To enable management of information covered by heterogeneous policies such as in this example, it has become standard for data-management systems from major vendors to go beyond table-level or column-level access control, by providing means to support row-level or even cell-level access control. Representative products that provide such *fine-grained access control* include Oracle's Virtual Private Database [25] and DB2's label-based access control [9].

© Springer International Publishing Switzerland 2015
S. Madria and T. Hara (Eds.): DaWaK 2015, LNCS 9263, pp. 403–415, 2015.
DOI: 10.1007/978-3-319-22729-0_31

Unlike table-level or column-level access control, fine-grained access control is often content based. That is, whether a user is allowed access to some data depends on the content of the data, and often also on the properties of the user, as illustrated in the above example. Content-based access control is a powerful mechanism that offers flexibility to express nontrivial constraints on user data access, which is often a must for the enforcement of privacy and corporate policies. However, correct configuration of fine-grained access control is challenging, both in centralized systems and in the cloud. For instance, many database systems offer a programming-style interface for specifying access-control rules. In this clearly error-prone process, a simple programming error may result in serious data-access violations. Specifically, even if the access-control rules for each user are implemented correctly, sensitive information might still be derived by colluding malicious users from the data that they are permitted to access. Consider an illustration that introduces our running example.

Example 1. Suppose a database relation Emp stores the name, department, and salary information about employees of a company. Sensitive database information that should not be disclosed to unauthorized users could include the relationship between the employee names and salaries in the Emp relation. Suppose user X is permitted to access the name-department relationships in Emp, via a projection of that relation on the respective columns. Further, user Y is granted access to the department and salary columns of Emp in a similar way.

Consider an instance of the Emp relation on which the user X sees exactly one relationship between employee names and departments, namely the pair (JohnDoe, Sales). Suppose the user Y can access, from the same instance of Emp, also a single relationship pair, (Sales, $50000), and no other information. If X and Y collude, it is straightforward for them to infer from these two tuples that employee JohnDoe's salary must be $50000 in the Emp relation. Thus, the sensitive name-salary relationship can be obtained in this case by unauthorized users just by comparing the data that they are permitted to see.

This example illustrates how content-based access control may be exploited by malicious users to derive sensitive information. The derivation in the example is *deterministic*. That is, the presence of the sensitive relationship (JohnDoe, $50000) in the back-end database is guaranteed by the information that these users are authorized to access. We call the problem of discovering whether it is possible to make such deterministic derivations of sensitive information the *information-leakage problem*. Our specific focus is as follows: Given a set of access-control rules for a database, and given the current database status, will a user or a set of colluding users be able to deterministically derive some secret information from their explicitly accessible data, and possibly from their knowledge of integrity constraints that must hold on the database? In this paper, we formalize and study this information-leakage problem in the context of fine-grained access control on relational data, with secret queries and access views expressed as SQL select-project-join *(SPJ)* queries with equality.

Our specific contributions are as follows:

- We formalize the information-leakage problem for access control on relational data, for a single secret query and one or more access-control views, possibly in presence of integrity constraints on the back-end database.
- We perform a theoretical study of this problem (in the absence of integrity constraints) in the setting where the secret query and each access-control policy is a SQL select-project-join (SPJ) query with equalities. As such queries are also called *conjunctive (CQ) queries* [11], we will refer to this setting as *the CQ setting*. Specifically, we develop an algorithm for information-leak disclosure in the CQ setting, and show its correctness.

In this paper we use some formal definitions, as our correctness results would not be provable without a formalization of the problem. At the same time, the presentation focuses on discussions, illustrations, and on outlining the results at the intuitive level. (A full detailed version of this paper [13] is available online.)

The rest of the paper is organized as follows. Section 2 formalizes the problem of information-leak disclosure. Sections 3 and 4 discuss two approaches to solving the problem: a natural approach based on data exchange (Sect. 3) and a more elaborate version, view-verified data exchange, with completeness guarantees (Sect. 4). Section 5 discusses related work, and Sect. 6 concludes.

2   The Problem Statement

In this section we formalize the problem of information-leak disclosure. As an overview, the problem is as follows: Given a set of answers, MV, to *access-control queries* V (which we call *views* in the rest of the paper), on an (unavailable) relational data set of interest I, and given another query Q, which we call *the secret query* in the sense that its answers must not be disclosed to unauthorized users: Which of the answer tuples to Q on I are "deterministically assured" by the contents of MV? That is, which tuples $\bar{t}$ must necessarily be present in the answer to Q on I, based on the information in V, in MV, and (optionally) in the set Σ of *integrity constraints* that hold on I? We say that there is *information leakage* of Q via MV and V if and only if at least one such tuple $\bar{t}$ exists. Note that the data set I is not available for making the determination.

Formally, suppose we are given a relational schema $\mathbf{P}$, set of dependencies Σ on $\mathbf{P}$, set V of views on $\mathbf{P}$, query Q over $\mathbf{P}$, and a (Σ-valid) set MV of view answers for V. (A set MV is Σ-*valid for* V whenever there exists a data set I that satisfies Σ and produces exactly MV as the set of answers to V. We use the notation "$V \Rightarrow_{I,\Sigma} MV$" to refer to such data sets I.) Then we call the tuple $\mathcal{D}_s = (\mathbf{P}, \Sigma, V, Q, MV)$ a *(valid) instance of the information-leak problem*.

We say that a tuple $\bar{t}$ is a *potential information leak* in instance $\mathcal{D}_s$ if for all I such that $V \Rightarrow_{I,\Sigma} MV$ we have $\bar{t} \in Q(I)$. Further, we say that *there is a disclosure of an information leak* in $\mathcal{D}_s$ iff there exists a tuple $\bar{t}$, such that one can show deterministically, using only the information in $\mathcal{D}_s$, that $\bar{t}$ is a potential information leak in $\mathcal{D}_s$. In this case, such a $\bar{t}$ is called a *witness* of the disclosure.

The intuition here is as follows. Suppose attackers are examining the available set of view answers MV, trying to find answers to the secret query Q on the back-end data set, call it I, of schema $\mathbf{P}$ such that I has actually given rise to MV. (That is, I is an actual instance of the organizational data, such that $\mathcal{V} \Rightarrow_{I,\Sigma} MV$ and such that the set MV was obtained by applying to I the queries for $\mathcal{V}$.) However, assuming that the attackers have access only to the information in $\mathcal{D}_s$, they have no way of determining which data set J, with the property $\mathcal{V} \Rightarrow_{J,\Sigma} MV$, is that actual data set I behind the set MV. Thus, their best bet is to find all the tuples $\bar{t}$ that are in the answer to Q on *all* data sets J as above.

Definition 1. *(Problem of Information-Leak Disclosure).* Given a valid instance $\mathcal{D}_s$ of the information-leak problem, the *problem of information-leak disclosure for $\mathcal{D}_s$* is to determine whether there is a potential information leak in $\mathcal{D}_s$.

The following example, which formalizes our running example from Sect. 1, provides an illustration of the definitions.

Example 2. Suppose a relation Emp stores information about employees of a company. The attributes of Emp are Name, Dept, and Salary, with self-explanatory names. A "secret query" Q (in SQL) asks for the salaries of all the employees:

(Q): SELECT DISTINCT Name, Salary FROM Emp;

Consider two views, V and W, that are defined for some class(es) of users, in SQL on the schema $\mathbf{P}$. The view V returns the department name for each employee, and the view W returns the salaries in each department:

(V): DEFINE VIEW V(Name, Dept) AS SELECT DISTINCT Name, Dept FROM Emp;
(W): DEFINE VIEW W(Dept, Salary) AS SELECT DISTINCT Dept, Salary FROM Emp;

Suppose some user(s) are authorized to see the answers to V and W. At some point the user(s) can see the following set MV of answers to these views:

$MV = \{$ V(JohnDoe, Sales), W(Sales, \$50000) $\}$.

Consider a tuple $\bar{t} = $ (JohnDoe, \$50000). Assume that these users are interested in finding out whether $\bar{t}$ is in the answer to the query Q on the "back-end" data set, I. Using the approach introduced in this paper, we can show that the relation Emp in such a data set I is uniquely determined by V, W, and MV:

$$\text{Emp } in \text{ } I \text{ } is \{ \text{ Emp(JohnDoe, Sales, \$50000) } \}. \tag{1}$$

The only answer to Q on this data set I is the above tuple $\bar{t}$. Thus, the presence of $\bar{t}$ in the answer to Q on the data set of interest is deterministically assured by the information that these users are authorized to access.

3 Using Data Exchange for Information-Leak Disclosure

In this paper, we address the problem of information-leak disclosure by developing an approach that outputs the set of disclosed information leaks for each problem instance $\mathcal{D}_s$ where $\Sigma = \emptyset$. Clearly, only those approaches that are sound and complete algorithms (for certain classes of inputs) can be used to solve correctly the problem stated in Definition 1 of Sect. 2. Here, *completeness* means that for each potential information-leak tuple $\bar{t}$ for a valid input $\mathcal{D}_s$, the given approach outputs $\bar{t}$. Further, *soundness* means that for each $\bar{t}$ that the approach outputs for such a $\mathcal{D}_s$, the tuple $\bar{t}$ is a potential information leak for $\mathcal{D}_s$.

We now begin the development of a sound and complete approach, by showing how data exchange [6] can be used for information-leak disclosure. The ideas in this section parallel those of [10,26]. These ideas are not a contribution of this paper, as they are sound but not complete in our context (see end of Sect. 3). The main contribution of this paper is an approach that builds on but does not stop at data exchange; we present that approach in Sect. 4. The approach of Sect. 4 is sound and complete for information-leak disclosure for CQ access-control policies and secret queries, for the case where where $\Sigma = \emptyset$. We now begin laying down the framework for that approach. (Please see [13] for the details.)

3.1 Reviewing Data Exchange

A *data-exchange setting* $\mathcal{M}$ is a triple $(\mathbf{S}, \mathbf{T}, \Sigma_{st} \cup \Sigma_t)$, where $\mathbf{S}$ and $\mathbf{T}$ are disjoint schemas, Σ_{st} is a finite set of integrity constraints applying from $\mathbf{S}$ to $\mathbf{T}$,[1] and Σ_t is a finite set of integrity constraints over $\mathbf{T}$. Instances of $\mathbf{S}$ are called *source* data sets and are always ground data sets (i.e., never include "null values," also called "nulls"). Instances of $\mathbf{T}$ are *target* data sets.

Given a source data set I, we say that a target data set J is a *solution for I* (*under $\mathcal{M}$*) if (I, J) satisfies $\Sigma_{st} \cup \Sigma_t$, denoted "$(I, J) \models \Sigma_{st} \cup \Sigma_t$." Solutions are not always unique; *universal solutions* are, intuitively, "the most general" possible solutions. Constructing a universal solution for source data set I can be done by chasing [18] I with $\Sigma_{st} \cup \Sigma_t$. The chase may not terminate or may fail; in the latter case, no solution exists [18]. If the chase does not fail and terminates, then the resulting target data set is guaranteed to be a universal solution for I.

While checking for existence of solutions is undecidable [6], a positive result is known for the case where Σ_t is a set of "weakly acyclic" [18] dependencies:[2]

Theorem 1. [18] *Let $\mathcal{M} = (\mathbf{S}, \mathbf{T}, \Sigma_{st} \cup \Sigma_t)$ be a fixed data-exchange setting, with Σ_t weakly acyclic. Then there is a polynomial-time algorithm, which decides for every source data set I whether a solution for I exists. Whenever a solution for I exists, the algorithm computes a universal solution in polynomial time.*

---

[1] The intuition is that tuple patterns occuring over $\mathbf{S}$ constrain tuple patterns over $\mathbf{T}$.

[2] Weakly acyclic dependencies [6] are types of tuple- and equality-generating integrity constraints that commonly occur in practice and have nice formal properties.

The universal solution of Theorem 1, called the *canonical* universal solution [18], is the result of the chase.

Query Answering: Assume that a user poses a query Q over the target schema **T**, and I is a given source data set. Then the usual semantics for the query answering is that of *certain answers* [6], which intuitively are tuples that occur in the answer to Q on all possible solutions for I in $\mathcal{M}$:

$$certain_{\mathcal{M}}(Q, I) = \bigcap \{ Q(J) \mid J \text{ is a solution for } I \}.$$

Computing certain answers for arbitrary first-order queries is an undecidable problem. For unions of CQ queries *(UCQ queries)*, which is a query language including all CQ queries, we have the following positive result:

Theorem 2. [18] *Let* $\mathcal{M} = (\mathbf{S}, \mathbf{T}, \Sigma_{st} \cup \Sigma_t)$ *be a data-exchange setting with* Σ_t *a weakly acyclic set, and let* Q *be a UCQ query. Then the problem of computing certain answers for* Q *under* $\mathcal{M}$ *can be solved in polynomial time.*

To compute the certain answers to a UCQ query Q with respect to a source data set I, we check whether a solution for I exists. If so, compute any universal solution J for I, and compute the set $Q_{\downarrow}(J)$ of all the tuples in $Q(J)$ that do not contain nulls. It can be shown that $Q_{\downarrow}(J) = certain_{\mathcal{M}}(Q, I)$.

3.2 Data Exchange in Information-Leak Disclosure

We now reformulate the problem of finding certain query answers in data-exchange settings into our information-leak disclosure problem of Definition 1.

Suppose that attackers are given a valid conjunctive (CQ) instance $\mathcal{D}_s = (\mathbf{P}, \Sigma, \mathcal{V}, Q, MV)$ of the information-leak problem, with Σ a set of weakly acyclic dependencies. The attackers are interested in finding tuples $\bar{t}$, such that for all data sets I with $\mathcal{V} \Rightarrow_{I, \Sigma} MV$, we have $\bar{t} \in Q(I)$.

We now show how a straightforward reformulation of $\mathcal{D}_s$ turns the above problem into an instance of the problem of computing certain answers in data exchange. We first construct a set Σ_{st} of tuple-generating dependencies *(tgds)* [6], as follows. For a view V in the set $\mathcal{V}$ in $\mathcal{D}_s$, consider the CQ query $V(\bar{X}) \leftarrow body_{(V)}(\bar{X}, \bar{Y})$ for V. We associate with this $V \in \mathcal{V}$ the tgd $\sigma_V : V(\bar{X}) \rightarrow \exists \bar{Y} \, body_{(V)}(\bar{X}, \bar{Y})$. We then define the set Σ_{st} to be the set of tgds σ_V for all $V \in \mathcal{V}$. Then $\mathcal{D}_s$ can be reformulated into the following data-exchange setting:

$$\mathcal{M}^{(de)}(\mathcal{D}_s) = (\mathcal{V}, \mathbf{P}, \Sigma_{st} \cup \Sigma),$$

with a source data set MV and a query Q on the target schema $\mathbf{P}$. We call the triple $(\mathcal{M}^{(de)}(\mathcal{D}_s), MV, Q)$ *the associated data-exchange instance for* $\mathcal{D}_s$.

For valid CQ instances $\mathcal{D}_s$, consider an algorithm, which we call the *data-exchange algorithm for disclosing information leaks*. (See [13] for the details.) First, compute the canonical universal solution, $J_{de}^{\mathcal{D}_s}$, for the source data set MV in the data-exchange setting $\mathcal{M}^{(de)}(\mathcal{D}_s)$. Then output, as a set of disclosed information leaks in $\mathcal{D}_s$, all those tuples in $Q(J_{de}^{\mathcal{D}_s})$ that do not contain nulls.

When we assume, similarly to [28], that everything in $\mathcal{D}_s$ is fixed except for MV and Q, then from Theorem 2 due to [18] we obtain immediately that this algorithm always terminates and runs in polynomial time. We have shown that this data-exchange algorithm is sound for valid CQ inputs:

Theorem 3. *The data-exchange algorithm is sound for valid CQ inputs.*

This algorithm is not complete. Indeed, it returns the empty set in Example 2 while the correct solution is the set of tuples {(JohnDoe, $50000)}.)

4   View-Verified Data Exchange

We now build on the intuition and results of Sect. 3, to introduce an approach that is provably sound and complete for information-leak disclosure for conjunctive access-control policies and secret queries, for the case where $\Sigma = \emptyset$.

4.1   The Intuition

Intuitively, the problem with the data-exchange algorithm of Sect. 3 is that its canonical universal solution, when turned into a ground data set,[3] may produce a *proper superset* of the given set of view answers MV.

Example 3. Recall that in Example 2 in Sect. 2, the relation Emp in the "back-end" data set I is uniquely determined by the problem inputs, see Eq. (1). Specifically, that relation Emp generates *exactly* the set MV of Example 2 as answers to the access-control views V and W in the example.

At the same time, the data-exchange algorithm of Sect. 3 produces, for the setting of Example 2, a canonical universal solution $J_{de}^{\mathcal{D}_s} = \{$ Emp(JohnDoe, Sales, NULL$_1$), Emp(NULL$_2$, Sales, $50000) $\}$. If we replace NULL$_1$ in $J_{de}^{\mathcal{D}_s}$ by a (new) value $70000, and NULL$_2$ by a (new) value JanePoe, then the answers to the views on the resulting instance would generate the tuples W(Sales, $70000) and V(JanePoe, Sales), in addition to the two MV tuples in Example 2.

Key Idea: We conclude that to turn the algorithm of Sect. 3 into a *complete* approach, we need to "tighten" the set of data sets that are described by the canonical universal solution $J_{de}^{\mathcal{D}_s}$ of that algorithm. The idea of the "tightening" is as follows: For all the NULL values in the data set $J_{de}^{\mathcal{D}_s}$ that "generate extra tuples" (in addition to MV) as answers to the given access-control views, we transform those NULLs in $J_{de}^{\mathcal{D}_s}$ into constants from MV, in such a way that the resulting answers to the access-control views form exactly the set MV.

Consider the following illustration of this key idea of our proposed *view-verified data-exchange algorithm*. The data-exchange solution $J_{de}^{\mathcal{D}_s}$ of Example 3 has two NULLs. In our proposed approach, we replace the first NULL by $50000, and replace the second NULL by JohnDoe. Then the resulting data set is exactly the relation Emp that attackers can infer using their knowledge, see Eq. (1).

---

[3] A *ground data set* is a data set without null values.

4.2 The Chase Formalism for View-Verified Data Exchange

We now formalize the key idea formulated at the end of Sect. 4.1. The main tool used in view-verified data exchange is *MV-induced dependencies*. Let $V(\bar{X}) \leftarrow \phi(\bar{X}, \bar{Y})$ be a CQ query in $\mathcal{V}$. Suppose the tuples of V in MV are $\{\bar{t}_1, \bar{t}_2, \ldots, \bar{t}_{mv}\}$. Then the *MV-induced generalized egd (MV-induced ged)* τ_V for V is

$$\tau_V : \phi(\bar{X}, \bar{Y}) \rightarrow \vee_{i=1}^{mv}(\bar{X} = \bar{t}_i). \tag{2}$$

Intuitively, the consequent of τ_V has a disjunction of "tuples in the answer to V in MV." Using this definition, we formalize chase of data sets with MV-induced dependencies, via a straightforward extension of the approach of [18]. (We omit here special cases such as zero tuples of V in MV, or Boolean views V; please see [13] for a detailed formal treatment.) Consider an illustration.

Example 4. Consider instance $\mathcal{D}_s$ with all the elements except MV as in Example 2 of Sect. 2. Let $MV' = \{ V(c, d), V(g, d), W(d, f) \}$. Here, we use c, d, f, and g as succinct abbreviations for `JohnDoe`, `Sales`, `$50000`, and `JanePoe`, respectively, and also abbreviate `Emp` to E. Thus, MV' can be interpreted as the result of adding the tuple `V(JanePoe, Sales)` to the MV of Example 2.

The data-exchange approach of Sect. 3 yields this canonical solution:

$$J_{de}^{\mathcal{D}_s} = \{ E(c, d, NULL_1), E(g, d, NULL_2), E(NULL_3, d, f) \}.$$

The set of answers without nulls to the query Q on $J_{de}^{\mathcal{D}_s}$ is empty. Thus, the data-exchange approach of Sect. 3 discovers no information leaks in $\mathcal{D}_s$.

In applying our view-verified data-exchange approach to the input $\mathcal{D}_s$, we first construct the MV'-induced gegds, τ_V and τ_W, one for each of V and W:

$$\tau_V : E(X, Y, Z) \rightarrow (X = c \wedge Y = d) \vee (X = g \wedge Y = d).$$
$$\tau_W : E(X, Y, Z) \rightarrow (Y = d \wedge Z = f).$$

Intuitively, each conjunct in the consequents of τ_V and τ_W corresponds to one tuple in MV'. For instance, "$(X = c \wedge Y = d)$" in τ_V represents $V(c, d)$ in MV'.

Now the canonical data-exchange solution $J_{de}^{\mathcal{D}_s}$ can be chased with τ_V, using the chase process similar to that in [18]. The chase turns $J_{de}^{\mathcal{D}_s}$ into *two* data sets J_1 and J_2. Intuitively, J_1 is $J_{de}^{\mathcal{D}_s}$ that is "constrained with tuple $V(c, d)$ of MV' using τ_V, and J_2 is $J_{de}^{\mathcal{D}_s}$ "constrained with $V(g, d)$ using τ_V.

$$J_1 = \{ E(c, d, NULL_1), E(g, d, NULL_2), E(c, d, f) \}.$$
$$J_2 = \{ E(c, d, NULL_1), E(g, d, NULL_2), E(g, d, f) \}.$$

(J_1 results from assigning $NULL_3 := c$, and J_2 from $NULL_3 := g$.)

We then use the same procedure to apply τ_W to each of J_1 and J_2. As a result, the following data set $J_{vv}^{\mathcal{D}_s}$ is obtained from each of J_1 and J_2:

$$J_{vv}^{\mathcal{D}_s} = \{ E(c, d, f), E(g, d, f) \}.$$

Unlike $J_{de}^{\mathcal{D}_s}$, the data set $J_{vv}^{\mathcal{D}_s}$ produces exactly MV' as the set of answers to V and W. Further, the answers to the secret query Q on $J_{vv}^{\mathcal{D}_s}$ are (c, f) and (g, f). These two tuples (intuitively, the salary \$50000 for each of JohnDoe and JanePoe) can be obtained by easy reasoning by human attackers who use V and W as access-control policies. That is, the view-verified data exchange discovers exactly the disclosed information leaks that can be inferred by human attackers.

4.3   View-Verified Data Exchange for Conjunctive Instances

We now introduce the view-verified data-exchange algorithm for disclosing information leaks for conjunctive (CQ) instances $\mathcal{D}_s$, for the case where $\Sigma = \emptyset$.

Let $\mathcal{D}_s$ be a CQ instance of information leak with $\Sigma = \emptyset$. We construct the set $\Sigma_{\mathcal{D}_s}^{(MV)}$ of its MV-induced dependencies. Then we compute (as in Sect. 3) the canonical universal solution $J_{de}^{\mathcal{D}_s}$ for the data set MV in the data-exchange setting $\mathcal{M}^{(de)}(\mathcal{D}_s)$. We then obtain all the results of chasing $J_{de}^{\mathcal{D}_s}$ with $\Sigma_{\mathcal{D}_s}^{(MV)}$. Denote by $\mathcal{J}_{vv}^{\mathcal{D}_s}$ the set of all these chase results. We call each $J \in \mathcal{J}_{vv}^{\mathcal{D}_s}$ a view-verified universal solution for $\mathcal{D}_s$. For each $J \in \mathcal{J}_{vv}^{\mathcal{D}_s}$, we compute the set $Q_\downarrow(J)$ of those tuples in $Q(J)$ that do not contain nulls. The final output is the set

$$\bigcap_{J \in \mathcal{J}_{vv}^{\mathcal{D}_s}} Q_\downarrow(J). \tag{3}$$

Algorithm 1. Algorithm for view-verified data exchange

Input: Valid CQ input $\mathcal{D}_s = (\mathbf{P}, \emptyset, \mathcal{V}, Q, MV)$;
Output: A set of disclosed information leaks in $\mathcal{D}_s$.
Compute the canonical universal solution $J_{de}^{\mathcal{D}_s}$;
if $J_{de}^{\mathcal{D}_s} = \emptyset$ **then**
$\quad\llcorner$ return $\emptyset$;
Let $\mathcal{T}$ be the finite chase tree of $J_{de}^{\mathcal{D}_s}$ with $\Sigma_{\mathcal{D}_s}^{(MV)}$;
$w \leftarrow \emptyset$;
$checked \leftarrow false$;
for $each \ nonempty \ leaf \ (i.e., \ final \ chase \ result) \ J \ in \ \mathcal{T}$ **do**
$\quad$ $tmp \leftarrow \emptyset$;
$\quad$ **for** $each \ t \in Q(J)$ **do**
$\quad\quad$ **if** $t \ does \ not \ contain \ nulls$ **then**
$\quad\quad\quad\llcorner$ $tmp \leftarrow tmp \cup \{t\}$;
$\quad$ **if** $checked = false$ **then**
$\quad\quad$ $w \leftarrow tmp$;
$\quad\quad$ $checked = true$;
$\quad$ **else**
$\quad\quad\llcorner$ $w \leftarrow w \cap tmp$;
return w;

View-verified data exchange addresses the shortcoming of the approach of Sect. 3. Recall that the canonical universal solution $J_{de}^{\mathcal{D}_s}$ of Sect. 3 might not cover "tightly enough" all the data sets of interest to the attackers. We address this problem here, by using our extension of the chase to generate from $J_{de}^{\mathcal{D}_s}$ a set $\mathcal{J}_{vv}^{\mathcal{D}_s}$ of data sets that are each "tighter" than $J_{de}^{\mathcal{D}_s}$ in this sense.

We have shown that the view-verified data-exchange approach is a sound and complete algorithm for the problem of disclosing information leaks for valid CQ inputs with $\Sigma = \emptyset$. In particular, the set $\mathcal{J}_{vv}^{\mathcal{D}_s}$ is well defined, in that the "chase tree" in view-verified data exchange is always finite and has at least one nonempty leaf (i.e., chase result). We have also shown that the set $\mathcal{J}_{vv}^{\mathcal{D}_s}$ is "just tight enough," in the following sense: Denote by $certain(Q, \mathcal{D}_s)$ the set of all the potential information leaks for $\mathcal{D}_s$. Then the expression in Eq. (3), which is the intersection of all the "certain-answer expressions" for Q and for the elements of the set $\mathcal{J}_{vv}^{\mathcal{D}_s}$, is exactly the set $certain(Q, \mathcal{D}_s)$. (Please see [13] for all the details.)

Theorem 4. *View-verified data exchange is a sound and complete algorithm for discovering potential information leaks for valid CQ instances with $\Sigma = \emptyset$.*

5 Related Work

The problem considered in this paper can be viewed as the problem of *inference control*, whose focus is on preventing unauthorized users from computing sensitive information. This problem has been studied extensively in database security, in both probabilistic (e.g., [23]) and deterministic versions. We consider the deterministic version of the problem. The reason is, when one can show that users can infer sensitive information in a conservative deterministic way, then the access-control rules are unsatisfactory and must be changed immediately. Work on deterministic inference control has been reported in [10,26], with solutions that can be made only sound but not complete for our information-leakage problem. To the best of our knowledge, none of the procedures in inference control yields sound and complete algorithms for the information-leak disclosure problem.

Generally, the literature on privacy-preserving query answering and data publishing is represented by work on data anonymization and on differential privacy; please see [12] for a survey. Most of that work focuses on probabilistic inference of private information (see, e.g., [3]), while in this paper we focus on the possibilistic situation, where an adversary can deterministically derive sensitive information. (An interesting direction of future work is to see whether the approach of [3] can be combined with that of this current paper to address the probabilistic situation.) Further, our model of sensitive information goes beyond associations between individuals and their private sensitive attributes.

Policy analysis has been studied for various types of systems, including operating systems, role-based access control, trust management, and firewalls [4,5,20,22,24]. Typically, two types of properties are studied. The first type is static properties: Given the current security setting (e.g., non-management privileges of users), can certain actions or events (e.g., separation of duty) happen?

Our analysis of information leakage falls into this category. What is different in our approach is that our policy model is much more elaborate, as we deal with policies defined by database query languages. The other type of properties in policy analysis is dynamic properties when a system evolves; that direction is not closely related to the topic of our paper.

The authors of [1] consider a problem whose statement is similar to ours. The results of [1] are different from ours, as we focus on developing constructive algorithms, rather than on doing complexity analysis as in [1]. Moreover, [1] use in their analysis a type of complexity metric that does not apply to data-access control. (The complexity metric that applies to our problem is that of [28].)

Our technical results build on the influential data-exchange framework of [18]. Our MV-induced dependencies of Sect. 4 resemble target-to-source dependencies Σ_{ts} introduced into (peer) data exchange in [19]. The difference is that Σ_{ts} are embedded dependencies defined at the schema level. In contrast, our MV-induced dependencies embody the given set of view answers MV.

6   Conclusion

Semantically seemingly benign access-control views may allow adversaries to infer sensitive information when considering specific answers to those views. We have seen that it may be easy enough for a human such as a database administrator (DBA) to see the leak of information. However, when given nontrivial security policies defined through views, the problem quickly becomes challenging. In particular, adversaries may obtain sensitive information through various inference methods. Known techniques for capturing specific inference methods often result in sound but incomplete solutions.

In this paper we studied the information-leakage problem that is fundamental in ensuring correct specification of data access-control policies, in centralized systems as well as in the cloud. We formalized the problem and investigated two approaches to detecting information leakage. We showed that the naturally arising approach based on data exchange [6] is efficient and sound, but is incomplete. We then introduced a sound and complete view-verified data-exchange approach, which can be used as a foundation of detection tools for DBAs.

This paper reports on our first effort in a formal study of information leakage in data. There are many important problems to investigate in our future work. Specifically, our current discussion assumes no integrity constraints on the data. It is important to develop techniques that would extend the view-verified data-exchange approach to detect inferences that exploit integrity constraints, and to study their complexity. Further, in this paper we focused on the problem setting where a set MV of answers to access-control views (policies) is given. In the future work it is also important to study the more general problem, which only considers integrity constraints and view definitions (policies) without a specific MV. It gives a much stronger security guarantee if we can show that information leakage cannot occur for all possible sets of view answers MV.

References

1. Abiteboul, S., Duschka, O.: Complexity of answering queries using materialized views. In: PODS, pp. 254–263 (1998)
2. Abiteboul, S., Hull, R., Vianu, V.: Foundations of Databases. Addison-Wesley, Reading (1995)
3. Agrawal, R., Bayardo Jr., R.J., Faloutsos, C., Kiernan, J., Rantzau, R., Srikant, R.: Auditing compliance with a hippocratic database. In: VLDB, pp. 516–527 (2004)
4. Al-Shaer, E., Hamed, H., Boutaba, R., Hasan, M.: Conflict classification and analysis of distributed firewall policies. IEEE JSAC **23**(10), 2069–2084 (2005)
5. Ammann, P., Sandhu, R.S.: Safety analysis for the extended schematic protection model. In: Proceedings of the IEEE Symposium on Security and Privacy, pp. 87–97 (1991)
6. Barcelo, P.: Logical foundations of relational data exchange. SIGMOD Rec. **38**(1), 49–58 (2009)
7. Bertino, E., Ghinita, G., Kamra, A.: Access control for databases: concepts and systems. Found. Trends Databases **3**(1–2), 1–148 (2011)
8. Biskup, J., Bonatti, P.A.: Controlled query evaluation for known policies by combining lying and refusal. Ann. Math. Artif. Intell. **40**(1–2), 37–62 (2004)
9. Bond, R., See, K.Y.-K., Wong, C.K.M., Chan, Y.-K.H.: Understanding DB2 9 Security. IBM Press, Indianapolis (2006)
10. Brodsky, A., Farkas, C., Jajodia, S.: Secure databases: constraints, inference channels, and monitoring disclosures. IEEE TKDE **12**(6), 900–919 (2000)
11. Chandra, A., Merlin, P.: Optimal implementation of conjunctive queries in relational data bases. In: STOC, pp. 77–90 (1977)
12. Chen, B.-C., Kifer, D., LeFevre, K., Machanavajjhala, A.: Privacy-preserving data publishing. Found. Trends Databases **2**(1–2), 1–167 (2009)
13. Chirkova, R., Yu, T.: Detecting information leakage in database access control with help from data exchange. Technical report (which is not a publication) TR-2013-1, NCSU (2013). http://www.csc.ncsu.edu/research/tech/reports.php
14. Deutsch, A.: XML query reformulation over mixed and redundant storage. Ph.D. thesis, Univ. Pennsylvania (2002)
15. Deutsch, A., Nash, A., Remmel, J.: The chase revisited. In: PODS, pp. 149–158 (2008)
16. Deutsch, A., Tannen, V.: Optimization properties for classes of conjunctive regular path queries. In: Ghelli, G., Grahne, G. (eds.) DBPL 2001. LNCS, vol. 2397, pp. 21–39. Springer, Heidelberg (2002)
17. Domingo-Ferrer, J. (ed.): Inference Control in Statistical Databases. LNCS, vol. 2316. Springer, Heidelberg (2002)
18. Fagin, R., Kolaitis, P., Miller, R., Popa, L.: Data exchange: semantics and query answering. Theor. Comput. Sci. **336**(1), 89–124 (2005)
19. Fuxman, A., Kolaitis, P.G., Miller, R.J., Tan, W.-C.: Peer data exchange. ACM TODS **31**(4), 1454–1498 (2006)
20. Harrison, M.A., Ruzzo, W.L., Ullman, J.D.: Protection in operating systems. Comm. ACM **19**, 461–471 (1976)
21. Kabra, G., Ramamurthy, R., Sudarshan, S.: Redundancy and information leakage in finite-grained access control. In: ACM SIGMOD Conference, pp. 133–144 (2006)
22. Li, N., Winsborough, W.H., Mitchell, J.C.: Beyond proof-of-compliance: safety and availability analysis in trust management. In: Proceedings of the IEEE Symposium on Security and Privacy, pp. 123–139 (2003)

23. Miklau, G., Suciu, D.: A formal analysis of information disclosure in data exchange. JCSS **73**(3), 507–534 (2007)
24. Motwani, R., Nabar, S., Thomas, D.: Auditing SQL queries. In: ICDE 2008 (2008)
25. The Virtual Private Database in Oracle9iR2. An Oracle White Paper (2002)
26. Stoffel, K., Studer, T.: Provable data privacy. In: Andersen, K.V., Debenham, J., Wagner, R. (eds.) DEXA 2005. LNCS, vol. 3588, pp. 324–332. Springer, Heidelberg (2005)
27. Zhang, X., Ozsoyoglu, M.: Implication and referential constraints: a new formal reasoning. IEEE TKDE **9**(6), 894–910 (1997)
28. Zhang, Z., Mendelzon, A.O.: Authorization views and conditional query containment. In: Eiter, T., Libkin, L. (eds.) ICDT 2005. LNCS, vol. 3363, pp. 259–273. Springer, Heidelberg (2005)

Author Index

Printed in the United States
By Bookmasters